SOME MILESTONES IN MOLECULAR BIOLOGY AND RECOMBINANT DNA TECHNOLOGY

1869 Miescher discovers DNA in trout sperm.

1882 Flemming describes chromosomes in dividing cells.

1917 Term *biotechnology* first used.

1943 Penicillin produced industrially.

1944 Avery, MacLeod, and McCarty demonstrate DNA is genetic material in bacteria.

1952 Hershey and Chase determine that T2 bacteriophage DNA is the material that enters host cells.

1953 Watson and Crick elucidate structure of DNA from Franklin and Wilkins' x-ray diffraction patterns.

1957 Kornberg isolates DNA polymerase I.

1958 The mode of DNA replication demonstrated experimentally.

1959 RNA polymerase discovered.

1960 mRNA and role in encoding information for amino acids discovered.

1961 Synthetic mRNA (UUU) used to begin to decipher the genetic code.

DNA renaturation discovered.

1962 Existence of restriction endonucleases in bacteria demonstrated.

Watson, Crick, and Wilkins awarded Nobel Prize for discovering structure of DNA.

1965 Discovery that plasmids encode antibiotic resistance.

1966 Genetic code completely elucidated.

1967 DNA ligase isolated.

1968 Nirenberg, Khorana, and Holley awarded Nobel Prize for elucidating genetic code.

1970 First restriction endonuclease isolated.

1972 DNA ligase joins two DNA fragments, creating first recombinant DNA molecules.

1973 DNA inserted into plasmid vector and transferred to host *E. coli* cell for propagation; cloning methods established in bacteria.

Potential hazards of recombinant DNA technology raise concerns.

1974 World moratorium on some recombinant DNA experiments.

1975 Development of experimental guidelines for recombinant DNA technology discussed at the Asilomar Conference on Recombinant DNA Molecules.

Southern blotting and hybridization developed for detecting specific DNA sequences.

Monoclonal antibodies produced.

1976 National Institutes of Health prepares first guidelines for physical and biological containment.

DNA sequencing methods developed.

1977 Genentech, the first biotechnology firm, established.

Introns discovered.

1978 Somatostatin, first human hormone produced by recombinant DNA technology.

Human insulin produced by recombinant DNA technology.

Arber, Smith, and Nathans awarded Nobel Prize for discovering restriction endonucleases.

1979 NIH guidelines relaxed.

1980 Landmark *Diamond v. Chakrabarty* case leads to first patent of genetically engineered microorganism.

Nobel Prize to Sanger and Gilbert for developing DNA sequencing methods; to Berg for first DNA cloning methods.

1981 First diagnostic kit based on monoclonal antibodies approved in U.S.

Transgenic mice and *Drosophila* fruit flies produced.

1982 First animal vaccine produced by recombinant DNA technology in Europe.

1983 Plants transformed by genetically modified Ti plasmids.

1985 Polymerase chain reaction developed for *in vitro* amplification of DNA.

DNA fingerprinting first developed by Alec Jeffreys.

1986 Research on the *in vitro* formation of tissues using degradable polymer scaffolds.

1988 Genetically engineered mouse that developed cancer

1990 First gene therapy trials approved in U.S.

Human Genome Project begins in U.S.

1993 Nobel Prize to Sharp and Roberts for discovering introns and RNA splicing; to Mullis and Smith for developing polymerase chain reaction.

1990s Automated DNA sequencing technology developed and refined.

1992 Development of automated DNA sequencing technologies.

1995 Completion of *Haemophilus influenzae* DNA sequence, the first free-living organism to be sequenced; the complete DNA sequence of *Mycoplasma genitalium;* after 1995 the complete genomic sequences have since been published or made available for *Methanococcus jannaschii, Synechocystis* sp., *Mycoplasma pneumoniae, Helicobacter pylori, Escherichia coli, Archaeoglobus fulgidis, Methanobacterium thermoautotrophicum,* and the yeast *Saccharomyces cerevisiae,* the first eukaryotic organism to be completely sequenced.

1996 Focus on the large-scale sequencing of the human genome.

1997 Nuclear transplantation experiment using nucleus from a differentiated cell produces a lamb called Dolly; expression of human proteins from sheep clones.

BIOTECHNOLOGY

An Introduction

Susan R. Barnum
Miami University

Wadsworth Publishing Company
I⊤P® An International Thomson Publishing Company

Belmont, CA • Albany, NY • Bonn • Boston • Cincinnati • Detroit • Johannesburg • London • Madrid • Melbourne
Mexico City • New York • Paris • Singapore • Tokyo • Toronto • Washington

Biology Editor: Gary Carlson and Jack Carey
Assistant Editor: Kristin Milotich
Editorial Assistant: Kerri Abdinoor and Marie Carigma-Sambilay
Developmental Editor: Mary Arbogast
Project Editor: Jerry Holloway
Print Buyer: Karen Hunt
Production Coordinator: Electronic Publishing Services Inc., NYC
Permissions Editor: Peggy Meehan / Veronica Oliva
Designer: Gary Head Design
Copy Editor: Kevin Gleason
Illustrators: Electronic Publishing Services Inc., NYC
Compositor: Electronic Publishing Services Inc., NYC
Printer: Webcom

I(T)P ® The ITP logo is a registered trademark under license.

Printed in Canada
1 2 3 4 5 6 7 8 9 10

For more information, contact Wadsworth Publishing Company, 10 Davis Drive, Belmont, CA 94002, or electronically at
http://www.thomson.com/wadsworth.html

International Thomson Publishing Europe
Berkshire House 168-173
High Holborn
London, WC1V 7AA, England

International Thomson Editorés
Campos Eliseos 385, Piso 7
Col. Polanco
11560 México D.F. México

Thomas Nelson Australia
102 Dodds Street
South Melbourne 3205
Victoria, Australia

International Thomson Publishing Asia
221 Henderson Road
#05-10 Henderson Building
Singapore 0315

Nelson Canada
1120 Birchmount Road
Scarborough, Ontario
Canada M1K 5G4

International Thomson Publishing Japan
Hirakawacho Kyowa Building, 3F
2-2-1 Hirakawacho
Chiyoda-ku, Tokyo 102, Japan

International Thomson Publishing GmbH
Konigswinterer Strasse 418
53227 Bonn, Germany

International Thomson Publishing Southern Africa
Building 18, Constantia Park
240 Old Pretoria Road
Halfway House, 1685 South Africa

Library of Congress Cataloging-in-Publication Data

Barnum, Susan.
    Biotechnology / Susan Barnum.
      p. cm.
    Includes index.
    ISBN 0-534-23436-4 (pbk.)
    1. Biotechnology.   I. Title.
TP248.2.B363   1998
660'.6—dc21   97-44429
                                CIP

CONTENTS

PREFACE

Biotechnology—the use of living organisms or their products to enhance our lives and our environment—is a broad and complex discipline that encompasses many specialized areas. Our understanding of living organisms has been immensely deepened by the contributions that recombinant DNA methods have made to basic scientific research. We can now isolate, analyze, and manipulate genes and ultimately alter the genetic makeup of living organisms. Indeed, biotechnology has already revolutionized agriculture and medicine.

I have been teaching college biotechnology courses since 1986, and have long been aware that students need a single book that offers a reasonably comprehensive introduction to biotechnology. That is what motivated me to write this book. However, the nature of modern biotechnology presents the writer of an introductory text with several challenges: The field is broad, diverse, and fast moving. Its breadth means that many areas must be addressed; its diversity means that no two teachers of what are nominally similar courses will necessarily address the same combination of topics; and its fast pace means that a book is truly up to date only when it is in proof stage. The last challenge can be addressed by a timely schedule of revisions and web site updates; the first and second challenges are addressed by integrating information from the many areas of biotechnology to give readers basic information on essential concepts and methods, and an understanding of how the field is evolving and what developments are on the horizon.

A brief tour of the book's chapters will indicate how this book meets these challenges in a number of ways. The first two chapters present in-depth historical reviews of the developments leading to modern biotechnology and of landmark experiments leading to recombinant DNA technology. Although the book is neither a methods manual nor a molecular biology text, Chapter 3 presents an overview of gene expres-

sion and Chapter 4 presents a review of the basic methods of recombinant DNA technology.

Each of chapters 5 through 8 focuses on a single broad area of research and development to which modern methods of biotechnology are being applied with significant results and intriguing potential for the future: microorganisms, plants, animals, and marine organisms—marine biotechnology is a developing area that is rarely covered in textbooks. In each chapter, examples of discoveries and new applications are described.

Chapter 9 describes the purposes and current achievements of the Human Genome Project; Chapter 10 reviews the contributions to date that biotechnology has made to medical science, while Chapter 11 describes its contributions to forensic science, giving particular attention to DNA profiling, or "fingerprinting." Finally, Chapter 12 focuses on regulatory and patent issues. At relevant points throughout the text the societal, legal, and environmental implications of developments are discussed.

In addition to breadth of coverage, other features enhance the book's usefulness:

- Several chapters are supplemented by appendices; these range from a detailed review of the immune response, to a summary of NIH guidelines for biological research.

- In several chapters—particularly in Chapter 9, The Human Genome Project—readers are referred to numerous World Wide Web sites for the most up to date information in rapidly moving areas. Because web sites are frequently changed, I recommend either looking addresses up on the book web site or conducting a keyword search to find the site.

- The generous use of cross-references to other chapters allows teachers and students to integrate information from different areas of biotechnology and different stages in its development to achieve a better understanding of this complex and broad field.

- Although no book can completely replace primary research papers and review articles in a rapidly advancing field, it can serve as a springboard for further study in a specific area or provide an overview for students not familiar with modern biotechnology. Therefore, at the end of each chapter I list several books and many articles—both general reviews and research papers—for further reading.

- A glossary gives definitions of key terms.

- End papers present a list of abbreviations and acronyms, a chronological listing of the historical development of molecular biology and recombinant DNA technology, and a year-by-year account of milestones in the Human Genome Project.

Who can use this book? Undergraduates and graduate students alike will find it valuable. Undergraduate science majors and nonmajors with some science background can use it as a stand-alone text for biotechnology courses, seminars, or special topics courses, or as a supplement in other courses such as cell biology, molecular biology, and genetics. Selected chapters can be used in courses geared more toward a particular group of organisms (for example, plant cell biology or microbial genetics) or in a general biology course for coverage of a particular topic. In graduate courses the book can provide a comprehensive overview of biotechnology, as well as background information, with primary papers and review articles assigned as additional reading along with each chapter. Medical students, professionals in industry and academia, and technical writers who are interested in learning about this exciting discipline will also find the book of value.

This book could not have been written without the help of many people. I am indebted to my editors, Jack Carey and Gary Carlson, for their encouragement, support, and guidance during the development of the book. I also thank Jack Carey for approaching me with the idea of writing a biotechnology book. I also acknowledge Jerry Holloway, senior project editor, for his helpful suggestions and encouragement; Kevin Gleason, developmental and copy editor, for his careful editing and constructive criticism of the manuscript; colleagues who carefully reviewed the manuscript and offered many suggestions for improving the content and accuracy; the many investigators who kindly provided summaries and published papers of their research and diagrams, photographs, or micrographs of their work; companies such as Mycogen, Calgene, and Monsanto, for providing photographs of their products and shared information about their work with recombinant organisms; and, finally, the many reviewers who contributed to the book's accuracy: James A. Bush, University of New Haven; Clare L. Chatot, Ball State University; William H. Harvey, Earlham College; Amy B. Mulnix, Earlham College; Dennis M. Smith, Wellesley College; Rod Anderson, Ohio Northern University; and Daniel R. Zeigler, The Ohio State University.

As an additional resource for instructors and students using this book, a web site may be found at

http://www.wadsworth.com/biology

This site contains links by chapter and sometimes by section to other web sites that will aid students in further exploring the exciting areas of biotechnology.

I dedicate this book to Dan and Emily,

who demonstrated unwavering patience

during the many hours

I spent completing it.

MODERN BIOTECHNOLOGY 1

WHAT IS BIOTECHNOLOGY?

We are witnessing a revolution in biotechnology that has far-reaching implications for medicine and agriculture and consequently for society beyond anything previously imaginable in the history of science. The Office of Technology Assessment of the United States Congress defines **biotechnology** as "any technique that uses living organisms or substances from those organisms, to make or modify a product, to improve plants or animals, or to develop microorganisms for specific uses." Thus, biotechnology encompasses tools and techniques, including those of **recombinant DNA** technology; the living organisms to be improved can be plants, animals, or **microorganisms**; and the products from these organisms can be new or rare—that is, not having existed before naturally, or being less abundant than we would like for certain needs or purposes. Biotechnology is multidisciplinary, involving a variety of the natural sciences—cell and molecular biology, microbiology, genetics, physiology, and biochemistry, to name only the major areas—as well as engineering and computer science. Biotechnology also encompasses recombinant DNA technology—indeed, the two terms are sometimes used synonymously. Recombinant DNA technology encompasses tools and techniques for genetically manipulating organisms ranging from bacteria and fungi to plants and animals. Scientists can change organisms and obtain products in ways not thought possible just forty years ago.

The many applications of biotechnology include the production of new and improved foods, industrial chemicals, pharmaceuticals, and livestock. Biotechnology has contributed developments such as virus-resistant crop plants and livestock; diagnostics for detecting genetic diseases, such as Huntington's chorea and hereditary cancers, and acquired diseases such as AIDS; therapies that use **genes** to cure diseases such as adenosine deaminase (ADA) deficiency; and recombinant vaccines to prevent diseases such as malaria.

Biotechnological strategies also offer hope for restoring the environment and protecting endangered species. Microorganisms are used to clean up toxic wastes from industrial and oil spills. Conservation biologists use genetic methods to identify particular populations of endangered or threatened species (they can even use a part of an animal—of a whale, for example, or of an elephant killed by poachers—for information that traces the animal to its population). By determining the genetic diversity of various plant and animal populations, genetic analysis can help zoos and field biologists improve conservation practices.

What events ushered in the era of modern biotechnology that began in the 1980s and plays such a significant role in our lives today? Note the word "modern" in the previous sentence. Biotechnology is broadly considered to be modern, but by strict definition—the manipulation of living organisms—biotechnology is in fact ancient (although only very recently known by the term). In this chapter, we briefly investigate the origins of biotechnology and review major contributory discoveries.

ANCIENT BIOTECHNOLOGY

History of Domestication and Agriculture

For two million years Paleolithic peoples lived in mobile camps and survived by hunting wild animals and collecting wild plants. The lives of these hunter-gatherers were dictated by the migratory habits of the animals they hunted and by the distribution of edible plants, especially of grasses such as wild wheat and barley. Then, some ten thousand years ago, something remarkable occurred that changed the course of civilization. Abandoning their nomadic ways, people settled and began domesticating plants and animals: They established agrarian societies. The archaeological record indicates that early farmers in the Near East (inclusive of present-day Iran, Iraq, Turkey, Syria, Jordan, Lebanon, and Israel) cultivated wheat, barley, and possibly rye. Sheep and goats provided milk, cheese, butter, and meat. Approximately seven thousand years ago in Africa, pastoralists roamed the Sahara (which was not then a desert) with large flocks of sheep, goats, and cattle. They also hunted and used grinding stones in food preparation. Exactly when these ancient people began to grow plants for food is not known. Archaeological evidence suggests that early farmers arrived in Egypt approximately six thousand years ago. They brought sheep, goats, and cattle and such crops as barley, emmer (an ancient wheat), flax, lentil, and chickpea. Archaeologists also have found ancient farming sites in the New World, the Far East, and Europe. Evidence suggests that agriculture developed independently in several areas of the world at approximately the same time; there appears to have been no one region of origin. We will never be certain what prompted this apparently sudden shift to a more sedentary lifestyle; it may have been a response to population increases and the increasing demand for food, to shifts in climate, or to the dwindling of the herds of migratory animals. Whatever the reasons, this change in lifestyle allowed civilizations to flourish and, with them, new technologies. Early farmers were able to control their environment in ways that nomadic people could not.

The origins of biotechnology thus date back some ten thousand years, to these early agrarian societies in which people collected the seeds of wild plants for cultivation and domesticated some species of wild animals living around them. Each year at harvest time, farmers collected the seeds of the plants with the most desirable traits and set them aside for planting the next year. Similarly, they bred only the most prized animals. By this practice of artificial selection, farmers gradually produced new varieties of plants and animals that retained the desirable traits found in the wild species but were modified in other ways that were beneficial to humans.

There is abundant evidence of ancient domestication and selection (Figure 1.1). The Babylonians and Egyptians left pictorial evidence that dogs, sheep, and cattle had been domesticated by 2000 B.C. (Figure 1.2). The Romans left written accounts of their selective breeding practices for improving livestock.

Ancient Plant Germplasm

For thousands of years, farmers at harvest time have selected seeds, cuttings, or tubers from superior plants to save for the next planting. Farmers often protected the stored seeds from insects by sealing them in clay pots or burying them in baskets covered with ash. They stored tubers in cool areas, and either replanted cuttings immediately or kept them in a dry location until the next planting time. Farmers thus saved their genetic stocks from season to season. They could exchange surplus stocks with neighbors or barter them in the local market, with the result that the plant variety became widespread in a region. (Large-scale organized seed production did not begin until the early 1900s.)

Ancient peoples began collecting plants several thousand years ago. Records show that around 2500 B.C. Sumerian collectors traveled west to Asia Minor to acquire vines, rose plants, and fig trees. The Egyptians collected plants such as East African incense trees, as was depicted on a temple built during the reign of Queen Hatshepsut (in the fifteenth century B.C.). By the 1500s A.D., plant-collecting expeditions became commonplace and collectors traversed the globe.

One of the most influential collectors of the twentieth century was Nikolai I. Vavilov (1887–1943), a Russian plant geneticist and agronomist who collected and cataloged thousands of ancient crop plants and their wild relatives. Between 1923 and 1931, he traveled extensively in the Soviet Union and in over fifty countries collecting economically important plant varieties—including beans, peas, chickpeas, maize, lentils, oats, rye, and wheat (26,000 varieties of wheat alone). Vavilov was instrumental in establishing, at the Institute of Plant Industry in Leningrad, one of the first gene banks for long-term storage of important plant **germplasm**. He demonstrated to scientists worldwide that germplasm collections had economic value and moreover were important to the disease resistance of crop plants in current use. As President of the Lenin Academy of Agri-

Figure 1.1 Corn cobs demonstrating the evolutionary changes from about 5000 B.C. to about A.D. 1500. The far left ear is wild, the others are early cultivated ears. The scale shows that the far right ear is over five times the length of the wild ear. All samples were found during excavations in the Tehuacan Valley, Mexico, under the auspices of the Robert S. Peabody Foundation for Archaeology.
Robert S. Peabody Museum of Archaeology, Phillips Academy, Andover, Massachusetts. All Rights Reserved.

cultural Sciences and Director of the Institute of Applied Botany (now the N.I. Vavilov Institute of Plant Industry), he initiated a comprehensive research and breeding program. His was the first organized, logical plan for crop genetic resource management.

Vavilov was arrested in 1940 on charges of espionage and died in prison from malnutrition in January 1943. Although charged with espionage, Vavilov had in fact become the victim of ideological forces in the Soviet Union. Under the encouragement of Trofim Lysenko (1898–1976), a rising leader in Soviet science under Stalin, the principles of Mendelian genetics were rejected in favor of Lamarckism: the theory that an organism can acquire physical traits in response to its environment and pass them on to its offspring, which also change in response to the environment. As opposed to Mendelian inheritance, these traits are acquired through, as it were, a desire or will to change. Lysenko argued that since plants acquire characteristics through environmental influence, selective breeding is not needed. Stalin banned experimentation and scientific inquiry, and Soviet geneticists could no longer continue their research.

As the Soviet government was suppressing Mendelian genetics, the United States was establishing centers for the preservation, study, and distribution of germplasm (in New York State in 1948 and in Georgia and Washington State in 1949). In 1959, the United States Department of Agriculture established a national gene bank, the National Seed Storage Laboratory, at Fort Collins, Colorado.

Much of the world's rich crop germplasm resides in the developing world because of its ancient agricultural past. Unfortunately, as farmers worldwide embrace modern agriculture with its high-yielding, "high-tech" crops, ancient varieties and their wild relatives are lost, and genetic diversity among crops has been eroded. Wild varieties, previously safe in remote, uncultivated areas are being destroyed by agricultural expansion and

Figure 1.2 Stone amulets from Iraq in the shape of sheep (pierced for suspension). Some amulets were used as stamp seals, and had designs drilled on the bottom. Protoliterate period, before 2900 B.C.

Table 1.1   Research Centers of the Consultative Group on International Research (CGIAR)

Center*	Location	Agricultural Programs	Focus Area
International Rice Research Institute (IRRI)	Philippines	Rice	Global
Centro Internacional de Mejoramiento de Maiz y Trigo (International Maize and Wheat Improvement Center) (CIMMYT)	Mexico	Wheat, maize, barley, triticale	Global
International Institute of Tropical Agriculture (IITA)	Nigeria	Maize, groundnut, sweet potato, yam, cassava, cowpea, soybean, plantain, agroforestry species	Tropical Africa Global
Centro Internacional de Agricultura Tropical (International Center for Tropical Agriculture) (CIAT)	Columbia	Cassava and beans Tropical pastures	Global Central and South America
West Africa Rice Development Association (WARDA)	Côte d'Ivoire	Rice	West Africa
Centro Internacional de la Papa (International Potato Center) (CIP)	Peru	Potato, sweet potato	Global Latin America
International Center for Agricultural Research in Dry Areas (ICARDA)	Syria	Barley, lentil, chickpea, fava bean, bread wheat, durum wheat, forage crops	Dry regions of North Africa and West Asia
International Crops Research Institute for Semi-Arid Tropics (ICRISAT)	India	Pearl millet, other millets, sorghum, groundnut, chickpea, pigeonpea	Global, semi-arid regions
International Network for the Improvement of Banana and Plantain (INIBAP)	France (germplasm stored in Belgium)	Banana and plantain	Global
International Plant Genetic Resources Institute (IPGRI)	Italy	Plant genetic resources General conservation of plant germplasm	Global
International Livestock Center for Africa (ILCA)	Ethiopia	Livestock production systems Forage germplasm	Tropical Africa
International Council for Research in Agroforestry (ICRAF)	Kenya	Agroforestry	Africa
International Service for International Agricultural Research (ISNIAR)	Netherlands	National agricultural research	Global
International Food Policy Research Institutes (IFPRI)	United States	Food policy	Global
Center for International Forestry Research (CIFOR)	Indonesia	Agroforestry Tropical trees	Global

*Additional nonplant research centers:
International Irrigation Management Institute (IIMI), Sri Lanka
International Center for Living Aquatic Resources Management (ICLARM), Philippines
International Laboratory for Research on Animal Diseases (ILRAD), Kenya

the widespread use of herbicides. There is now a world-wide effort to salvage remaining germplasm for preservation in gene banks.

The establishment of a global network of plant gene banks has been a significant achievement for world agriculture. Since 1971 this network has been coordinated by the Consultative Group on International Agricultural Research (CGIAR), a broadly based international consortium that works to strengthen national agricultural research programs in developing countries. The CGIAR supports 17 international agricultural research centers distributed around the world (Table 1.1), and ensures the conservation of plant species in approximately 450 non-CGIAR institutions in more than 90 countries. Germplasm storage banks house plant material such as seeds, plant cuttings, and tubers for future use and study. Their holdings are divided among short-, intermediate-, and long-term collections. Plant material in short-term storage at ambient temperatures must be used. Seeds in intermediate-term storage at −5 to 0°C can last up to 30 years before they must be grown out. Dried seeds stored in sealed containers at −20°C can last 100 years or longer. Periodic germination tests must be conducted to determine viability. Other types of material such as cuttings and tubers can be preserved in fields associated with the gene banks or even grown in tissue culture (described in Chapter 6, Plant Biotechnology). The research at CGIAR centers helps to preserve the world's plant genetic resources, increase the supply of basic foods to developing countries, increase productivity through genetic improvements of plants and livestock, and strengthen research programs in developing countries. CGIAR supports research that ranges from the development and introduction of integrated pest-management programs and biological control methods that encourage farmers to reduce the use of chemical pesticides, to the production of disease-resistant, high-yielding varieties of potato, beans, sorghum, plantain, cassava, rice, and wheat and forages, to name just a few. In addition, CGIAR is committed to strengthening national agricultural research in developing countries. Expert consultation and training programs provide thousands of scientists in developing countries with the technical expertise required to improve plants and livestock and to preserve their genetic resources.

History of Fermented Foods and Beverages

Fermented Foods   Once people settled in villages, their diets became more varied with the introduction of new foods. Many of these new foods were produced quite by accident. Microorganisms have affected human food for thousands of years. Bread, yogurt, cheese, wine, and beer are produced by **fermentation**—that is, a microbial process in which enzymatically controlled transformations of organic compounds occur. ("Fermentation" comes from the Latin *fervere*, "to boil": The addition of yeast to fruit juices during wine making or to cereal grains during bread baking or beer brewing produces a bubbling, from the production of carbon dioxide.) The aroma of baking bread comes from the alcohol that is produced, and trapped carbon dioxide causes the bread to rise. (In beer, the carbon dioxide forms the frothy "head," and the alcohol is present in the malt beverage.)

The existence of microorganisms and their role in contaminating foods are quite recent discoveries, dating back only some two hundred years. For millennia, the fermentation of foods and beverages was an art, practiced without scientific knowledge for guidance. Before methods of preserving food were devised, moist foods often became contaminated with bacteria, yeast, and molds. People learned to live with microbe-infected foods and encouraged contamination when they discovered that flavor and texture were improved.

Bread was one of the earliest foods; it predates the earliest agriculture. At some point, wild cereal grains were found to be edible. They could be dried in the sun and stored for years without spoiling. Early humans probably chewed the raw grains and only later produced flour and dough for baking. Egyptian models and paintings found in tombs show that grain first was ground in a mortar and then on a sloping stone (saddle quern) with a rubbing stone (Figure 1.3). The flour was sifted with sieves made of rushes. The milled flour was mixed with water to form a paste and, after salt was added, molded into loaves. The earliest loaves, being unleavened, were flat and dense (like pita bread); they were cooked on a flat stone over a fire or baked in a clay oven. Early bread was made from an ancient cultivated wheat called emmer (for example, *Triticum turgidum subspecies dicoccum*, a principal wheat of the Old World

Figure 1.3   A model of a bakery, Asyut, Egypt, Middle Kingdom (2040–1782 B.C.).

used in beer making), from einchorn (e.g., *Triticum monococcum*), or from barley (*Hordeum vulgare*).

Fermented dough was almost certainly discovered quite by accident when some dough was not baked immediately, underwent spontaneous fermentation, and, when baked, produced a lighter, expanded, and more palatable bread. Some time around 1800 B.C., the Egyptians and Babylonians learned that old, uncooked fermented dough could be used to ferment a new batch of dough. Bakers no longer had to depend on chance contamination. Some flour from a batch was removed from the mixing vessel and added to fresh flour; the resulting paste, or "starter," was used in the next day's dough. Records such as tomb paintings and reliefs and evidence such as carbonized remains of breads in tombs indicate that the Egyptians used fermentation to make bread. A wide variety of loaf shapes have been found—ovals, indented squares, triangles, cones, animal shapes. Analysis of bread (and beer) samples has demonstrated that the Egyptians were using a pure yeast strain, *Saccharomyces winlocki*, as early as approximately 1500 B.C. Since at that time bakeries and breweries were usually set up in close proximity, the bakers probably obtained *S. winlocki* from the brewery barm (liquid yeast skimmed from the surface of fermented beer).

Bread making was exported from Egypt and Mesopotamia to Greece and Rome. The Romans improved the technology, producing a lighter, leavened bread by using yeast skimmed from grain-malt wort (a liquid prepared with malt that when fermented produces alcohol). This method was used for nearly two thousand years, although dough "starters" were also still in use. Bread making was considered an art, and Roman society held bakers in high regard. The Romans took their bread making methods to the lands they conquered, where they were readily adopted. A bread-specific yeast was unavailable from Roman times through the Middle Ages. In fact, not until Pasteur's experiments between 1857 and 1863 was the connection made between the role of yeast and fermentation. Finally, between 1915 and 1920 the modern production of **baker's yeast** began. Today, a pure strain of *Saccharomyces cervasiae* is used almost universally and is provided in a dried and often compressed form.

Bread was not the only fermented food in the ancient world. By 4000 B.C., the Chinese were using lactic-acid-producing bacteria for making yogurt, molds for making cheese, and acetic acid bacteria for making wine vinegar. They also made soy sauce and other sauces by fermentation; the oldest known reference to soybean is Shen Nan's *Materia Medica* (2838 B.C.). Fermented rice was eaten not only in the Orient but in the Andes Mountains region of what is now Ecuador. Vegetables and fruits were fermented for preservation by being packed in vessels with salt or brine. This process probably originated in the Orient, in prehistoric times, and is still used today;

olives, pickles, and sauerkraut are examples of fermented produce, and in the Orient beets, turnips, lettuce, radishes, and vegetable mixtures are fermented.

Milk from domesticated animals has been a dietary staple probably since the earliest times of animal domestication. Rock drawings in the Libyan desert, dating back to almost 9000 B.C., depict a cow being milked. Mammalian milk, an emulsion of oil and water, is a stable product because the surface of fat globules adsorbs phospholipids and **proteins**. However, in the absence of preservation methods, milk readily becomes contaminated with lactic acid bacteria and is soured by heat.

Milk undergoes fermentation when bacterial action causes casein, the main protein in milk, to coagulate into a curd that separates from the thin, watery whey. Cheese curd from milk was first made between five thousand and nine thousand years ago. Early cheese makers discovered that using milk from a variety of animals—cows, goats, sheep, etc.—yielded cheeses with different textures, aromas, and flavors. Milk was probably first stored in animal skins or in bladders made from animal stomachs. **Enzymes** in the stomach worked with the natural bacteria in the milk to cause the casein to precipitate and form curds, which were subsequently dried. Early records report that cream, buttermilk, yogurt, sour cream, and butter also were consumed. As in early bread making, results were probably inconsistent, since people relied on spontaneous souring. The process was refined and improved through much experimentation.

Modern cheese makers inoculate the milk with lactic acid bacteria (for lactic acid fermentation) and add enzymes (such as rennet) to curdle the casein (Figure 1.4). (Rennet is found in the gastric juices produced in the fourth stomach of nursing calves and other animals.) After the whey is separated from the curd, salt is added to the curd prior to ripening. The essential steps in modern cheese manufacture are heating, separating the curd from the whey, draining the whey (which is mostly water with a little protein, lipid, lactose, and lactic acid), salting, pressing the curd, and ripening. New cheeses have continued to be discovered, often by accident. Camembert and Roquefort are two French cheeses that were created by accident. In the late eighteenth century, Camembert was first made when curd was inoculated with a mold (now known as *Penicillium camemberti*). At the end of the 1800s, Camembert was produced commercially, and its flavor was refined by use of a variety of yeasts for fermentation. Roquefort was first produced when curd was contaminated by a mold now called *Penicillium glaucum roqueforti*, found in French caves. The flavor and aroma particular to a variety of cheese result from the action of different microorganisms.

Fermented Beverages Beer making may have begun in Egypt between 6000 and 5000 B.C. (both bread and beer

Figure 1.4 Curdled milk is being stirred for cheese making.

a connection made between yeast cell activity and alcoholic fermentation, although the findings, derived from microscopic examination of various fermenting liquids, were subject to criticism. The renowned French chemist Louis Pasteur eventually established that yeast and other microbes are directly linked to fermentation. In two published manuscripts, *Études sur le Vin* (1866) and *Études sur la Bière* (1876), he described the experiments that led him to conclude that during the process of fermentation, yeast converts sugar into ethanol and carbon dioxide in the absence of air.

Wine was almost certainly first made accidentally when the juices from grapes became contaminated with yeast and other microbes and fermented naturally. Some authorities believe that wine making originated in the valley of the River Tigris, in present-day Iraq. However, the exact date is unknown. The ancient Egyptians, the Greeks, and the Romans made wine. Pottery wine jars, often labeled, have been found at archaeological sites. Today, wine is aged in large wooden barrels.

were Egyptian dietary staples). Early Babylonian records recommend the use of particular varieties of barley for making particular beers. Archaeological evidence indicates that the earliest brewers made beer by partially baking dough from barley just long enough to dry it and form a crust but not long enough to break down the enzymes involved in fermentation. The dry dough was soaked in water until fermentation was complete. The resulting acidic beer was strained and poured into jars for storage (Figure 1.5). Barley was placed in earthenware vessels that were buried until the barley germinated. This germinated barley, called malt, was removed, crushed, and made into a brewer's dough. Malt contains starch and fermentable sugars, coloring, and aromatic compounds that give the beverage a particular taste and aroma. In other regions, a variety of cereal grains, such as sorghum, corn, rice, millet, and wheat, were used in ancient brewing. Yeast sediments found in ancient beer urns from somewhat later periods indicate that brewing methods improved with time.

Brewing was considered an art until the fourteenth century A.D., when it was recognized as a separate trade using specialized skills. For several centuries, monasteries were the major brewers. Although brewers refined and improved their techniques, they knew nothing about the microbial basis of fermentation. In 1680 the Dutch biologist and microscopist Antony van Leeuwenhoek examined samples of fermenting beer through a microscope. He submitted a description and drawings of the yeast cells he observed to the Royal Society of London, but his discovery was forgotten. Not until 1837 was

Figure 1.5 Servant bottling beer and sealing the pottery jar with clay. Model found in a tomb, Giza, Egypt, fifth or sixth dynasty of the Old Kingdom (2498–2181 B.C.).

CLASSICAL BIOTECHNOLOGY

The term "classical biotechnology" can be used to describe the course of development that fermentation has taken from ancient times (that is, ancient biotechnology) to the present. The accumulation of scientific and applied knowledge from experiments and discoveries of the past has provided a solid foundation for the many industrial processes that today provide us with a plethora of products and services. From the mid-nineteenth century to the present, classical biotechnology has exploited our knowledge of cell processes to refine fermentation technology, just as have developments in modern biotechnology (described in subsequent chapters). Knowledge of fermentation has increased to such a level that a large number of important industrial compounds can be readily produced.

Brewers began producing alcohol on a large scale in the early 1700s. To produce what are known as the English, Dutch, Belgian, and red beers, brewers introduced top fermentation, a process in which the yeast rises to the surface of the liquid as fermentation progresses. In 1833, they introduced bottom fermentation; in this process, the yeast remains at the bottom of the vat. Most beers in the United States and Europe, as well as pale ales, are made by bottom fermentation. By the 1800s, brewers had accumulated enough knowledge to begin using pure yeast cultures in the fermentation process. The equipment designed by E.C. Hansen in 1886 for producing brewer's yeast is still in use today. In 1911, brewers adopted a method for measuring the amount of acid during mashing in order to better control the quality of their beers.

Vinegar is another fermentation product that illustrates progress in expertise and equipment. Wine was allowed to sit in shallow barrels until it was oxidized to vinegar by the action of microorganisms. Early vinegars were most likely the product of accident, but eventually producers realized that air (that is, oxygen) enhanced the transformation. Experimentation improved vinegar production by leading to the use of a fermentation chamber packed with a material like charcoal, through which the wine or some other alcohol slowly moved for aeration; modern vinegar production uses **fermenters**. By the turn of the century, vinegar producers had mastered variables in the control of vinegar quality. For example, they prepare a starter solution by inoculating the alcohol with vinegar (Figure 1.6).

From 1900 to 1940, the number of commercial fermentation products expanded to include glycerol, acetone, butanol, lactic acid, citric acid, and yeast biomass for baker's yeast. Indeed, industrial fermentation was established during World War I because Germany needed large quantities of the fermentation product glycerol for explosives. The Germans produced these quantities of glycerol by adding sodium bisulfite to the **substrates** of alcoholic fermentation. Acetone and butanol also were produced by fermentation for explosives during World

War I. Later, an anaerobic acetanobutylic fermentation using pure cell culture was developed to eliminate contamination by unwanted microorganisms.

The fermentation of organic solvents began during World War I, and by the 1940s had been improved by the establishment of aseptic techniques and use of fermenters that could be steam-sterilized to prevent microbial contamination. Improvements in fermenter design and in control of nutrients and aeration, methods of introducing and maintaining sterility, and methods of product purification and isolation have made it possible to produce rare and valuable chemicals.

World War II ushered in the age of the modern fermenter, or bioreactor, and antibiotics. **Antibiotics** were the first compounds to be produced primarily because of a need for drugs to combat bacteria during the war. Fermentation technology could not have been adapted to commercial antibiotic production without the development of strain isolation methods. The antibiotic penicillin was produced by fermentation of cultured *Penicillium*. Penicillin-producing strains were improved, and effective methods of sterile aeration and culture mixing were introduced. Large-scale penicillin recovery from cultures also was developed. The fermentation of other antibiotics quickly followed (Table 1.2).

Classical biotechnology has introduced chemical transformations that yield products with important therapeutic value. In the 1950s, biotransformation technology was developed to convert cholesterol to other steroids such as cortisone and sex hormones. Microorganisms that are "fed" the proper substrate transform

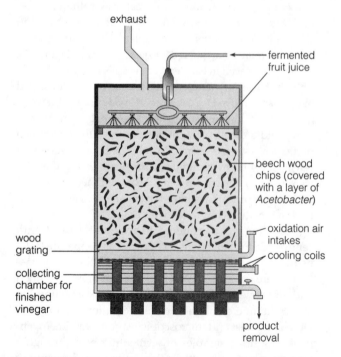

Figure 1.6 Large quantities of vinegar are produced by *Acetobacter* on a substrate of wood chips. Fermented fruit juice is introduced at the top of the column and the column is oxygenated from the bottom.

Table 1.2 Some Commercially Produced Antibiotics

Antibiotic	Producing Microorganism	Class
Produced by fungi		
Cephalosporin	*Cephalosporium acremonium*	Broad-spectrum
Griseofulvin	*Penicillium griseofulvum*	Fungi
Penicillin	*Penicillium chrysogenum*	Gram-positive bacteria
Produced by Gram-positive, Spore-Forming Bacteria		
Bacitracin	*Bacillus subtilis*	Gram-positive bacteria
Polymyxin B	*Bacillus polymyxa*	Gram-negative bacteria
Produced by Gram-Positive Bacterium, Actinomycete		
Amphotericin B	*Streptomyces nodosus*	Fungi
Chloramphenicol	*Streptomyces venezuelae* (now chemical synthesis)	Broad-spectrum
Cycloheximide	*Streptomyces griseus*	Pathogenic yeasts
Cycloserine	*Streptomyces orchidaceus*	Broad-spectrum
Erythromycin	*Streptomyces erythreus*	Mostly Gram-positive bacteria
Kanamycin	*Streptomyces kanomyceticus*	Gram-positive bacteria
Lincomycin	*Streptomyces lincolnensis*	Gram-positive bacteria
Neomycin	*Streptomyces fradiae*	Broad-spectrum
Nystatin	*Streptomyces noursei*	Fungi
Streptomycin	*Streptomyces griseus*	Gram-negative bacteria (*Mycobacterium tuberculosis*)
Tetracycline	*Streptomyces rimosus*	Broad-spectrum

J. Ingraham and C. Ingraham, *Introduction to Microbiology*, Table 29.4. Copyright © 1995 Wadsworth Publishing Co.

it into the desired compound (see Chapter 5, Microbial Biotechnology). For example, cholesterol can be converted to steroids such as estrogen and progesterone by a microbial hydroxylation reaction (that is, addition of an OH group to the cholesterol ring). (Microorganisms can readily carry out the hydroxylations and dehydroxylations that are essential steps in the conversion of steroids.) Plant steroids can even be used in the chemical transformation as substrates. This ability of microorganisms to synthesize a wide variety of compounds and use unusual substrates has been exploited commercially. By the mid-1950s, **amino acids** and other primary **metabolites** (those used for cell growth) were being produced. Large amounts of a specific metabolite can be produced by shifting the direction of cell metabolism during the fermentation process. Other fermentation products such as enzymes and vitamins were also produced at this time.

By the 1960s, microbial cells were being produced on a large scale as a source of protein. Aeration methods were developed to new levels of sophistication, and continuous culturing replaced batch culturing. Eventually computers were used to control fermenter operations. In the 1960s and 1970s, many secondary metabolites (those compounds that are not used for cell reproduction and growth) began to be produced by fermentation and screened for therapeutic activity.

Today, primary metabolites such as amino acids (Table 1.3), pharmaceutical compounds, and a variety of

Table 1.3 Produced Amino Acids and Their Uses

Amino Acid	Use
Alanine	Added to fruit juice to improve taste
Aspartate	Added to fruit juice to improve taste
Cysteine	Added to bread and fruit juice to enhance flavor
Glutamate (MSG)	Added to many foods to enhance flavor
Glycine	Enhances flavor of sweetened foods
Histidine + tryptophan	Prevents rancidity in various foods
Lysine	Used in Japan to make bread a more complete protein
Methionine	Makes soybean products a more complete protein

J. Ingraham and C. Ingraham, *Introduction to Microbiology*, Table 29.5. Copyright © 1995 Wadsworth Publishing Co.

Table 1.4 Commercially Important Enzymes Produced by Microorganisms

Enzyme	Activity	Producing Microorganism	Use
Cellulase	Hydrolyzes cellulose	*Trichoderma konigi*	Digestive aid
Collagenase	Hydrolyzes collagen	*Clostridium histolyticum*	Promotes wound/burn healing
Diastase	Hydrolyzes starch	*Aspergillus oryzae*	Digestive aid
Glucose isomerase	Converts glucose to fructose	*Streptomyces phaeochromogenes*	Converts glucose from hydrolyzed cornstarch to a sweetener
Invertase	Hydrolyzes sucrose	*Saccharomyces cerevisiae*	Candy manufacture
Lipase	Hydrolyzes lipids	*Rhizopus* spp.	Digestive aid
Pectinase	Hydrolyzes pectin	*Sclerotina libertina*	Clarifies fruit juice
Protease	Hydrolyzes protein	*Bacillus subtilis*	Used in detergents

J. Ingraham and C. Ingraham, *Introduction to Microbiology*, Table 29.6. Copyright © 1995 Wadsworth Publishing Co.

chemicals, hormones, and pigments are produced by industrial fermentation for commercial use. Many antibiotics are commercially produced: Antibiotic-producing microorganisms are grown in large fermenters, and the antibiotic end product is collected. Enzymes with a variety of uses (Table 1.4) have been commercially produced from microorganisms and from animal and plant cells. Biomass was first used for food in Germany during World War I, and commercial production for animal and human consumption (such as **single-cell protein**) continues today, as does the mass production of baker's yeast, which began early in the twentieth century.

THE FOUNDATIONS OF MODERN BIOTECHNOLOGY

The ability to manipulate living organisms with today's precision requires intricate knowledge of cell structure, of the biochemical reactions that take place within the cell, and of the genetic makeup of cells. Modern, multidisciplinary biotechnology is the result of scientific discoveries and technological developments that span the more than three hundred years from the first microscopes to the first molecular cloning experiments. What were the major discoveries that led to today's sophisticated biotechnology?

Early Microscopy and Observations

The microscope revolutionized science. Until its advent, natural scientists could describe only what they could see with the naked eye. In 1590, the Dutch spectacle maker Zacharias Janssen made the first compound microscope (that is, one having more than one lens to magnify the image). This primitive two-lens microscope magnified an image 30 times. Microscopes were widely used by 1665, when Robert Hooke, a physicist and Curator of Instru-

ments for the Royal Society of London, examined the structure of thinly sliced cork under the microscope. In his treatise *Micrographia*, he described small rectangular compartments, which he called "cellulae" (Latin for "small chambers"); Figure 1.7 shows his microscope and his drawing of what he observed. These structures were actually the walls of dead cells. A Dutch shopkeeper, Antony van Leeuwenhoek, ground lenses as a hobby; his single lenses magnified an image 200 times. In 1676, he

Figure 1.7 Robert Hooke's compound microscope and the cork tissue he sketched.

examined pond water samples and saw small living organisms (protozoa and fungi), which he called "animalcules" (for "small animals"), that were visible only through his lenses (Figure 1.8). In 1683, he saw still smaller creatures—bacteria. Nevertheless, the low resolution and magnification of these crudely constructed microscopes severely limited the understanding of cells.

Development of Cell Theory

As their tools of inquiry improved, scientists began to realize that tissues were composed of cells, and that these cells could divide to generate more cells. Therefore, each cell was a living, functioning unit. In 1838, the German botanist Matthias Schleiden determined that all living plant tissue was composed of cells, and that each plant arose from a single cell. A year later, the German physiologist Theodor Schwann came to a similar conclusion for animals. The cell theory was refined by the German pathologist Rudolf Virchow, who concluded in 1858 that "all cells arise from cells" and that the cell is the basic unit of life. Until this time, the prevailing theory was "vitalism": Only the complete organism, rather than its individual parts, possessed life. After the cell theory was introduced, vitalism gradually fell from favor. By the early 1880s, with the improvement of microscopes, tissue preservation techniques, and stains, scientists made significant advances in the understanding of cell structure and function.

The Role of Biochemistry and Genetics in Elucidating Cell Function

While the secrets of cell structure, organization, and reproduction were unfolding, scientists were elucidating the biochemical and genetic nature of organisms. Most early nineteenth century researchers believed that the organic and inorganic worlds (that is, the living and non-living) were distinctly separate; the laws of chemistry applied only to the inorganic world and not to the biochemical processes of living organisms. Therefore, only living tissues could synthesize organic molecules. However, in 1828, the German chemist Friedrich Wohler obtained crystallized urea, a waste product in mammalian urine, from ammonium cyanate in the laboratory. This was an important discovery: The experiment demonstrated that an organic compound made by a living organism could be synthesized from inorganic compounds in the laboratory. This chemical synthesis encouraged chemists to synthesize other organic compounds.

Between 1850 and 1880, Louis Pasteur made significant contributions to our knowledge of living processes. Observing that wine sometimes became sour as it aged, Pasteur discovered that yeast cells present in the wine contributed to spoilage. He determined that wine was preserved if heated during the interval after the alcohol was made and before lactic acid was produced. This heating, called pasteurization, plays an important role in food preservation today.

Scientists had been engaged in an ongoing controversy about whether life could arise spontaneously. Experiments performed in 1668 and later in 1768 suggested that life must come from life. Although these experiments provided convincing evidence that higher life forms did not arise spontaneously, scientists thought that perhaps microorganisms could so arise. In 1860 Pasteur conducted an experiment to demonstrate that spontaneous generation of microorganisms indeed did not occur. He placed sterilized broth of boiled meat extracts in a swan-necked flask and a straight-necked flask (Figure 1.9). After some time only the broth in the straight-neck flask became contaminated. Pasteur

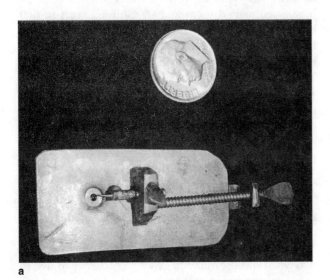

a

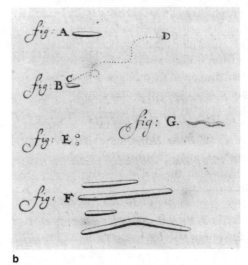

b

Figure 1.8 (a) Small handheld microscope used by Antony van Leeuwenhoek and (b) animalcules from his drawings published in 1684.

Figure 1.9 Swan-necked flask used by Louis Pasteur to demonstrate that microorganisms did not spontaneously form in sterilized broth.

argued that air-borne microbes inoculated the broth after entering through the straight neck but were trapped in the trough made by the bend in the swan neck. Therefore, microorganisms gave rise to microorganisms. These results also supported the cell theory proposed earlier in the century. By the end of the nineteenth century, the cell theory was widely accepted, knowledge about the biochemical basis of organisms was rapidly accumulating, and the biochemical analysis of cell components was well underway.

In 1896, Eduard Buchner converted sugar to ethyl alcohol using yeast extracts rather than intact yeast cells. These "ferments" (as he called the extracts) were eventually found to consist of enzymes ("enzyme" means "in yeast") or biological catalysts. Now these important chemical transformations could be conducted outside the cell.

The chemical structure of protein was of great interest; scientists believed that the proteins held the key to heredity and biochemical processes, and they were intrigued by the complexity and variety of proteins. In the 1920s and 1930s, the biochemical reactions of many important metabolic pathways were elucidated. By 1935 all twenty amino acids were isolated. Researchers were greatly aided by advances in instrumentation. In the late 1920s the ultracentrifuge was developed, and by the early 1940s ultracentrifugation methods had been perfected for separating cell organelles and macromolecules by size, shape, and density. In 1932, the German electrical engineer Ernst Ruska built the first electron microscope (400× magnification). Although primitive at first, the instrument was greatly improved in resolution and magnification, and in the 1940s and 1950s was used routinely to elucidate cell ultrastructure. Now researchers could integrate what they learned about cell organization with their knowledge of biochemical processes within the cell in order to create a more complete picture of the cell.

The study of the genetic nature of living organisms, which helped integrate prior knowledge of cells with what was known of the principles of heredity, is rooted in the mid-nineteenth century. In 1857, Austrian botanist and Augustinian monk, Gregor Mendel, began to experiment with peas grown in the monastery garden. He systematically cross-pollinated plants to examine the patterns of inheritance of seven traits (petal color, seed color, and seed texture among them) that existed in two alternative forms such as green vs. yellow seeds, smooth vs. wrinkled seeds, tall vs. dwarf plants. Mendel determined that each parent pea plant contributed to its progeny one unit of heredity for each trait, in either a dominant or recessive form. From his experiments with peas, Mendel formulated the principles of inheritance that later came to be known as Mendelian genetics. However, his findings, which he published in 1865, were not understood and remained in obscurity until 1900, when three botanists independently rediscovered his paper and recognized the significance of his work.

In 1869, Johann Friedrich Miescher, a Swiss biochemist, isolated from the nuclei of white blood cells a substance that he called "nuclein" and that is now called nucleic acid. At the time, nuclein was not known to be the hereditary material, but shortly afterwards an important series of observations led to the identification of chromosomes as the carriers of genetic material. In 1882, the German cytologist, Walter Flemming, described threadlike bodies that were visible during cell division and the equal distribution of their material to daughter cells. Although he did not know the significance of what he saw, the threadlike bodies were chromosomes (*chromosome* means "colored body," because chromosomes stain intensely; the term was coined in 1888 by W. Waldeyer), and sister **chromatids** were being equally distributed to the daughter cells. Shortly after the rediscovery of Mendel's experiments and conclusions, Walter Sutton, an American cytologist, determined in 1903 that chromosomes were the carriers of Mendel's units of heredity—or "genes," as the Danish botanist Wilhelm Johannsen named them in 1909. Sutton observed that during meiosis (the reductive division that produces **haploid** egg and sperm cells), the gametes produced receive only one chromosome of each morphologic type. Thus, he reasoned that meiosis was the mechanism by which the hereditary units are distributed. Through meiosis, many different combinations of traits can be obtained according to how the different chromosomes are oriented during distribution to the **germ cells**.

THE NATURE OF THE GENE

Between 1930 and 1952, researchers focused much effort on determining the relationship of genes and proteins. They established an important connection between the two when specific mutated genes were demonstrated to produce changes in certain enzymes. The first experiments were conducted by George W. Beadle and Boris Euphrussi with mutants in the fruit fly *Drosophila* and by

Beadle and Edward L. Tatum with the bread mold *Neurospora*. Charles Yanofsky and others conducted experiments with the bacterium *Escherichia coli*. Yanofsky demonstrated that there was a direct relationship, or colinearity, between the ordering of mutant sites within a gene and the linear sequence of amino acids in a protein. He and his colleagues determined the colinearity by introducing **mutations** along the tryptophan synthetase gene (*trpA*) and correlating the locations with changes in specific amino acids of the protein. Thus, genes determine the structure of proteins.

Several elegant experiments conducted in the 1940s increased our understanding of the chemical nature of the gene. Two strains of *Streptococcus pneumoniae*, a bacterium responsible for a form of pneumonia, were used: a **virulent** "smooth" strain (S) with a gelatinous coat, and a less virulent "rough" strain (R) that lacks the coat. In 1928, the British physician Fred Griffith had hypothesized that a "transforming principle" from the S strain of bacterial cells was responsible for the conversion of R bacterial colonies to the S form. The S strain was lethal to mice, but R cells were not (Figure 1.10). However, when Griffith injected mice with a mixture of heat-killed S and live R bacteria, the mice died, and the bacterial colonies that were isolated and plated were of the S type. In 1944, Avery, MacLeod, and McCarty extended Griffith's investigations to identify the transforming principle. They mixed the R strain with **DNA (deoxyribonucleic acid)** extracted from S-type bacteria and isolated S colonies after plating onto media (Figure 1.11). When added to the mixture, the enzyme deoxyribonuclease (DNase), which digests or breaks down DNA, abolished this transfor-

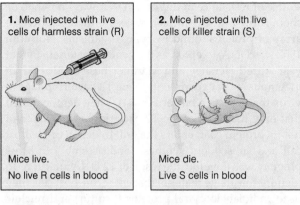
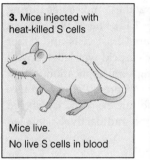
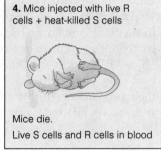

1. Mice injected with live cells of harmless strain (R)

Mice live.
No live R cells in blood

2. Mice injected with live cells of killer strain (S)

Mice die.
Live S cells in blood

3. Mice injected with heat-killed S cells

Mice live.
No live S cells in blood

4. Mice injected with live R cells + heat-killed S cells

Mice die.
Live S cells and R cells in blood

Figure 1.10 Results of Griffith's experiments with lethal and nonlethal strains of the bacterium *Streptococcus pneumoniae*.

mation of R to S colonies. Proteases (enzymes that digest proteins) did not prevent transformation. Thus, DNA was determined to be Griffith's transforming principle.

Unfortunately, such an unexpected conclusion often meets resistance, despite the evidence. What was the

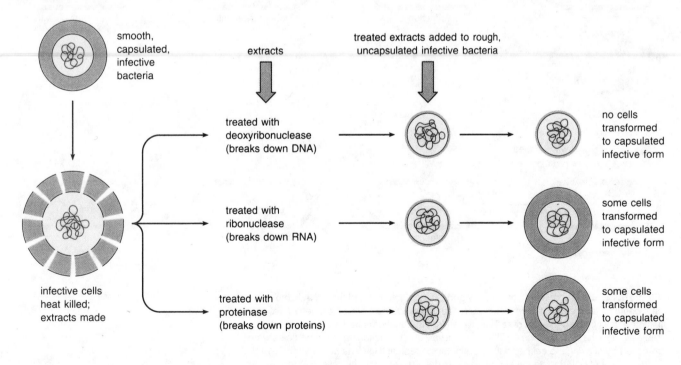

smooth, capsulated, infective bacteria

extracts

treated extracts added to rough, uncapsulated infective bacteria

treated with deoxyribonuclease (breaks down DNA)

no cells transformed to capsulated infective form

treated with ribonuclease (breaks down RNA)

some cells transformed to capsulated infective form

infective cells heat killed; extracts made

treated with proteinase (breaks down proteins)

some cells transformed to capsulated infective form

Figure 1.11 Experiments by Avery, MacLeod, and McCarty demonstrated that DNA was the "transforming principle" responsible for the transformation of *Streptococcus pneumoniae* from nonvirulent to virulent.

source of resistance in this case? DNA was known to be chemically a much simpler molecule than the diverse array of proteins found in living organisms. Not until 1952 was DNA accepted as the genetic material. In that year Alfred Hershey and Martha Chase conducted an ingenious set of experiments using T2 **bacteriophage** (a virus that infects a bacterial host). They demonstrated that only the DNA of the T2 bacteriophage DNA, and not its protein coat, enters the cell. Infection of the host cell and subsequent synthesis and assembly of virus progeny was therefore the result of T2 DNA infection. To identify the phage material that enters the host cell, they used radiolabeled viral proteins (^{35}sulfur) and **nucleic acids** (^{32}phosphorus) to follow these molecules during viral infection of *E. coli* (Figure 1.12). Phages were grown in medium containing bacterial host cells and either bacteriophage protein or nucleic acid was labeled with ^{35}S or ^{32}P, respectively. During viral replication within host cells, phage progeny incorporated either ^{35}S into their proteins or ^{32}P into the nucleic acid. Unlabeled cells were infected with the radioactively labeled

phages. After infection, cells were put in a blender to remove the attached phage particles. Analysis of the host cells and culture medium showed that the ^{32}P-labeled nucleic acid was inside the bacterial cells and was used to synthesize new viral progeny. The ^{35}S-labeled proteins remained in the medium. This was strong experimental evidence that DNA was the genetic material.

The final threshold to modern molecular biology was crossed when James Watson and Francis Crick elucidated the structure of DNA, in 1953. Rosalind Franklin, an expert x-ray crystallographer, and Maurice Wilkins (who shared the 1962 Nobel Prize in chemistry with Watson and Crick) had produced x-ray diffraction patterns of DNA, and Erwin Chargaff had established the DNA base ratios (**pyrimidine** and **purine** bases were in a ratio of 1:1; adenine equaled thymine and cytosine equaled guanine). By studying the x-rays with the knowledge of Chargaff's findings, Watson and Crick were able to develop a structural model of DNA. Their model described the width of the molecule, the number of bases per helical turn, and the spacing and location of the

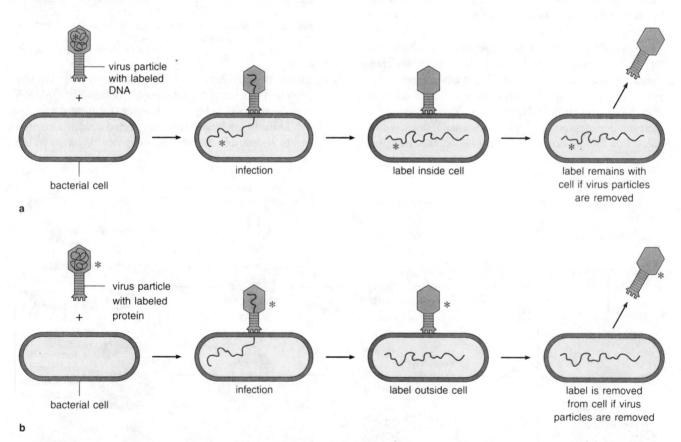

Figure 1.12 The experiments of Hershey and Chase demonstrated that DNA is the material that contains the genetic information required for the production of viral progeny. Growing infected cells in the presence of radioactive sulfur or phosphorus had the effect of radiolabeling the viral proteins or DNA, respectively. After infection, the viral particles were removed by shaking the cells and the location of the radioactivity was determined. (a) Radioactivity of DNA-labeled viral particles was detected in bacterial cells. (b) Radioactivity of protein-labeled viral particles was removed when bacterial cells were shaken.

bases, as well as how the molecule could replicate according to complementary base-pairing rules (Figure 1.13).

Experiments by many other scientists rapidly followed to determine how the information in the gene is decoded and how gene expression is regulated in both prokaryotic and eukaryotic organisms. The scientific foundation of modern biotechnology was provided not only by the discovery and understanding of nucleic acids, but also by the elucidation of the enzymatic tools required for DNA manipulation *in vitro*. By studying **DNA replication**, **DNA repair**, and the immunity of cells to viral infection, scientists purified the enzymes that later allowed them to synthesize DNA, to cut DNA into specific fragments, and to rejoin them in different combinations *in vitro*.

Many connections between structures and functions were made years later by insightful scientists who reexamined the conclusions of early experiments. The accumulated knowledge of cell structure, biochemistry, and heredity opened the door to modern molecular biology and biotechnology. Today scientists are applying what is known about cellular and molecular processes of living organisms to enhance our quality of life.

By the mid-1970s the scientific foundation was in place. However, many people were unprepared for the DNA revolution that followed and the important scientific achievements that were made during this time. The road was not a smooth one in the beginning.

Figure 1.13   James D. Watson (left) and Francis H. C. Crick, demonstrating their model of DNA structure deduced from x-ray diffraction data obtained by Wilkins and Franklin.

General Readings

C.W. Cowan and P.J. Watson, eds. 1992. *The Origins of Agriculture: An International Perspective.* Smithsonian Institution Press, Washington, D.C.

R.S. MacNeish. 1992. *The Origins of Agriculture and Settled Life.* University of Oklahoma Press, Norman, Oklahoma.

E. Oura, H. Suomalainen, and R. Viskari. 1982. Breadmaking. In A.H. Rose, ed. *Fermented Foods.* Academic Press, New York, pp. 87–146.

H. Wilson. 1988. *Egyptian Food and Drink.* Shire Egyptology, Bucks, United Kingdom.

D. Zohary and M. Hopf. 1993. *Domestication of Plants in the Old World,* 2nd ed. Clarendon Press, Oxford, England.

Additional Readings

O.T. Avery, C.M. MacLeod, and M. McCarty. 1944. Studies on the chemical nature of the substance inducing transformation of *Pneumococcal* types. *J. Exp. Med.* 79:137-158.

G.W. Beadle and E.L. Tatum. 1941. Genetic control of biochemical reactions in *Neurospora. Proc. Natl. Acad. Sci. USA.* 27:499-506.

B. Bracegirdle. 1989. Microscopy and comprehension: The development of understanding of the nature of the cell. *Trends Biochem. Sci.* 14:464-468.

A. Claude. 1975. The coming of age of the cell. *Science* 189:433-435.

J.S. Fruton. 1976. The emergence of biochemistry. *Science* 192:327-334.

F. Griffith. 1928. Significance of pneumococcal types. *J. Hygiene* 27:113-159.

J.R. Harlan. 1971. Agriculture origins: Centers and noncenters. *Science* 174:468-474.

J.R. Harlan and J.M.J. De Wet. 1973. On the quality of evidence for origin and dispersal of cultivated plants. *Curr. Archaeol.* 14:51-62.

J.R. Harlan and D. Zohary. 1966. Distribution of wild wheats and barleys. *Science* 153:1074-1080.

A.D. Hershey, A. and M. Chase. 1952. Independent functions of viral protein and nucleic acid in growth of bacteriophage. *J. Gen. Physiol.* 36:39-56.

G.C. Hillman and M.S. Davies. 1990. Measured domestication rates in wild wheats and barley under primitive cultivation and their archaeological implications. *J. World Prehist.* 4:157-222.

B.A. Law. 1982. Cheeses. In A.H. Rose, ed. *Fermented Foods.* Academic Press, New York.

M. McCarty and O.T. Avery. 1946. Studies on the chemical nature of the substance inducing transformation of pneumococcal types. II. Effect of deoxyribonuclease on the biological activity of the transforming substance. *J. Exp. Med.* 83:89-96.

J.H. Quastel. 1984. The development of biochemistry in the 20th century. *Can. J. Cell Biol.* 62:1103-1110.

D. Rindos. 1980. Symbiosis, instability, and the origins and spread of agriculture: A new model. *Curr. Anthropol.* 21:751-772.

W.S. Sutton. 1903. The chromosomes in heredity. *Biol. Bull.* 4:231-251.

A. Tannahill. 1973. *Food in History.* Stein and Day, New York.

E.R. Vedamuthu. 1982. Fermented Milks. In A.H. Rose, ed. *Fermented Foods.* Academic Press, New York, pp. 199–226.

J.D. Watson and F.H.C. Crick. 1953. A structure for deoxyribose nucleic acid. *Nature* 171:737-738.

J.D. Watson and F.H.C. Crick. 1953. General implications of the structure of deoxyribonucleic acid. *Nature* 171:964-967.

THE DNA REVOLUTION: PROMISE AND CONTROVERSY 2

Rapid advances in the field of **molecular biology** have brought revolutionary change in human and veterinary medicine, agronomy, animal science, environmental science, and even patent law. We can anticipate still more new treatments for diseases, improved crops and livestock that are disease resistant and more productive, and a cleaner, safer environment.

Techniques for manipulating DNA developed out of basic biological research directed toward understanding how genes are expressed and regulated in living organisms. These techniques were developed when the conventional genetic and biochemical methods that had been so successful for studying bacteria proved inadequate for studying complex **eukaryotic cells** and their **genomes**. Further progress in understanding the molecular genetics of higher organisms required new techniques.

Recombinant DNA technology revolutionized the study of molecular biology. Genes can be isolated, amplified, sequenced, and expressed in a different host cell. This process yields large quantities of a gene product. Specific DNA fragments are routinely cut from the **chromosome**, inserted into a vehicle called a **vector** DNA, and placed in a host cell. When the host cell divides, this recombinant DNA molecule replicates, producing a clone (a genetically identical copy). Recombinant DNA technology, also known as **gene cloning** and genetic engineering, is described in Chapter 4, Basic Principles of Recombinant DNA Technology.

THE EARLY YEARS OF MOLECULAR BIOLOGY

During the early years after the emergence of molecular biology as a discipline, in the early 1950s, basic research focused on elucidating the molecular basis of life. Questions focused on how a fertilized egg stores information for development, how the information is organized and retrieved when needed, and how genes control the metabolic and developmental characteristics of an organism. Answers to these profound questions provided the foundation for powerful new technologies, and the revelations of life processes gave rise to many new applications.

The field of molecular biology arose from important discoveries about the cell (recounted in Chapter 1). Genes were found to be made of DNA, and not protein as previously believed. Thus, the genetic information is encoded in the relatively simple chemical and molecular structure of DNA. The double-stranded DNA molecule employs an efficient way to produce duplicate copies of itself during cell division by unzipping the strands and making copies of the template (parental) strands (see Chapter 3, Gene Expression). These discoveries inspired researchers to seek the answers to questions about the nature of genetic information, how hereditary information in DNA is stored, and how it is retrieved and utilized.

In the 1950s and 1960s research focused on two main questions: (1) How does the sequence of bases in the gene relate to the sequence of amino acids in the protein? (2) What is the cell decoding process that produces a protein from the information encoded by the gene? After Watson and Crick elucidated the structure of DNA in 1953, progress was rapid. Genetic experiments in 1956 supported the hypothesis that the sequence of base pairs in DNA specified the information, or messages, of that DNA. Matthew Meselson and Frank Stahl, in 1957, demonstrated how DNA is replicated by separating the complementary strands of the double helix. In that same year, Francis Crick and colleagues hypothesized that the DNA bases specify the linear amino acid sequence of a protein, and that each amino acid is specified by a triplet of bases. There was speculation that, in eukaryotes, a **ribonucleic acid** (RNA) served as a messenger between DNA in the nucleus and **ribosomes** (where proteins are synthesized) in the cytoplasm. That is, genetic information must flow from DNA to a **messenger RNA** (mRNA) to protein. The existence of messenger RNA was confirmed in 1960.

In 1961, Marshall Nirenberg and J. H. Matthei made the first attempt to break the **genetic code** by using synthetic mRNA. For their *in vitro* experiment with *Escherichia coli*, they made a mixture of cell components—*E. coli* ribosomes, amino acid-**transfer RNAs** (tRNAs), GTP (guanosine triphosphate), magnesium, and translation factors for initiation, translocation, and release—but they omitted nucleic acids and the cell membrane. Thus, the test tube contained all the components necessary for protein synthesis except the mRNA (see Chapter 3, Gene Expression). They identified the **codons** (that is, the code words, or triplet ribonucleotides) for the individual amino acids incorporated into protein from simple synthetic messages. The first codon identified was UUU (called poly-U mRNA) for phenylalanine. Nirenberg and Severo Ochoa continued this work using artificial mRNAs such as AAA for lysine, CCC for proline, and GGG for glycine. By mixing nucleotides in specific ratios (for example, uracil and guanine in a ratio of 5U:1G), they determined the amino acids coded by the synthetic mRNA, but they could not determine the order of the bases. For example, although they established that leucine contains two U's and one G, they did not know that the order was UGU.

In 1964, Nirenberg and Phillip Leder developed an assay that enabled them to determine which codons in the mRNA specified which amino acids. They synthesized specific mRNA sequences comprising different codon combinations and used these sequences in a binding assay designed to identify the aminoacyl-tRNA (a transfer RNA with its bound amino acid) that was bound to a particular codon-ribosome complex. A synthetic mRNA was added to *E. coli* ribosomes and radiolabeled amino acids bound to their appropriate tRNAs. The tRNAs (adapter molecules that properly position amino acids during protein synthesis) bound to the appropriate codon of the mRNA-ribosome complex by complementary base pairing between the mRNA codon and the anticodon of the tRNA (Figure 2.1). Those aminoacyl tRNAs that bound to the ribosome complex were trapped on a filter. However, tRNAs with noncomplementary anticodons passed through, and no radioactivity was detected on the filter. The amino acid bound to the tRNA was identified and associated with a specific codon. Not long after the development of this binding assay, all the codons that encoded amino acids and stops (that is, stop codons that terminated protein synthesis) were elucidated. By 1966, the complete 64-triplet code used by all living organisms had been elucidated (Figure 2.2).

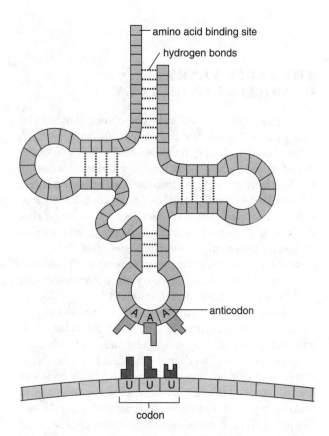

Figure 2.1 The folded shape of aminoacyl transfer RNA resembles a cloverleaf. The two functional regions are the three-base anticodon region that hydrogen bonds to the codon of the messenger RNA, and the end of the molecule that binds to the amino acid specified by the anticodon.

THE FIRST RECOMBINANT DNA EXPERIMENTS

During much of the 1950s and 1960s, bacteriophages were the focus of research in molecular biology. Much effort concentrated on viral genome replication and the generation of progeny. Researchers in the late 1960s,

First Base	Second Base				Third Base
	U	C	A	G	
U	phenylalanine	serine	tyrosine	cysteine	U
	phenylalanine	serine	tyrosine	cysteine	C
	leucine	serine	**stop**	**stop**	A
	leucine	serine	**stop**	**tryptophan**	G
C	leucine	proline	histidine	arginine	U
	leucine	proline	histidine	arginine	C
	leucine	proline	glutamine	arginine	A
	leucine	proline	glutamine	arginine	G
A	isoleucine	threonine	asparagine	serine	U
	isoleucine	threonine	asparagine	serine	C
	isoleucine	threonine	lysine	arginine	A
	(start) methionine	threonine	lysine	arginine	G
G	valine	alanine	aspartate	glycine	U
	valine	alanine	aspartate	glycine	C
	valine	alanine	glutamate	glycine	A
	valine	alanine	glutamate	glycine	G

Figure 2.2 The genetic code includes three stop codons and 61 amino acid-coding triplets.

although lacking the requisite methods, were nevertheless discussing the mixing of unrelated DNAs and the moving of hybrid molecules from one species to another. By the 1970s, they were using **plasmids** (autonomously replicating, extrachromosomal circular DNAs found in bacteria) and viruses as vehicles to transfer DNA into cells. Numerous publications reported successful results of experiments using transformation or **transduction** of bacterial cells with DNA. Some enzymes to manipulate DNA had been isolated; for example, **DNA polymerase**, **DNA ligase**, and **restriction endonucleases**. The implications of recombinant DNA technology were obvious to those scientists who had knowledge of the work: This technology showed promise not only for studying gene structure and regulation but also for producing mammalian proteins in bacterial cells for commercial use. Many scientists at that time also saw the potential for human gene therapy once methods had been developed for transferring DNA into mammalian cells.

The experiments of Paul Berg, Herbert Boyer, Stanley Cohen, Janet Mertz, and Ronald Davis and their colleagues, published over 20 years ago, ushered in the era of modern biotechnology and genetic engineering. Their findings dramatically changed the way molecular biology and recombinant DNA experiments were conducted. Recombinant DNA could be produced by recombining two DNA molecules from different sources and transferring the recombinant molecule into a host cell.

Cohen extended his research to the replication of *E. coli* plasmids. In late 1971, a medical student in Cohen's laboratory found that bacteria treated with calcium chloride took up plasmid DNA harboring antibiotic resistance genes. The progeny plasmids were clones of the plasmids used in the transformation experiments. Cohen then was able to use plasmids to transfer genes into host bacterial cells. He knew from discoveries in the mid-1950s that bacteria could become resistant to antibiotic drugs from plasmid-encoded resistance genes and that antibiotic resistance could be transferred to related bacteria. Cohen studied how plasmids reproduce and are transferred to other host cells. He cut plasmids into fragments to identify essential DNA regions for replication and antibiotic resistance.

Two significant studies hinted at the exciting future potential of recombinant DNA technology. In the first study, at Stanford University in 1971, Paul Berg and his postdoctoral fellows, David Jackson and Robert Symons, were able to construct circular dimers of SV40 DNA (simian virus 40 DNA) and insert the *E. coli* galactose **operon** and a few genes from lambda phage (a type of bacteriophage) for replication in *E. coli*. Their method involved multiple steps: They linearized SV40 DNA with *Eco*RI (a restriction enzyme—restriction endonuclease—that cuts double-stranded DNA at a specific sequence within the molecule), modified the DNA ends with lambda exonuclease (an enzyme that removes nucleotides from the free ends of DNA), added a string of adenine nucleotides to the end of one DNA molecule and thymine nucleotides to the other DNA with the enzyme terminal transferase (that is, added complementary A's and T's for annealing), filled gaps generated in the DNA with DNA polymerase, and finally ligated the two types of DNAs with DNA ligase (Figure 2.3). This work demonstrated that two unrelated DNA molecules could be joined to one another. Berg and his colleagues speculated that mammalian cells could be transformed with these recombinant DNAs and then tested for expression of foreign genes.

In the other study, published in 1972, Janet Mertz and Ronald Davis, at Stanford University, demonstrated that *Eco*RI in one step generated 3' **cohesive** (or staggered) **ends** rather than blunt ends (Figure 2.4). The cohesive ends of linearized SV40, cut with *Eco*RI, spontaneously annealed through hydrogen bonding and

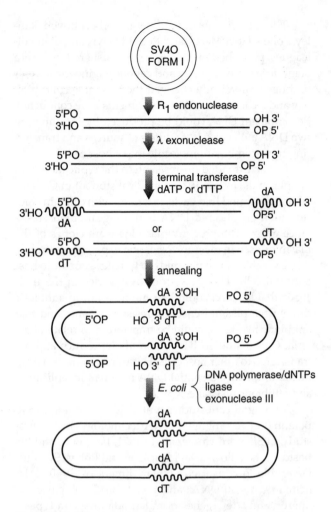

Figure 2.3 The method used by Berg and his colleagues to produce covalently closed, circular SV40 dimer molecules.

Information about how to cut plasmid DNA at a specific site and insert a new fragment was presented in November 1972 at a meeting in Honolulu, Hawaii. Mertz and Davis had just published their work showing that *Eco*RI produces cohesive ends, and Cohen presented his method of transferring plasmid DNA to a bacterial host. At the same meeting, Herbert Boyer, of the University of California, San Francisco, presented a paper about the purification of *Eco*RI from *Escherichia coli*. Boyer's restriction enzyme research developed from his studies of the molecular basis of conjugation using bacterial plasmids and host-controlled restriction/modification of bacteriophage. Boyer and his colleagues hoped to use restriction enzymes to generate specific plasmid fragments for experimental purposes. In Hawaii, Cohen and Boyer discussed a collaboration in which *Eco*RI would be used to generate DNA fragments for insertion into Cohen's plasmids. These recombinant molecules would be used to transform an *E. coli* host. At the time no one knew if molecules engineered in the laboratory could be propagated in living organisms.

After Hawaii, Boyer went to Cold Spring Harbor Laboratories, in New York, and learned how molecular biologists were using **gel electrophoresis** and **ethidium bromide** staining to visualize DNA cut with *Eco*RI (see Chapter 4, Basic Principles of Recombinant DNA Technology). Boyer immediately realized that gel electrophoresis was a powerful analytical tool for monitoring DNA digestions with restriction enzymes.

Herbert Boyer, Robert Helling, Stanley Cohen, and Annie Chang began a collaborative research project to ligate specific DNA into a vector and transform an *E. coli* host (results were published in 1973). Each investigator pursued a complementary area of research to provide valuable information to the collaborative effort. Boyer was studying the restriction and modification (methylation) process by which bacteria protect themselves from foreign DNA while protecting their own DNA from restriction. Helling, at the University of Michigan,

were covalently bonded by DNA ligase. Mertz and Davis had found an easy method of *in vitro* recombination for generating specifically oriented hybrid DNAs and allowing the original DNA molecules to be retrieved by cutting with *Eco*RI.

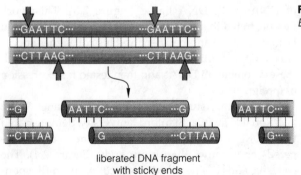

liberated DNA fragment
with sticky ends

Figure 2.4 The generation of cohesive ends by the restriction enzyme *Eco*RI.

was investigating the process of transduction by which bacteriophages (bacterial viruses) transfer genes between bacteria. At Stanford University, Cohen and his technician Chang were examining the process of plasmid transfer and replication as well as how bacteria establish drug resistance.

This collaborative group used a small bacterial plasmid, pSC101 (with a tetracycline resistance gene and one *Eco*RI restriction site), to make other useful cloning vectors (for example, pSC102, pSC105, pSC109) with different antibiotic-resistance genes. By March 1973, the cloning experiment showed much promise. *Eco*RI produced fragments of a large plasmid and linearized the cloning vector (harboring tetracycline and kanamycin antibiotic-resistance genes). The cloning vector and plasmid fragments were mixed together. The enzyme DNA ligase was added to covalently join the ends (Figure 2.5), and bacteria were transformed with the DNA solution. The bacteria were plated onto solidified agar medium supplemented with tetracycline and kanamycin antibiotics to select for those bacteria transformed with recombinant DNA molecules (also described in Chapter 4). The experiments were successful: Recombinant DNA could be isolated from the bacteria hosts. This method of cutting and rejoining DNAs to produce a recombinant DNA that could replicate in a host cell became known as DNA cloning. The research culminated in a paper, "Construction of Biologically Functional Bacterial Plasmids *in Vitro*," published in the *Proceedings of the National Academy of Sciences*. This publication formed the basis for a patent ("Process for Reproducing Biologically Functional Molecular Chimeras"), of which Stanford University was the beneficiary. The method provided the groundwork for much of modern biotechnology. The practical applications of the method and its implications were immediately apparent.

Scientists thought that restriction and modification enzymes in foreign bacterial hosts might be barriers to transformation of these cells. However, Cohen and Chang discovered that they could transfer bacterial DNA to unrelated bacteria. In a 1974 issue of the *Proceedings of the National Academy of Sciences*, they demonstrated that DNA from *Salmonella* and *Staphylococcus* could be transferred and expressed in *E. coli*. By August 1973, Cohen, Boyer, Berg, Helling, Chang, Howard Goodman (Stanford University) and John Morrow, Berg's graduate student, successfully transferred RNA genes from the African clawed frog, *Xenopus laevis*, to *E. coli*. The eukaryotic genes were transcribed into RNA.

Figure 2.5 The production of a recombinant DNA molecule by restriction endonuclease digestion and the ligation of complementary ends.

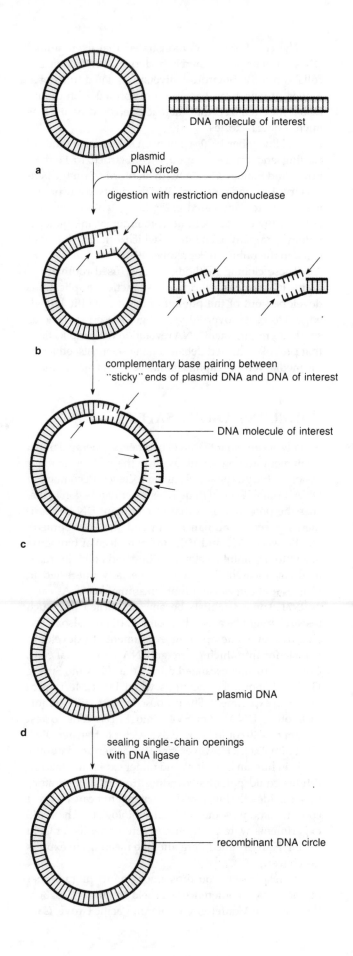

a plasmid DNA circle

DNA molecule of interest

digestion with restriction endonuclease

b

complementary base pairing between "sticky" ends of plasmid DNA and DNA of interest

c DNA molecule of interest

d plasmid DNA

sealing single-chain openings with DNA ligase

recombinant DNA circle

The collaborators demonstrated that even animal DNA could be propagated in bacteria. When Cohen and collaborators at Stanford University isolated functional mouse protein from bacteria transformed with mouse DNA, they realized that the bacteria could be used as microscopic factories.

In November 1980, a patent on the basic methods of cloning and transformation was awarded to Herbert Boyer and Stanley Cohen, with Stanford again the beneficiary. A second patent granted the rights to any organisms that are engineered using the patented methods. Herbert Boyer also became cofounder of the biotechnology company Genentech, which offered shares of stock to the public in September 1980.

These cloning methods are now used routinely in many laboratories worldwide—practical applications developed out of the pursuit of basic scientific knowledge. However, over 20 years ago, these experiments sparked a recombinant DNA revolution with implications that provoked heated debate among scientists, ethicists, the media, venture capitalists, lawyers, and many others.

CONCERNS ABOUT SAFETY

Initially, recombinant DNA experiments generated much excitement among scientists since the method made it easier to study a cell's molecular processes. Recombinant DNA became a social issue when scientists began to discuss the potential risks that recombinant DNA experiments posed to human health and the environment.

Between 1971 and 1973, many molecular biologists began to use animal cells in culture and animal viruses in their research. They were especially interested in viruses that can cause tumors in animal hosts. As early as 1971, some scientists became concerned that such research might be a health hazard. Paul Berg also raised concerns by contemplating experiments to develop a vehicle for introducing foreign DNA into animal cells. Concerns further escalated during the 1971 Cold Spring Harbor Laboratory meetings when Janet Mertz proposed an experiment: She proposed to introduce tumor-promoting DNA from SV40 into *E. coli* cells to test whether SV40 could be used as a vector to transfer DNA into animal cells. However, since the bacterium naturally exists in human intestines and under experimental conditions could potentially induce tumors in laboratory rats, the idea that it would be engineered caused great concern among some molecular biologists. The SV40 experiments were postponed when a scientist at Cold Spring Harbor called Berg after the meetings to express his concern.

In response to concerns among a growing number of scientists, a conference was held in January 1973, at the Asilomar Conference Center in Pacific Grove, California, to specifically address the biohazards of working with animal viruses, and particularly those that promote tumors. The conference resulted in a book, published by Cold Spring Harbor Laboratory, entitled *Biohazards in Biological Research*, but it attracted very little attention. The presentations at the Gordon Conference on Nucleic Acids, in June 1973, focused on the use of restriction enzymes for recombining DNAs from unrelated species. The end-of-meeting discussions were primarily responsible for sparking a national DNA debate that eventually became international. Conference participants decided that the concerns should be announced publicly; accordingly Dr. Maxine Singer of the National Institutes of Health (NIH) and Dr. Dieter Soll of Yale University stated the concerns in a letter to the President of the National Academy of Sciences and to the journal *Science*. They requested that the academy establish a committee to investigate the potential consequences of recombinant DNA and to formulate laboratory guidelines. The academy established the Committee on Recombinant DNA Molecules, and in July 1974, its members—Paul Berg and other well-respected scientists—published a letter in *Science* and in the British scientific journal *Nature*. In this letter, entitled "Potential Biohazards of Recombinant DNA Molecules," they urged that an international meeting be convened to discuss guidelines, that potential biohazards be evaluated, and that experiments be terminated until biohazards had been evaluated. In short, they proposed a moratorium.

The published letters made headlines in U.S. newspapers, and focused discussions on biological containment and the possibility of a voluntary moratorium. The NIH responded in October 1974 by establishing the Recombinant DNA Molecule Program Advisory Committee (RAC). The mandate of this committee was to evaluate the risks in using recombinant DNA, to establish guidelines for recombinant DNA research, and to develop methods for preventing the accidental release of molecules to humans and the environment. The committee's guidelines are described after the following account of the second Asilomar conference.

The Asilomar Conference

A second international Asilomar Conference on Recombinant DNA Molecules was held February 24–27, 1975 (see Watson and Tooze, *The DNA Story*, Document 2.2, pp. 41–42 for the conference program). The conference was sponsored by the National Academy of Sciences, funded by the NIH and the National Science Foundation, and organized by Paul Berg. It was attended by over a hundred molecular biologists, reporters, and an assortment of lawyers and other interested people, meeting to discuss the progress made in recombinant DNA technology and to evaluate the risks and potential dangers posed by recombinant DNA molecules. The recommendations that emerged from the conference were to be incorporated into the NIH guidelines. Participants voted to impose guidelines that would help ensure the safety of those working with potentially hazardous recombinant molecules. They also proposed biological containment through the use of disabled host-vector systems that would require special laboratory conditions for their survival. This measure would eliminate any chance of release outside the laboratory.

Members of the organizing committee spent the last night of the conference drafting guidelines that included laboratory safety or containment measures for a variety of recombinant DNAs, including host cells and vectors (Table 2.1). The committee had hoped for a vote supporting the guidelines before the conference

ended, but agreement was difficult to reach. Those who believed there was very little risk strongly opposed the guidelines and moratorium, while others believed that no one could predict with certainty what the biohazards might be. However, the majority of the participants did agree that there was some cause for concern and that there might be some risk; thus, guidelines were approved. The committee's report was quickly drafted, and the guidelines appeared in *Science* on June 6, 1975.

Most scientists believed that they would be able to continue their recombinant DNA research without further delay, and were unprepared for the debate that ensued. Their research became a focal point for controversy. Research scientists, public interest groups, the U.S. Congress, corporations, and private citizens became engaged in a heated debate that would last almost four

Table 2.1   Classification of Experiments in the Provisional Statement of the Asilomar Conference

Containment Level	Techniques/Practices	Types of Experiments
Low containment	Good microbiological techniques, encompassing use of lab coats, use of mechanical pipettes, a prohibition on eating in the lab, and use of biological safety cabinets for procedures that generate large aerosols.	Experiments involving organisms that normally exchange genetic information. Most experiments involving DNA from prokaryotes, lower eukaryotes, plants, invertebrates, and cold-blooded vertebrates.
Moderate containment	Physical containment required for handling moderate-risk oncogenic viruses, encompassing use of gloves and lab coats, use of biological safety cabinets for transfer operations, vacuum lines protected by filters, negative pressure that is maintained in limited-access laboratories.	Experiments with bacterial genes that affect pathogenicity or antibiotic resistance. Experiments with viral DNA involving the linkage of viral genomes in whole or in part to prokaryotic vectors and their introduction into prokaryotic cells (if "disarmed" vectors are used). Experiments with segments of oncogenic viral DNA if these segments are nontransforming. Experiments with segments of nononcogenic viruses. Experiments with DNA from warm-blooded vertebrates. Experiments with any DNA and animal vectors.
High containment	Use of facilities that are isolated from other areas by air locks, clothing exchanges, and shower rooms and that have treatment systems to inactivate or remove biological agents that may be contaminant in exhaust air and liquid and solid wastes. All agents are handled in biological safety cabinets. Personnel wear protective clothing and shower on leaving. Facility is maintained under negative air pressure.	Experiments with high-risk viruses. Experiments linking eukaryotic or prokaryotic DNA to prokaryotic vectors when the resulting organism might express a toxic or pharmacologically active product.

Originally adapted from "Provisional Statement of the Conference and Proceedings," RDHC, Feb. 27, 1975 and used in S. Wright, *Molecular Politics*, University of Chicago Press, 1994. Used by permission of the publisher.

more years. Questions about risk could not be clearly answered since data were unavailable. Molecular biologists remained in disagreement among themselves about the safety of recombinant DNA technology and the restrictions to be placed on experimentation.

The DNA debates became an international issue. The government of Great Britain, for example, became involved and in the summer of 1974 established The Working Party on the Experimental Manipulation of the Genetic Composition of Microorganisms. Most likely, a fatal accident with the smallpox virus at the London School of Hygiene and Tropical Medicine in 1973 prompted this action.

DRAFTING THE NIH GUIDELINES

The RAC, established by the NIH just prior to the second Asilomar conference, had three tasks: (1) Evaluate the potential biological and ecological hazards of DNA recombination, (2) Develop procedures to minimize the spread of recombinant molecules in animal populations, and (3) Devise guidelines for researchers using potentially dangerous recombinant molecules and organisms. In the spring and summer of 1975, RAC drafted guidelines incorporating some of the recommendations made at Asilomar. RAC classified the many types of recombinant DNA experiments and the DNA of pathogenic organisms according to potential hazard. Both physical containment guidelines (P1–P4) and biological containment guidelines (EK1–EK3) were established on the basis of hypothetical hazards (Table 2.2).

The NIH guidelines recognized four levels of risk, from minimal to high, and devised four corresponding containment levels, P1 to P4. P1 laboratories were to take simple precautions such as prohibiting workers from pipeting by mouth, while P4, the highest level of containment, was reserved for those experiments that used the most lethal and pathogenic microorganisms such as anthrax bacilli and smallpox viruses, as well as tumor viruses and toxin genes. Such work required a laboratory with negative-pressure air locks and the use of extreme caution in handling biological material. All experiments were to be conducted in laminar-flow hoods with filtered or incinerated exhaust air.

Biological containment addressed the use of host cells and vectors that could not reproduce outside the laboratory if they escaped. Organisms would be disabled so that they would die outside the specialized laboratory environment. An example that occurs in nature is the bacterium *E. coli* strain EK2, which cannot survive outside the human gut because mutations prevent the cells from synthesizing thymine, necessary for DNA replication. This mutation would be lethal since the cell could not replicate its DNA unless provided with thymine. Modified vectors also were to be used for recombinant DNA research so that the DNA could not be transferred accidentally from one host cell to other cells.

Levels of biological containment ranged from the lowest, EK1, to the highest, EK3. Initially, only EK1 host-vector systems were available for research. An EK2 host-vector system was not developed and approved until April 1976. Although special containment laboratories had been used in the past by the military for biological warfare investigations or for the handling of life-threatening pathogens, the NIH and RAC had not approved these facilities for P4 containment. Thus, no P4 facilities were available.

These NIH guidelines were more stringent than those proposed at Asilomar. They were completed in December 1975 and released publicly on June 23, 1976 (see *Federal Register* vol. 41 No. 131, July 7, 1976 and Appendix A in this book), one and a half years after Asilomar. At the time of their release, EK3 host-vector systems

---

Table 2.2 Biological and Physical Containment Measures for Laboratory Safety

Biological Containment	Physical Containment
EK1 Standard laboratory *E. coli* K12 host strain and compatible plasmid and virus vectors	P1 Standard microbiological procedures
EK2 Genetically altered, disabled *E. coli* that cannot survive outside controlled laboratory conditions	P2 P1 standards plus KEEP OUT notice displayed at the laboratory entrance
EK3 An extremely weakened *E. coli* strain that is 10^6 times less likely to escape from the laboratory	P3 P2 standards plus negative-pressurized laboratory or the use of negative-pressure cabinets
	P4 Used for extremely dangerous host-vector systems; P3 plus investigators must change clothes and shower before exiting the facility; air locks must be used

Note: Both physical and biological containment would be matched to provide maximum laboratory safety. For example, an experiment might be classified as P2 + EK1 or P3 + EK2. Therefore, if mammalian DNA were used, especially if the investigator was working with primate DNA, a P3 + EK3 or P4 + EK2 level of containment might be required.

Adapted from an article by Nicholas Wade, *Science* vol. 190 (November 21, 1975).

had not yet been developed and P4 levels of physical containment existed only in special military facilities or at medical laboratories where lethal pathogens were studied. Therefore no molecular biologists had access to these facilities. Since the NIH guidelines were relaxed in the years to come, an EK3 system never was developed to meet the most stringent requirements.

Because of speculation that the closer the phylogenetic relationship between humans and the species used in recombinant experiments, the greater the risk to humans, experiments that required the use of DNA from mammals or viruses were terminated and DNA samples were destroyed. Unfortunately, this measure precluded using the new technology for research on cancer or using bacteria as hosts for human genes to isolate important human proteins such as insulin and growth hormone. Only the DNA of fish, amphibians, or invertebrates could be used in experiments.

However, the NIH was enforcing the guidelines only at institutions and laboratories receiving federal funding. Compliance by facilities in the private sector, such as industrial laboratories, was voluntary. Individuals working in industry would have proprietary rights or patentable information to protect, and disclosure of experiments might preclude protection. Many research scientists complied voluntarily, and destroyed recombinant DNA molecules when experiments did not meet NIH guidelines. In some cities, additional guidelines were adopted, such as those specified by the Cambridge Experimentation Review Board. Anyone conducting recombinant DNA research in Cambridge, Massachusetts, had to follow both the NIH and board restrictions. Federal legislation was even proposed to regulate recombinant DNA research. Some scientists expressed fears that legislation would restrict intellectual freedom and impede important scientific progress.

Revision of the NIH Guidelines

In 1977, RAC began to reevaluate the NIH guidelines for revision. As part of the review process, experts assessed the biology and ecology of *E. coli*, the primary host organism for recombinant DNA technology. Reviewers concluded that the laboratory strain *E. coli* K12 is too defective genetically to survive outside the laboratory. In response to new information and additional discussion, RAC proposed revised NIH guidelines that reduced the stringency of some of the containment specifications for many types of experiments. The proposed guidelines were published in the *Federal Register* for public comment on September 27, 1977, but were further revised before the final version was published on December 22, 1978 and adopted on January 2, 1979.

In 1978, at the same time that research in laboratories across the nation was terminated or suspended, Genentech announced that *E. coli* could synthesize the

mammalian hormone somatostatin. This was the first time a functional mammalian protein had been produced by recombinant DNA technology from a gene that had been synthesized chemically.

In the 1979 NIH guidelines, only a few types of experiments required P4-level containment, and *E. coli* K12 was subjected to only P1-level physical containment. Additional revised guidelines published in the *Federal Register* on January 29, 1980 further relaxed containment requirements. Other changes included granting experimental review to institutional biosafety committees established by organizations receiving federal funds and others voluntarily observing guidelines. These committees would be informed of *E. coli* K12 experiments either before or after initiation of experiments. This change represented a significant relaxation of regulations. In addition, although industrial compliance was voluntary, a code of practice that protected proprietary rights was included. In November 1980, the RAC recommended further that the review of recombinant DNA experiments covered by the NIH guidelines be left entirely to local institutional biosafety committees (Appendix B).

CURRENT AND FUTURE CONCERNS

As experiments demonstrated that recombinant DNA and host cells did not pose any risk to human health or the environment, concerns about the safety of the technology subsided. Concerns today focus on applications and ethical implications. For example, in the medical field, gene therapy experiments raise the question of eugenics (artificial human selection) as well as testing for the presence of diseases with no cure (such as Huntington's chorea). In agriculture, there is some concern that the spread of genes from transgenic crop plants to weeds or insect pests may cause problems (such as herbicide-resistant weeds), or that engineered microorganisms released into the environment may adversely affect the environment.

One of the first occasions for public objection to the release of an organism into the environment arose from the intended use of "ice-minus" bacteria, an engineered bacterium, *Pseudomonas syringae*, designed to protect crop plants against frost. In August 1984, social activists filed suit to block an attempt to use "ice-minus" bacteria and a federal judge issued an injunction against the tests. In April 1987, after a costly delay, testing was able to begin on strawberry plants. Initially those spraying the plants with bacteria took the precaution of wearing protective clothing (Figure 2.6). These engineered frost-resistant bacteria proved to be quite safe. However, naturally occurring "ice-minus" bacteria have recently been isolated and are being marketed instead of the engi-

Figure 2.6 A "moon suit" was used to protect the sprayer against unknown biohazards when applying "ice-minus" bacteria to strawberry plants.

neered version. Field testing of other genetically engineered microorganisms and plants followed. For example, agronomically important virus-resistant crop plants have been engineered. These plants harbor a specific viral coat protein gene that enables the plants to synthesize a protein that confers resistance to infection by the virus. Virus-resistant plants were first grown in the greenhouse and then transferred to the field for trials beginning in 1988.

Progress continues in many areas. Today, as a result of recombinant DNA technology, hundreds of genetically engineered disease-, pest-, and herbicide-resistant plants are awaiting field testing; rapid progress is being made in identifying genes involved in human disease; and new medical treatments are being developed. The legal and regulatory maze has impeded progress, and industry must still grapple with cumbersome regulatory committees (see Chapter 12, Regulation, Patents, and Society), but the positive impacts on society are slowly being realized.

General Readings

J.D. Watson and J. Tooze. 1981. *The DNA Story.* W.H. Freeman and Company, San Francisco.

S. Wright. 1994. *Molecular Politics: Developing American and British Regulatory Policy for Genetic Engineering, 1972-1982.* The University of Chicago Press, Chicago.

Additional Readings

A.C.Y. Chang and S.N. Cohen. 1974. Genome construction between bacterial species *in vitro*: Replication and expression of *Staphylococcus* plasmid genes in *Escherichia coli*. *Proc. Natl. Acad. Sci. USA* 71:1030-1034.

S.N. Cohen, A.C.Y. Chang, H.W. Boyer, and R.B. Helling. 1973. Construction of biologically functional bacterial plasmids *in vitro*. *Proc. Natl. Acad. Sci. USA* 70:3240-3244.

J. Hedgpeth, H.M Goodman, and H.W. Boyer. 1972. DNA nucleotide sequence restricted by the RI endonuclease. *Proc. Natl. Acad. Sci. USA* 69:3448-3452.

D.A. Jackson, R.H. Symons, and P. Berg. 1972. Biochemical method for inserting new genetic information into DNA of Simian Virus 40: Circular SV40 DNA molecules containing lambda phage genes and the galactose operon of *Escherichia coli*. *Proc. Natl. Acad. Sci. USA* 69:2904-2909.

J.L. Marx. 1976. Molecular cloning: Powerful tool for studying genes. *Science* 191:1160-1162.

J.E. Mertz and R.W. Davis. 1972. Cleavage of DNA by RI restriction endonuclease generates cohesive ends. *Proc. Natl. Acad. Sci. USA* 69:3370-3374.

J.F. Morrow, S.N. Cohen, A.C.Y. Chang, H.W. Boyer, H.M. Goodman, and R.B. Helling. 1974. Replication and transcription of eukaryotic DNA in *Escherichia coli*. *Proc. Natl. Acad. Sci. USA* 71:1743-1747.

C. Norman. 1976. Genetic manipulation: Guidelines issued. *Nature* 262:2-4.

H. Schmeck, Jr. 1986. Recombinant DNA controversy: The right to know—and to worry. In R.A Zalinskas and B.K. Zimmerman, eds. *The Gene-Splicing Wars: Reflections on the Recombinant DNA Controversy*. Macmillan, New York. pp 93-106.

M. Singer and D. Soll. 1973. Guidelines for DNA hybrid molecules. *Science* 181:1114.

K. Struhl, J.R. Cameron, and R.W. Davis. 1976. Functional genetic expression of eukaryotic DNA in *Escherichia coli*. *Proc. Natl. Acad. Sci. USA* 73:1471-1475.

B.K. Zimmerman. 1986. Asilomar and the formation of public policy. In R.A Zalinskas and B.K. Zimmerman, eds. *The Gene-Splicing Wars: Reflections on the Recombinant DNA Controversy*. Macmillan, New York. pp 3-10.

N.D. Zinder. 1986. A personal view of the media's role in the recombinant DNA war. In R.A Zalinskas and B.K. Zimmerman, eds. *The Gene-Splicing Wars: Reflections on the Recombinant DNA Controversy*. Macmillan, New York. pp 109-118.

To manipulate living cells and organisms, scientists must have a comprehensive understanding of gene structure, gene function, and the regulation of gene expression. In addition to knowing the cell- and tissue-specific genes that encode important traits, they must have also identified and studied the regulatory sequences that control gene activity. This chapter presents a brief review of the basic principles of molecular biology, paying particular attention to DNA structure, the replication of DNA, the transmission of information from DNA for protein synthesis, and the regulation of product synthesis. For a more detailed review of genes and their regulation, consult a comprehensive molecular biology textbook; a few are listed in the Readings at the end of the chapter.

DNA STRUCTURE

Deoxyribonucleic acid (DNA) is a long polymer consisting of repeating units called deoxyribonucleotides. A deoxyribonucleotide has three components: (1) a pentose sugar or deoxyribose, (2) a phosphate group, and (3) one of four nitrogen-containing bases. Figure 3.1 shows the four bases: adenine (A), guanine (G), thymine (T), and cytosine (C). DNA stores its genetic information in the four nitrogen-containing bases; adenine and guanine are double-ring structures called purines, while thymine and cytosine are single-ring structures called pyrimidines. The bases project inward from the sugar–phosphate backbone, and hydrogen bonding between opposite bases (that is, one on each DNA strand) holds the two strands of the DNA molecule together. The deoxyribonucleotides are linked by 5′–3′ phosphodiester bonds; that is, the phosphate group attached to the number-5 carbon atom (5′ carbon) of the deoxyribose of one deoxyribonucleotide is connected to the number-3 carbon (3′ carbon) of the adjacent deoxyribonucleotide **nucleotide** (Figure 3.2). Alternating pentose sugar and phosphate make up the backbone of DNA. The end of the nucleic acid strand that terminates with the 5′ carbon atom is called the 5′ end and the end with the 3′ carbon atom is the 3′ end.

Complementary Base Pairing

Two DNA strands base pair with each other to form a double-stranded molecule, or double helix. The **x-ray diffraction pattern** of B-DNA (the most common conformation) shows a helical configuration, with both strands of the molecule winding around a common central axis to form a spiral (like a spiral staircase). Two

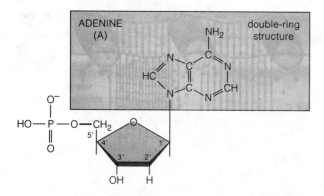

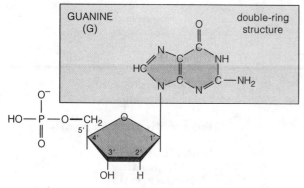

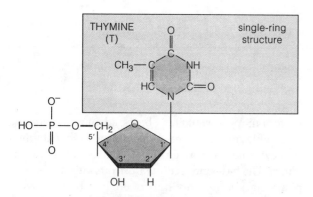

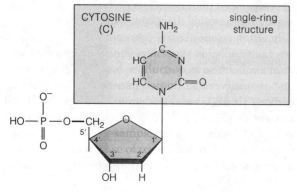

Figure 3.1 The four nucleotides of DNA. The numbers 1'–5' refer to the carbon atoms on the pentose sugar to which other groups are attached.

grooves are formed by this conformation: a minor groove, and a wider, major groove (Figure 3.3). In this most common DNA conformation, a deoxyribonucleotide is located every 3.4 Å, with approximately 10.5

nucleotides for every helical turn of 34 Å. Each B-DNA double helix is 20 Å in diameter. Other isomers of DNA also exist; however, only B- and Z-DNAs exist in cells. Table 3.1 shows the structural differences in the DNA types.

Nucleotides base pair with one another through hydrogen bonding of the bases (Figure 3.2). To maintain a constant helix diameter, a purine always pairs with a pyrimidine; adenine always base pairs with thymine through two hydrogen bonds, and cytosine always base pairs with guanine through three hydrogen bonds. Because each strand of the helix is a complement of the other, such base pairing is called complementary. For hydrogen bonding of the bases, the strands must be antiparallel to one another so that one strand goes from 5' to 3' and the other 3' to 5'. Thus, the DNA molecule has polarity: the ends differ from one another.

The two strands do not spontaneously separate under physiological conditions because the many hydrogen bonds keep the base pairs together. However, high temperatures (near boiling) or pH extremes (pH < 3 or pH > 10) can break hydrogen bonds so that the two strands separate, or are denatured. If the temperature is gradually lowered (for example, 65°C), the complementary strands can recombine or reanneal.

DNA REPLICATION

When a cell divides to yield two daughter cells, the genetic material must be reproduced accurately, or replicated, so that each daughter cell contains identical DNA copies. Accuracy in replicating is essential, since the DNA stores the genetic information that the cell uses. During mitosis, the two DNA molecules move to opposite sides of the dividing cell, and cell division produces two daughter cells, each containing identical double-stranded DNA molecules.

To replicate, the two strands of the double helix must separate. Once they are separated, the base-pairing rules allow each single-stranded molecule to act, during replication, as a template for the formation of a new complementary strand. Each of the two identical double-stranded molecules thus formed consists of one original, parent strand and one new, daughter strand. Because half of the new molecule is original material, this mode of replication is called semiconservative (Figure 3.4).

Enzymes catalyze the successive stages of the replication process. First, two enzymes, **DNA helicase** and **DNA topoisomerases**, initiate the separation, or "unzipping," of the two strands of the DNA molecule. The enzymes bind to a specific site on the molecule called the origin of replication. The two separated strands form a **replication fork** (Figure 3.5). DNA helicase unwinds the DNA, and topoisomerases relax the supercoils produced by the unwinding of the double helix. Single-strand binding proteins attach to the exposed bases of each separated strand to stabilize them. The enzyme

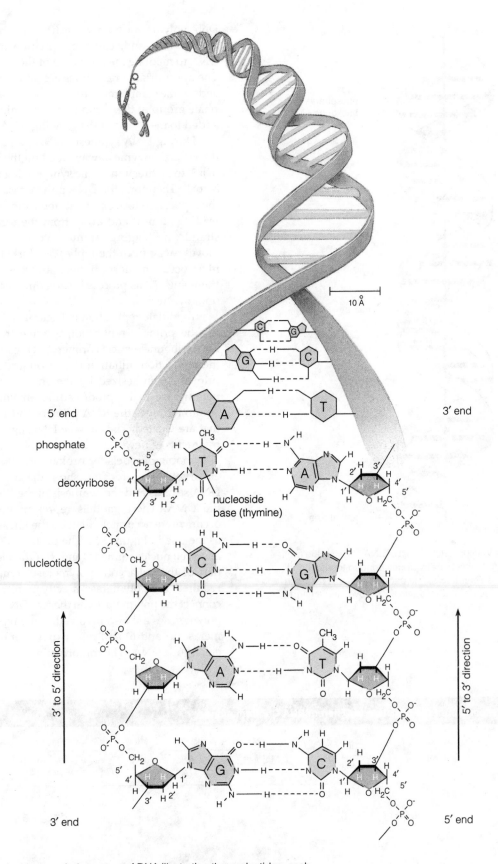

Figure 3.2 The double-stranded structure of DNA illustrating the nucleotides, each composed of a deoxyribose sugar, a phosphate, and a nitrogen-containing base. The two strands are held together by hydrogen bonds (dashed lines) between pairs of bases. Guanine (G) and cytosine (C) are held together by three hydrogen bonds, and adenine (A) and thymine (T) by two. The two strands are antiparallel, since one strand is 5′ to 3′ in one direction and the complementary strand is 3′ to 5′.

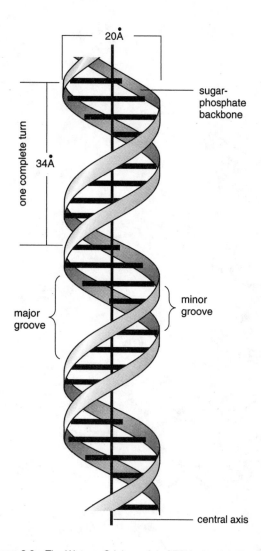

one complete turn

20Å

34Å

sugar-phosphate backbone

major groove

minor groove

central axis

Figure 3.3 The Watson-Crick model of DNA showing the two strands of a DNA double helix hydrogen bonded together and spiraling around a central axis. The packing of atoms results in two grooves, a wide one called the major groove, and a narrow one called the minor groove.

DNA polymerase binds to each single strand and moves along the strand, using information in the template DNA to mediate the formation of the new DNA strand. The appropriate nucleotides are added to the growing ends of the complementary daughter strands. The phosphate group of the 5′ end of the incoming nucleotide is added to the 3′ end of the growing DNA strand.

During DNA synthesis, DNA polymerase is a unidirectional enzyme, always reading the DNA template in a 3′ to 5′ direction, while synthesizing the new strand 5′ to 3′. Therefore, the DNA polymerase moves toward the replication fork on one strand (sometimes called the **leading strand**) and away from the fork on the other strand (the **lagging strand**). When DNA polymerase moves away from the replication fork, DNA synthesis must occur in short, discrete stretches called Okazaki fragments. (This process is sometimes called discontinuous synthesis.)

To initiate replication in bacteria, DNA polymerase III (the primary replication enzyme) requires a short RNA (ribonucleic acid) primer that is complementary to the replication initiation region (Figure 3.5). The RNA primer is synthesized by the enzyme primase. DNA polymerase I, the proofreading enzyme involved in repair, removes the RNA primers and fills in the gaps that are created. The enzyme DNA ligase connects the fragments of newly synthesized DNA.

Looping models for **prokaryotes** (Figure 3.6a) and eukaryotes (Figure 3.6b) show the unwinding of the DNA strands and the synthesis of the RNA primer and the DNA. These models represent how replication occurs more accurately than do the linear models (such as the one in Figure 3.5). The proteins PCNA (proliferating cell nuclear antigen) and RF-C (a release factor) are part of the eukaryotic polymerase complex; PCNA is required for polymerase δ activation, and RF-C may connect the polymerases on the two DNA strands. DNA polymerases are also involved in the replication (polymerases α and δ) and repair (polymerases β and ε) of eukaryotic DNA. The conformation in the model shows

| Conformation | Base Pairs per Turn | Tilt of Base Pairs | Groove Width | | Groove Depth | | Approximate Diameter (Å) |
			Major (Å)	Minor (Å)	Major (Å)	Minor (Å)	
A	11	+20°	10.9	3.7	2.8	13.5	23
B	10.1–10.6	−6°	5.7	11.7	7.5	8.5	20
C	9.3	−8°	4.8	10.5	7.9	7.5	19
D	8	−16°	1.3	8.9	6.7	5.8	
E	7.5						
Z	12	−7°	8.8	2.0	3.7	13.8	18

Table 3.1 Comparison of the Structure of Different DNA Conformations

Adapted from S. L. Wolfe, *Molecular and Cellular Biology*, Table 14.1. Copyright © 1993 Wadsworth, Inc.

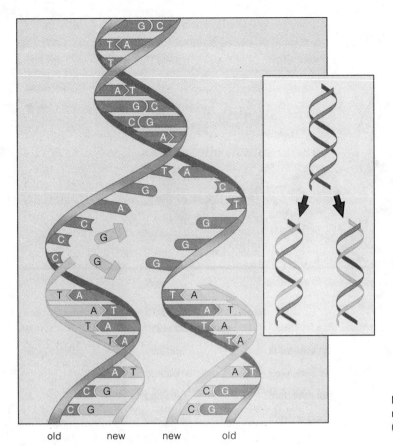

old new new old

Figure 3.4 Semiconservative replication of DNA. Each newly replicated DNA molecule is composed of one parent (old) and one daughter (new) strand.

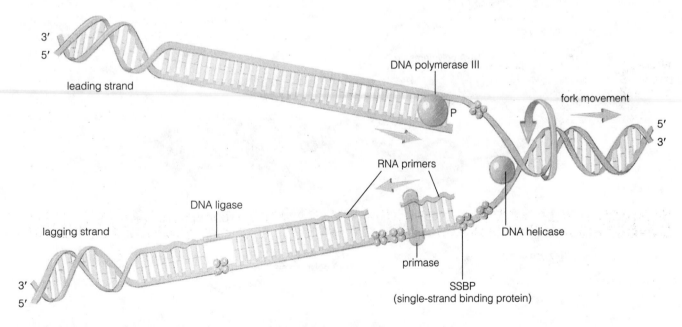

3′
5′

leading strand

DNA polymerase III

P

fork movement

5′
3′

RNA primers

DNA ligase

primase

DNA helicase

SSBP
(single-strand binding protein)

lagging strand

3′
5′

Figure 3.5 Replication of DNA at a replication fork in bacteria. The template is denatured, DNA helicase unwinds the DNA strands in front of the replication fork, and single-stranded binding proteins bind to the single strands and keep them denatured. One strand serves as a template for discontinuous synthesis of the "lagging" strand and continuous DNA synthesis of the "leading" strand. Replication of the DNA strands proceeds in the opposite direction with leading strand replication moving in the direction of the fork. Primase adds short RNA primers to the lagging strand to "prime" replication, and DNA ligase produces a continuous DNA strand from discontinuous replication.

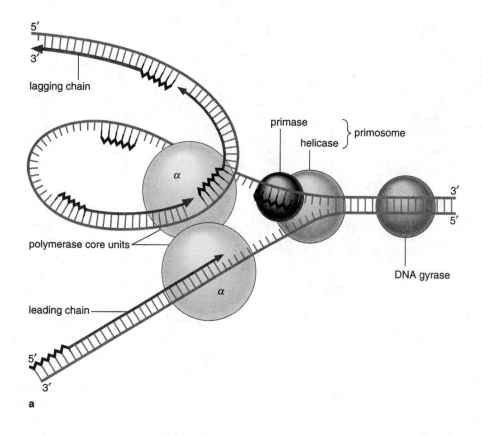

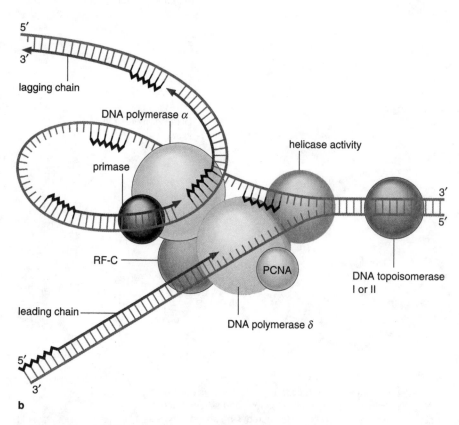

Figure 3.6 Looping models for replication in (a) prokaryotes and (b) eukaryotes allow DNA polymerases to move in the same direction on the leading and lagging strands. In prokaryotic replication, the core DNA polymerase III is composed of two α-subunits. Additional proteins, PCNA and RF-C are involved in the eukaryotic replication complex.

that the lagging strand is looped in the same orientation as the leading strand. Thus, the DNA polymerases on both strands move in the same direction. Most polymerases have exonuclease activity that enables them to remove nucleotides one at a time. This proofreading process ensures that mispaired nucleotides can be removed and replaced with the correct ones.

THE GENE

A gene is a discrete stretch of nucleotide bases on a strand of DNA that serves as a unit of information. The gene can reside on either strand, and can range in size from a few hundred to many thousand of the nucleotide bases A, T, G, and C—the chemical units of information storage. (Noncoding regions separate the genes on a strand.) The number and sequence of the bases within a particular gene determine the information that the gene carries—that is, the amino acid sequence of the specific **polypeptide** or protein it will encode. The information in a gene, however, cannot be translated directly into a polypeptide. The process takes two steps—**transcription** and **translation**—and requires three intermediaries, all of them ribonucleic acids. The process can be summarized:

$$\text{DNA} \xrightarrow{\text{transcription}} \text{RNA} \xrightarrow{\text{translation}} \text{polypeptide}$$

We look at the two steps after briefly reviewing ribonucleic acids.

RIBONUCLEIC ACID

RNA resembles DNA in all ways except the following structural features:

1. The base uracil (U) substitutes for thymine (T), and pairs with adenine (A);

2. The pentose sugars of the backbone are ribose molecules instead of deoxyribose (Figure 3.7); and

3. The RNA molecule is single stranded.

There are, however, exceptions to the last statement: Sometimes short nucleotide sequences that are complementary to each other within the RNA molecule base pair to form short double-stranded regions. This intramolecular base pairing may be needed to maintain the integrity of the molecule or to permit some RNA functions (such as those of transfer RNA, described below). Most RNA molecules are much shorter than DNA molecules, and while DNA is always present within the cell, RNA molecules are present only transiently, and are degraded after a specific period of time.

Different genes encode functionally distinct types of RNAs. Some genes encode messenger RNA (mRNA); others encode transfer RNA (tRNA) and **ribosomal RNA** (rRNA). Although mRNA encodes the amino acid sequence of the protein or polypeptide to be synthesized, both tRNA and rRNA molecules (which do not code for protein) are required for protein synthesis. tRNA, a small nucleic acid of approximately 75 ribonucleotides, folds into a secondary structure that resembles a cloverleaf (Figure 3.8). In the cell, tRNAs serve as adapter molecules by carrying the appropriate amino acid to the site of protein synthesis.

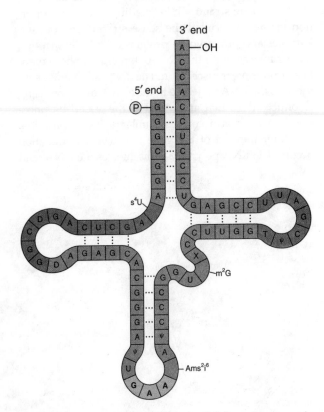

Figure 3.8 A mature transfer RNA showing the anticodon and 3' amino acid binding regions.

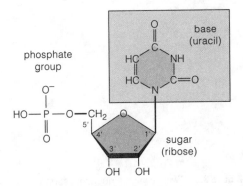

Figure 3.7 Nucleotide structure of RNA with uracil as the base and a ribose as the pentose sugar.

There are several lengths, or sizes, of ribosomal RNA (rRNA). These are given by their "S" value or Svedberg unit—their relative sedimentation rates during centrifugation: 5S, 16S, and 23S in prokaryotes, and 5S, 5.8S, 18S, and 28S in eukaryotes. rRNAs combine with a large number of proteins to form large cytoplasmic structures called ribosomes. A ribosome is the workbench of protein synthesis, where amino acids are covalently bonded together to form the polypeptide. Ribosomes consist of two subunits, a large and a small, which come together in the cytoplasm to form the mature ribosome (Figure 3.9).

TRANSCRIPTION

RNA is synthesized, or transcribed, from a DNA template. Through transcription, the genetic information stored in the DNA is used to make an RNA that is complementary to the DNA. The enzyme **RNA polymerase** recognizes nucleotide sequences upstream, just before the start of the coding region, or gene (usually denoted by the number of nucleotides from the start of transcription). These sequences, or **promoters**, allow the RNA polymerase to be placed correctly on the DNA strand near the gene to be transcribed. As Figure 3.10 shows, the RNA polymerase binds to the promoter, the DNA strands separate, and the polymerase moves along the DNA template strand while "reading" the coding portion (sometimes called the sense strand). In both eukaryotes and prokaryotes, proteins called transcription factors interact with the RNA polymerase to help regulate or promote transcription. The RNA polymerase synthesizes the RNA molecule in the 5' to 3' direction using ribonucleotides that are added to the 3' end by phosphodiester bonds in a process similar to DNA replication.

After the end of the gene is reached, termination occurs: The RNA polymerase disengages the DNA, and the new RNA molecule is released. At some termination sites in bacteria, specific transcription termination proteins, or rho proteins are involved; at other sites, termination is signaled by the formation of a hairpin loop or intramolecular base-paired stem and open loop structure.

Promoters

In both prokaryotes and eukaryotes, most promoters define the beginning of a gene and signal the start of transcription. Exceptions do exist; some promoters are within the gene, and, in prokaryotes, several genes can be part of a polycistronic operon with the promoter located at the beginning of the transcription unit.

In bacteria, one type of RNA polymerase is present within the cell and associates with a protein factor or sigma factor (σ). Sigma factors provide the binding specificity of RNA polymerase to specific promoters. A single sigma factor binds to the RNA polymerase before transcription of a gene is initiated, and is released immediately after transcription begins. In *E. coli*, the most common sigma factor that recognizes most of the promoters in the cell is σ^{70}. The most common promoter is composed of two regions of sequences approximately 10 and 35 nucleotides upstream from the transcriptional start (Figure 3.11). These segments are called the –10 region (also called the Pribnow box) and –35 region. The designation +1 in the figure identifies the first nucleotide at the transcriptional start site. The *E. coli* RNA polymerase most likely recognizes and binds to the –35 region. The polymerase also binds to the –10 region, which then denatures and becomes single-stranded so that transcription can begin. Comparison among promoter regions of genes reveals common sequence motifs. A **consensus sequence** represents the bases most likely to be found at each particular site within the sequence being examined. The promoters of many genes were examined to determine the consensus sequence of TTGACA (–35) and TATAAT (–10), which is recognized by the major sigma

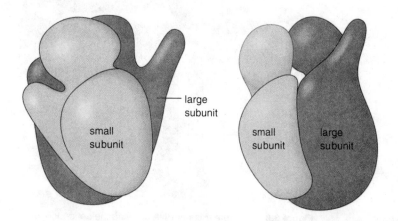

Figure 3.9 Two views of a ribosome showing the conformations of the large and small subunits.

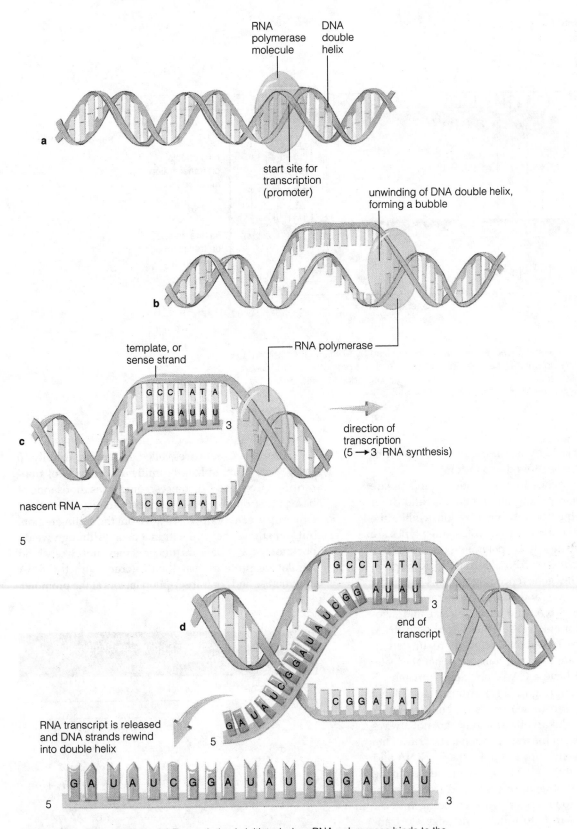

Figure 3.10   RNA synthesis. (a) Transcription is initiated when RNA polymerase binds to the DNA at the promoter region. (b) The double-stranded DNA unwinds. (c) As the RNA polymerase travels along the DNA template, nucleotides are added to the growing RNA strand. (d) When RNA polymerase reaches a terminator, the RNA transcript is released and transcription is terminated.

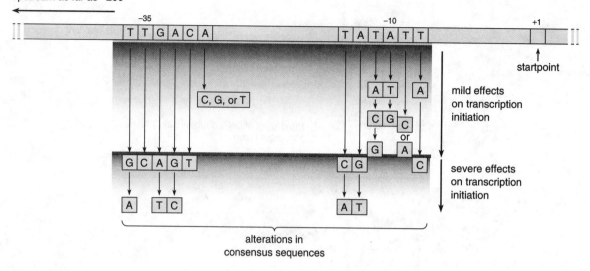

other sequences with regulatory effects on transcription located upstream as far as –200

mild effects on transcription initiation

severe effects on transcription initiation

startpoint

alterations in consensus sequences

Figure 3.11 Bacterial promoter sequences that are recognized by sigma factor σ^{70} linked to RNA polymerase. RNA polymerase has the highest affinity for those promoters with TTGACA and TATATT consensus sequences. Deviations in the consensus sequence weaken the binding as shown by the length of the downward arrows.

factor, σ^{70} (see Table 3.2, which also shows the consensus sequences of promoters for other σ factors).

Promoters of genes in eukaryotic organisms are more complex than prokaryotic promoters, and differ from them in sequence and organization. Unlike the single RNA polymerase in prokaryotes, three types are present in eukaryotes. RNA polymerase I transcribes ribosomal RNA genes for 5.8S, 18S, and 28S; RNA polymerase II transcribes genes encoding proteins (and most small nuclear RNAs); and RNA polymerase III transcribes tRNA, 5S rRNA, and small nuclear and cytoplasmic RNAs. For eukaryotic genes encoding mRNA, the promoter is the TATA box, located approximately 25 base pairs upstream from the transcriptional start site. Many genes also have a CAAT box approximately 75 base pairs upstream (Figure 3.12). Animals and plants have different consensus sequences for both the TATA and CAAT boxes. Eukaryotic promoters are not always located just in front of the transcriptional start site of the gene coding sequence (for example, the 5S rRNA and tRNA genes have internal promoters). In addition to the RNA polymerase, other proteins called transcription factors bind to promoters of genes encoding mRNA, tRNA, and rRNA to help initiate and regulate transcription. Gene-specific regulatory proteins also are involved, and act in combination to increase the binding affinity of RNA polymerase for the promoter. These regulatory proteins can modulate transcription by interacting either directly with RNA polymerase or with one of the transcription factors.

Short DNA sequences called **enhancers**, usually 50 to 100 base pairs in length, influence the level of transcription from great distances, sometimes thousands of base pairs from the gene. Enhancers are usually located either upstream or downstream from the coding region, but sometimes they are within a gene. Although we do not know how enhancers function, they most likely bind regulatory proteins that then interact with the RNA polymerase and/or transcription factors at the promoter.

Table 3.2   E.coli Promoter Consensus Sequences for Some Sigma Factors

Sigma Factor	Consensus Sequences
σ^{70}	TTGACA----------------------TATAAT –35 –10
σ^{54}	GTGGC----------------TTGCA –26 –14
σ^{32}	CCCCC----------------------TATAATTA –39 –16
σ^{26}	TAAA-------------GCCGATAA –35 –10
σ^{23}	TATAATA –15

Modified from an original by M. S. Z. Horwitz and L. A. Loeb, *Prog. Nucleic Acids Res. Mol. Biol.* 38:137 (1990) and used in S. L. Wolfe, *Molecular and Cellular Biology*, Table 15.3. Copyright © 1993 Wadsworth, Inc.

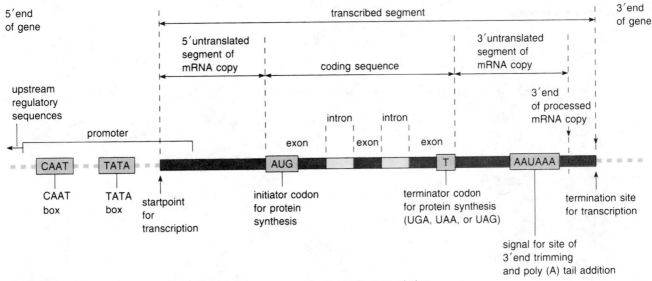

Figure 3.12 Eukaryotic promoter elements and other sequences important for transcription and translation. Introns also may be present within the gene coding sequence. The transcribed segment is presented as RNA (with uracil in the sequence). The CAAT and TATA boxes are shown as the DNA missense sequence (the sequence complementary to the DNA templates).

The DNA must loop out to enable distant DNA regions to interact with the enhancer. This looping out may result from the interaction between the regulatory protein and proteins associated with the promoter.

RNA Processing

In eukaryotic cells, most newly synthesized RNA—primary transcripts called **heterogeneous nuclear RNA (hnRNA)**—must be modified before it is fully functional, mature messenger RNA. Often a string of adenine ribonucleotides, called a poly(A) tail, is added to the 3′ end of eukaryotic mRNA. This process is called polyadenylation. A "cap" structure (**5′ cap**) of a modified (usually methylated) guanine base is often observed at the 5′ end. The 5′ and 3′ modifications may facilitate transport out of the nucleus, protect the mRNA from degradation in the cytoplasm, and maintain stability for translation.

In the primary transcript of most eukaryotic genes, noncoding sequences, called **introns**, intervene between the coding sequences, the exons. The introns are spliced out of the newly synthesized RNA so that the exons are adjacent to one another. Figure 3.12 reviews the organization of sequence elements in a typical eukaryotic gene encoding a messenger RNA.

Primary transcripts of ribosomal and transfer RNAs undergo similar processes to become mature RNAs. Introns are spliced from tRNAs, and both the 5′ and 3′ ends are processed, including the addition at the 3′ end of CCA where the amino acid binds. In eukary-

otes, a large rRNA transcript is synthesized and cut in a stepwise process to generate 28S, 18S, and 5.8S rRNAs.

PROTEINS

Proteins are large organic compounds that are major cellular determinants of an organism's characteristics. The many types of proteins found in cells include enzymes that catalyze cellular reactions for both biosynthetic and catabolic activities, hormones that help regulate cellular metabolic activities, antibodies involved in the immune response, transcription factors that help control transcriptional activity, and structural proteins such as tubulin and actin that determine the shape of cells and allow cells and organisms to be motile. Protein-encoding sequences make up only about 10% of the human genome; the rest consists of introns (noncoding regions within genes), short repetitive sequences, sequences for rRNAs and tRNAs, and regulatory or control sequences. The specific functions of many of repetitive sequences are unknown.

Proteins are composed of amino acids joined by covalent bonds. Figure 3.13 shows the 20 amino acids that make up the polypeptides. The great diversity of proteins in organisms is due to the variability of polypeptide length, the many sequence combinations (or linear order) possible from 20 different amino acids, and the three-dimensional conformations after folding of the molecules. Some functional proteins are com-

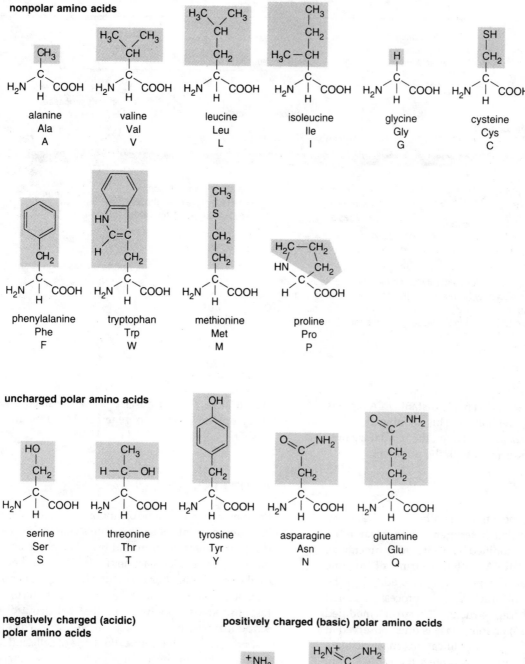

Figure 3.13 The 20 amino acids used in protein synthesis. The shaded boxes highlight the side chains that determine the chemical characteristics of protein. Both the three- and one-letter abbreviations are shown.

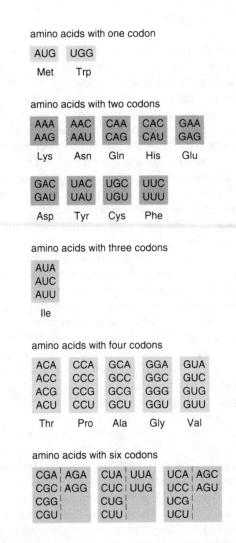

Figure 3.14 Peptide bond formation between two amino acids. "R" represents the side chains of each amino acid.

posed of a single polypeptide chain (for example, myoglobin), while others are made up of many (for example, hemoglobin).

Each amino acid has the same basic backbone; its unique side group (R) determines its chemical characteristics. The chemical properties of an amino acid's side group determine which of four main classes the amino acid is assigned to: nonpolar, uncharged polar, negatively charged (acidic) polar, and positively charged (basic) polar. Amino acids are joined by covalent bonds (**peptide bonds**) between the carbon atom in the carboxyl group of one amino acid and the nitrogen atom of the amino group of the adjacent amino acid (Figure 3.14). The resulting polypeptide has directionality: One end terminates with a free amino group (–NH$_2$), the amino terminus, and the other terminates with a free carboxyl (–COOH), the carboxyl terminus. The combined chemical properties of the amino acid side groups determine the chemical nature of the protein.

THE GENETIC CODE

The nucleotide sequence of DNA and mRNA specifies the amino acid sequence of a polypeptide. Although there are 20 amino acids, the four bases are sufficient to encode the amino acid sequence of a protein. Years ago, scientists discovered that three bases are required to encode the information for a single, specific amino acid. A simple calculation demonstrates that two nucleotides used together are not enough to encode all 20 amino acids (four nucleotides used two at a time: $4^2 = 4 \times 4 = 16$). However, nucleotides used three at a time can generate 64 combinations ($4^3 = 4 \times 4 \times 4 = 64$). Thus, a triplet code is sufficient to specify all 20 amino acids. Insertion and deletion mutagenesis experiments demonstrated a triplet code by determining the **reading frame** of the sequence. One, two, and three nucleotides were inserted or deleted to determine the effect on the reading frame of the encoded information. Addition or deletion of one or two nucleotides produced a subsequent change in the reading frame. The addition or deletion of three nucleotides close together restored the original reading frame. Therefore, nucleotides are "read" in threes for each amino acid. These triplets in the messenger RNA are called codons.

The genetic code is degenerate, because more than one codon can specify a particular amino acid (Figure

3.15 shows the amino acids arranged according to the extent of degeneracy). The protein synthesis (that is, translation) machinery "reads" these codons so that the appropriate amino acids are bonded together to generate the correct protein encoded by the messenger RNA. Three stop codons (UAA, UAG, UGA) in the genetic code signal the end of the polypeptide and terminate translation. A start codon signifies the beginning of the polypeptide in translation. Methionine, specified by the codon AUG, is usually the first amino acid, but it is not the only possibility; other amino acids can be used at the beginning of the protein.

amino acids with one codon

AUG	UGG
Met	Trp

amino acids with two codons

AAA	AAC	CAA	CAC	GAA
AAG	AAU	CAG	CAU	GAG
Lys	Asn	Gln	His	Glu

GAC	UAC	UGC	UUC
GAU	UAU	UGU	UUU
Asp	Tyr	Cys	Phe

amino acids with three codons

AUA
AUC
AUU
Ile

amino acids with four codons

ACA	CCA	GCA	GGA	GUA
ACC	CCC	GCC	GGC	GUC
ACG	CCG	GCG	GGG	GUG
ACU	CCU	GCU	GGU	GUU
Thr	Pro	Ala	Gly	Val

amino acids with six codons

CGA	AGA	CUA	UUA	UCA	AGC
CGC	AGG	CUC	UUG	UCC	AGU
CGG		CUG		UCG	
CGU		CUU		UCU	

Figure 3.15 The amino acids arranged according to the extent of degeneracy.

The correct protein sequence is synthesized only if the message is read in the appropriate reading frame. In-frame translation can occur only if the translational apparatus (i.e., ribosome/aminoacyl-tRNA complex) initially binds to the correct codon at the translational start site. For example, suppose the following three-letter words represent codons:

THE BIG RED DOG WAS SAD

When read in frame, the message makes sense. If reading is shifted one letter (or nucleotide), it becomes meaningless:

HEB IGR EDD OGW ASS AD

Thus, to generate the correct amino acid sequence for each polypeptide synthesized, the translation machinery must initiate translation at the correct start site and continue reading triplets in frame.

TRANSLATION

In eukaryotes, transcription occurs in the nucleus where newly synthesized RNA is processed, but translation occurs after the RNA is transported to the cytoplasm, where the ribosomes are assembled. Translation, in brief, is the conversion of information encoded in the mRNA sequence into a sequence of amino acids forming a polypeptide chain. It is a four-step process: initiation, elongation, translocation, and termination. During initiation, the small ribosomal subunit first binds to the initiator tRNA (the first aminoacyl-tRNA) and its amino acid, and this complex binds to the end of the mRNA near the 5' cap site. The small subunit then scans the mRNA until the initiator AUG (encoding the amino acid methionine) is recognized. The large ribosomal subunit is then added. Numerous initiation protein factors play important roles in the initiation process.

The translation process in prokaryotes (for example, *Escherichia coli*) is similar to that of eukaryotes, although the process begins while the gene is being transcribed, fewer initiation factors are involved, and a special sequence in the 5' region of the mRNA (Shine-Delgarno sequence) is complementary to a sequence of the 16S rRNA portion of the ribosome, which establishes the correct alignment of the ribosome on the mRNA. The placement of the ribosome on the mRNA determines which bases will be read in triplets as codons. This important step ensures that the message will be read in-frame for synthesis of the correct protein. The **anticodon** of the initiator tRNA with the amino acid—usually methionine in eukaryotes and formylmethionine in prokaryotes (Figure 3.16)—base pairs with the codon AUG of the mRNA. The anticodon and codon base pairing occurs at a specific site on the ribosome. Therefore,

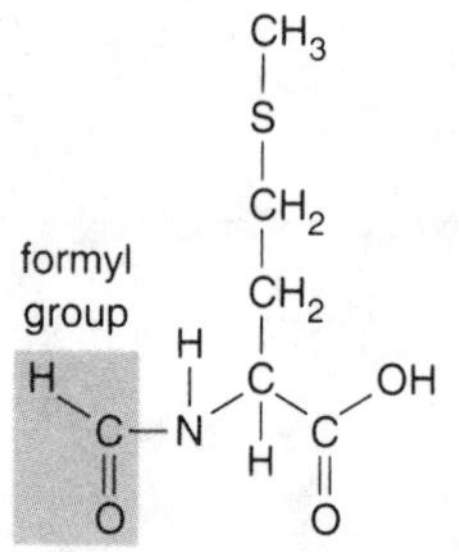

Figure 3.16 The structure of formylmethionine (fMet). The formyl group is shown in the shaded box.

tRNAs serve as adapter molecules or interpreters that convert the RNA code or codons into the language of amino acids. Attached to the tRNA is the amino acid that is specified by the codon to which it has base paired. In this way, the tRNA carries the appropriate amino acid to the site of protein synthesis.

Nucleic acids remain antiparallel in codon–anticodon base pairing. The first two bases in an anticodon–codon pairing are specific and exactly follow base-pairing rules. However, the third position is not so restrictive. Here nonstandard base pairing, or "wobble" pairing, can occur (Table 3.3). Thus, there is not always a specific type of tRNA for each of the 61 codons that correspond to amino acids. Many tRNAs recognize more than one codon for a particular amino acid.

The ribosome accommodates one aminoacyl tRNA at each of the special sites A and P. Once translation is initiated and the ribosome and initiator tRNA with its amino acid are in place on the mRNA, the polypeptide is synthesized by the process of elongation: Amino acids are added stepwise to the first amino acid. The amino acid specified by the next codon of the mRNA is brought to the ribosome by a second tRNA whose anticodon is complementary to the second codon (Figure 3.17). If complementary base pairing occurs, the amino acid attached

Table 3.3 Codon-Anticodon Pairing at the Third Base of the Codon According to the Wobble Hypothesis	
Anticodon	Codon
U (or ψ)	U, G, or A
C	G
A	U
G	U or C
I	U, C, or A

ψ= the modified base pseudouridine; I = the modified base inosine.

From F. H. C. Crick, J. Mol. Biol. 19:548 (1966) as used in S. L. Wolfe, *Molecular and Cellular Biology*, Table 16.3. Copyright © 1993 Wadsworth, Inc.

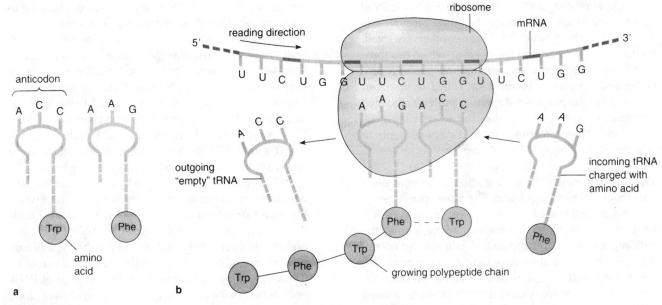

Figure 3.17 The steps of polypeptide synthesis. (a) The tRNAs with their bound amino acids. (b) At the ribosome, the tRNAs transfer their amino acids to the growing polypeptide chain. The bases of the anticodons of the tRNAs form transient hydrogen bonds with the bases of the complementary codons of the mRNA. Here, the ACC anticodon of the tRNA carrying tryptophan base pairs with the codon for tryptophan, UGG. From the energy released by the hydrolysis of ATP, the polypeptide is transferred to the tryptophan bound to its tRNA. A peptide bond is established and the tRNA that lost the polypeptide is released from the ribosome. The ribosome moves to the next codon on the mRNA, UUC, which codes for phenylalanine, and the process continues with the anticodon AAG of tRNA-phenylalanine base pairing with the codon.

to the first tRNA is transferred to the amino acid on the second aminoacyl-tRNA and the amino acid carried by the second transfer RNA is added to the growing polypeptide. A peptide bond is made between, for example, methionine and the second amino acid. The first tRNA, now without its amino acid, is released from the ribosome and mRNA. The ribosome complex moves one codon on the mRNA, in a process known as translocation. A new codon is exposed, and another tRNA can transfer the next specified amino acid to the growing polypeptide. Thus, the polypeptide is elongated one amino acid at a time. The process requires several elongation factors, each with specific functions. Elongation of the polypeptide chain continues until a stop codon is reached (UAA, UAG, or UGA) and release factors interact with the stop codon to terminate translation. The ribosomal subunits dissociate, the messenger and transfer RNAs are released, and the newly synthesized protein is used by the cell.

REGULATION OF GENE EXPRESSION

In both prokaryotic and eukaryotic cells, complex mechanisms regulate the amount of particular proteins synthesized. The cell synthesizes only what is required. In eukaryotes, cellular control can occur at several levels, and may be quite complex (that is, regulation occurs

during and after transcription and during and after translation). Transcriptional control is especially common in prokaryotes; it includes gene activation and inactivation and regulation of the transcription initiation rate. Translational control includes regulating the rate at which specific messenger RNAs are translated as well as controlling the availability of specific mRNAs for translation. The stability of mRNA is also important, since the rate of mRNA degradation determines how long the message is available for translation. For example, the mRNA for the protein tubulin has a half-life of 4 to 12 hours, while the insulin receptor and the enzyme pyruvate kinase mRNAs have half-lives of 9 hours and 30 hours, respectively. Prokaryotes and eukaryotes use specific mechanisms of gene regulation that are quite different.

Prokaryotic Gene Expression

Microorganisms must respond rapidly to sudden fluctuations in the environment. Proteins or enzymes may be required for only a very brief time, and when conditions change, they might not be needed. Bacteria must rely on inducers that transmit signals so that genes can be turned on or off in response to environmental cues (such as nitrogen starvation, heat and salt stress, light intensity). In bacteria, a single promoter often controls several coding regions, called cistrons. This arrangement is called an operon (Figure 3.18). The genes that are part of the same operon have related functions within the cell. Some *E. coli* operons are large, such as the histidine biosynthesis operon (*his* operon), which has 11 genes; others are small, such as the lactose operon (*lac* operon), which has three genes. An operon consists of structural genes, a promoter, and a repressor binding site, called an **operator**, that overlaps the promoter. **Repressor proteins**, encoded by repressor genes, are synthesized to regulate gene expression. These proteins bind to the operator site to block transcription by RNA polymerase.

The *lac* Operon The well-studied *lac* operon illustrates bacterial gene regulation. The three operon genes, *lacZ* (β-galactosidase), *lacY* (permease), and *lacA* (acetylase), encode enzymes that transport and break down the milk sugar lactose into glucose and galactose for energy (Figure 3.19). In the absence of lactose, the enzymes are not needed, and the genes in the operon are not transcribed. In an inactive operon, the *lac* repressor, encoded by the *lacI* gene (not part of the operon), binds to the operator that overlaps the promoter region (Figure 3.20a). Transcription is prevented because the presence of the

repressor prevents the RNA polymerase from binding to the promoter. The lactose repressor has two binding sites: one for lactose compounds and the other for the DNA operator site. When lactose (sometimes called an inducer) is present, the operon is transcribed and the enzymes are synthesized. Lactose binds to the repressor to form a lactose–repressor complex that cannot bind to the operator region of the DNA. The promoter region is available to the RNA polymerase, which can bind and initiate transcription (Figure 3.20b). Ribosomes immediately attach to the newly synthesized RNA, and translation begins.

The regulation of an operon by a repressor is called negative control because genes are not transcribed when the repressor is bound. The *lac* operon can also be positively regulated so that the genes are transcribed at a higher rate than they would be without this type of regulation. Specific small molecules increase the rate of transcription by facilitating both the binding of RNA polymerase to the promoter and the separation of DNA strands. The preferred substrate for *E. coli* is not lactose but glucose; however, when glucose is absent, the concentration of **cyclic AMP** (cAMP) increases in the cell.

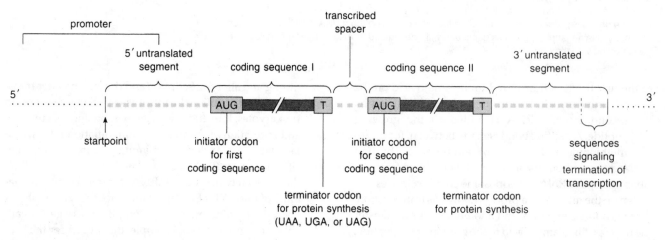

Figure 3.18 A prokaryotic operon.

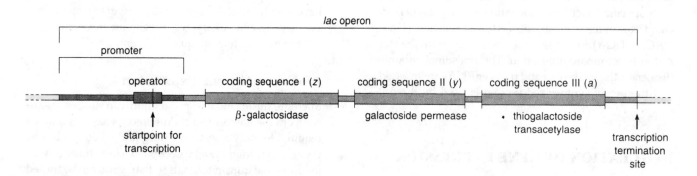

Figure 3.19 The *lac* operon showing the structural genes and promoter region. The repressor binds to the operator site on the DNA. β-galactosidase hydrolyzes lactose into glucose and galactose, and galactoside permease facilitates the transport of lactose and other sugars into the cell. The function of thiogalactoside transacetylase has not been completely elucidated.

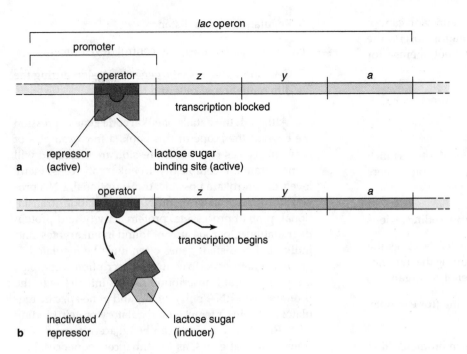

Figure 3.20 Regulation of the *lac* operon. (a) When the active repressor binds to the operator in the absence of lactose, RNA polymerase cannot bind to the promoter and transcription is blocked. (b) In the presence of lactose, a derivative of the sugar binds to the repressor and inactivates it so that the repressor can no longer bind to the operator, thus allowing RNA polymerase to bind to the promoter and transcription to proceed.

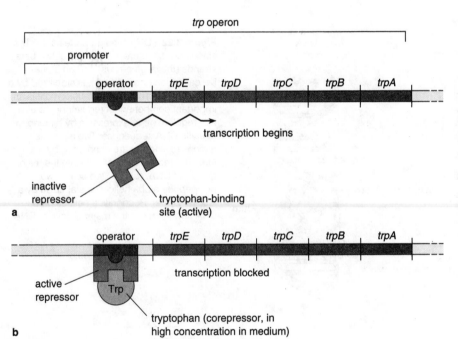

Figure 3.21 The *trp* operon showing the five structural genes and the promoter region. (a) In the absence of tryptophan, the repressor is inactive and cannot bind to the operator, thus allowing RNA polymerase to bind to the promoter and transcription to proceed. (b) When tryptophan is present it binds to and activates the repressor, thus allowing the activated repressor to bind to the operator and block transcription.

Cyclic AMP binds to a DNA-binding protein, **catabolite activator protein (CAP)**. The cAMP-CAP complex binds to a CAP binding site near the promoter region but not overlapping the region. As a result, *lac* operon transcription is greatly enhanced. When glucose is present, cAMP is low, and the cAMP-CAP complex does not form. Thus, the rate of transcription is not increased, or, if lactose is absent, the *lac* operon is not expressed.

The *trp* Operon The operon that regulates tryptophan biosynthesis is another example of negative control.

The synthesis of amino acids must be highly regulated to maintain the correct concentrations in the cell. The tryptophan (*trp*) operon consists of promoter and operator regions in addition to five genes encoding enzymes that catalyze the last steps of tryptophan biosynthesis.

In contrast to the repressor of the *lac* operon, the *trp* operon repressor is not active unless tryptophan binds to it. Tryptophan activates the repressor so that the repressor–tryptophan complex can bind to the operator region and block transcription (Figure 3.21). In this way, tryptophan acts as a corepressor to turn off transcription. Because an inactive repressor cannot bind to the operator when tryptophan is absent, the promoter remains available to RNA polymerase for transcription.

Eukaryotic Gene Expression

The regulation of gene expression in eukaryotes is much more complex and variable than in prokaryotes. There are several reasons for this complexity:

1. Larger genome size with extensive noncoding regions

2. Compartmentalization within the cell (such as the nucleus and other organelles), requiring that nuclear-encoded gene products be targeted to organelles

3. More extensive transcript processing (for example, introns)

4. Scattered genes each with their own promoter (that is, not organized into operons)

5. Regulation from a distance

6. Complex mechanisms to control development

7. Cell- and tissue-specific gene expression during the differentiation of cells and tissues

Although the details of eukaryotic gene expression are beyond the scope of this book, a few examples of how eukaryotes regulate expression are presented will demonstrate the complexity. In eukaryotic cells, many levels of control are possible: transcriptional, RNA processing, RNA transport, mRNA degradation, translational, protein processing, protein activity, and protein degradation. Operons are not found in eukaryotes, and individual or distant genes often must be regulated in groups or networks. To control transcription, a complex array of general transcription factors interact with the promoter and RNA polymerase, and gene-specific regulatory proteins interact with regulatory protein binding sites. Regulatory regions can be adjacent to or distant from structural genes, as are enhancer sequences. Proteins that bind to DNA sequences are called *trans*-acting

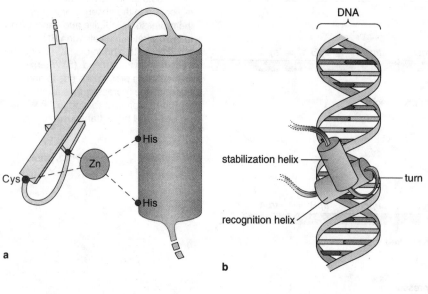

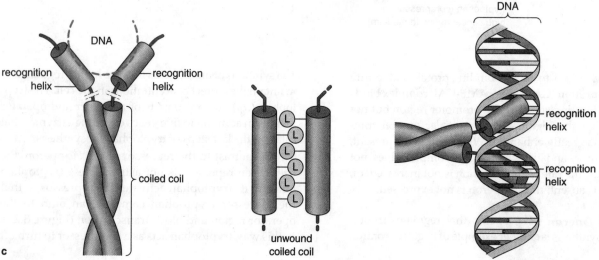

Figure 3.22 DNA-binding proteins. (a) The zinc-finger structure, composed of two beta strands (the large arrows) and an alpha-helical region (the cylinder), is stabilized by a zinc atom. (b) The helix-turn-helix motif is an important component of regulatory proteins that regulate gene expression by binding to specific DNA sequences. Two alpha-helical regions (cylinders) are connected by a short turn. The recognition helix fits into the major groove of DNA. (c) A coiled coil of two polypeptide alpha helices is stabilized by a leucine zipper. Recognition helices bind to DNA sequences in the major groove of DNA.

factors, and the binding sites on the DNA are sometimes called *cis* elements. Specific sequences or *cis* elements and their corresponding regulatory proteins or *trans*-acting factors are often gene-specific; for example, heat shock genes have special heat shock sequences that bind heat shock protein factors. Regulatory proteins must have specific three-dimensional conformations which enable them to interact closely with specific DNA regions. Protein binding specificity is determined by special sequences sometimes called DNA-binding structural motifs. Figure 3.22 shows three classes of protein motifs found in regulatory proteins: the **zinc-finger**, the helix-turn-helix, and the **leucine-zipper** motifs. Proteins encoded by the **homeotic genes** (which control major developmental pathways) are examples of regulatory proteins with a helix-turn-helix motif. Within the gene is a **homeobox** sequence that encodes the **homeodomain**, a region where the helix-turn-helix is located. This region of the protein binds specifically to a sequence adjacent to the 5' flanking region of all genes controlled by the homeotic genes (Figure 3.23). Table 3.4 lists just a few of the many other eukaryotic regulatory proteins that have been identified.

In addition to controlling transcription, eukaryotes regulate their genes by controlling mRNA processing and transport of mRNAs to the cytoplasm, the rate of translation and the availability of mRNA (controlled through stability and the masking of mRNA), and protein processing. Two different cell types may process the same RNA differently, by alternative splicing, to yield two different proteins. That is, different introns and exons (or portions of them) in some intron-containing primary RNA transcripts are selectively removed or retained. An example is the gene calcitonin which is expressed in both the thyroid and hypothalamus. As Figure 3.24 shows, the pre-mRNA contains five introns

separating six exons. The transcript can be processed in either of two ways to generate calcitonin mRNA in the thyroid cells or calcitonin gene-related polypeptide (CGRP) mRNA in the hypothalamus cells (Figure 3.24).

UNDERSTANDING AND EXPLOITING GENE EXPRESSION

In the early years of molecular biology (certainly through the 1950s), the complexity of eukaryotic gene expression precluded rapid progress in understanding gene control. In these early years, bacteria and viruses were used as models for gene regulation, although ultimately the complex mechanisms of eukaryotic gene expression would have to be studied in eukaryotic organisms. The advent of recombinant DNA technology provided the means for such study.

The scientist must have a thorough understanding of eukaryotic gene expression to be able to manipulate the eukaryotic organism or desired product. The use of eukaryotic genes poses special challenges for biotechnologists, especially when a prokaryotic organism is used as a host for expression. Eukaryotic and prokaryotic genes are regulated in distinctly different ways because of differences in, for example, genome organization, promoter sequences, transcription factors, and processing mechanisms. To enhance the expression of a gene, one must know host promoter sequences as well as codon usage. Sometimes promoter sequences must be substituted or modified, or the number of gene copies may also be increased to maximize gene expression. A comprehensive understanding of eukaryotic gene expression enables the scientist to modify genes and regulatory sequences, permitting a select protein to be synthesized in a particular host cell.

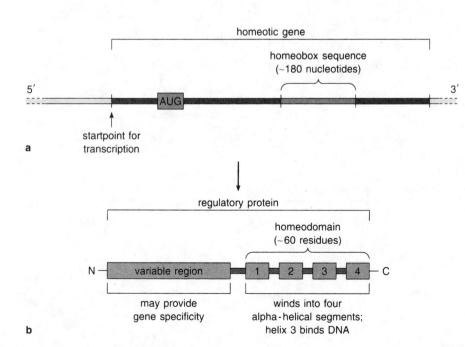

Figure 3.23 Homeotic genes contain a specific sequence called a homeobox (a) that encodes the homeodomain (b), a region of regulatory proteins. The homeodomain is composed of four alpha-helical regions, including a helix-turn-helix motif, one of which binds to a special sequence in the 5' region adjacent to those genes regulated by homeotic genes.

Table 3.4 Examples of Eukaryotic Regulatory Proteins

Regulatory Protein (*trans* element)	Sequence Recognized (*cis* element)	Source	Motif*	Genes Regulated
ACF		Mammals		Serum albumin
Antennapedia		*Drosophila*	HTH	Homeotic gene
AP-2	CCCCAGGC	Mammals		MHC class I, proenkephalin, metallothionein, others
AP-3		Mammals		SV40 viral enhancer
AP-4		Mammals		Proenkephalin, others
APF		Mammals		Serum albumin
CDF		Sea urchin		Histone H2B
C/EBP	TGTGGAAAG	Mammals	LZ	serum albumin, α-globin
CREB (ATF)	TGACGTCA	Mammals	LZ	Somatostatin, proenkephalin
Estrogen receptor	GGTCANNNTGACC	Birds, mammals	ZnF	Vitellogenin, ovalbumin, others
Fos/Jun (AP-1)	TGACTCA	Mammals	LZ	Growth regulation
GAL4		Yeast	ZnF	Enzymes of galactose metabolism
GCN4		Yeast	LZ	Genes encoding enzymes synthesizing amino acids
Glucocorticoid receptor	GAACANNNTCTTC	Mammals	ZnF	Prolactin, others
GT-1		Plants		RuBP carboxylase, chlorophyll-binding protein
HSF		Mammals, others		Heat shock proteins
ITF		Mammals		β-interferon
MAT proteins		Yeast	HTH	Mating type genes
NF-1 (CTF)	GCCAAT	Mammals		α-globin, β-globin, serum albumin, heat shock proteins
OCT-1 (OTF1)	ATTTGCAT	Mammals	HTH	snRNA, histone genes, others
OCT-2 (OTF2)	ATTTGCAT	Mammals	HTH	Antibody genes
Pit-1		Mammals	HTH	Growth hormone
SP1	GGGCGG	Vertebrates	ZnF	Wide variety
TUF		Yeast		Ribosomal proteins
Ultrabithorax		*Drosophila*	HTH	Homeotic gene

*HTH = helix-turn-helix; LZ = leucine zipper; ZnF = zinc finger.

Adapted from data presented in P. F. Johnson and S. L. McKnight, *Annu. Rev. Biochem*,. 58:799 (1989), C. Murre et al., *Cell* 56:777 (1989), and P. J. Mitchell and R. Tijan, *Science* 245:371 (1989) and used in S. L. Wolfe, *Introduction to Cell and Molecular Biology*, Table 17.1. Copyright © 1993 Wadsworth, Inc.

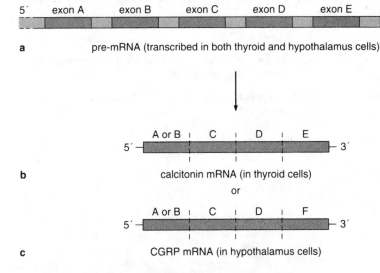

a pre-mRNA (transcribed in both thyroid and hypothalamus cells)

b calcitonin mRNA (in thyroid cells)

or

c CGRP mRNA (in hypothalamus cells)

Figure 3.24 An example of alternative splicing to yield two different mRNAs, calcitonin mRNA in thyroid cells and calcitonin gene related polypeptide (CGRP) mRNA in hypothalamus cells. (a) Pre-messenger RNA comprises six exons (A-F) and five introns. (b) A or B, C, D, and E exons are retained in the mRNA that encodes the hormone calcitonin while (c) exons A or B, C, D, and F are kept in CGRP hypothalamus mRNA, thought to play a role in the function of taste receptors. Exon F is an intron in calcitonin RNA and an exon in CGRP, while exon E is an exon for calcitonin RNA and an intron for CGRP. Exon A or B, located in the 5′ untranslated region, can be present in either protein.

General Readings

B. Alberts et al. 1994. *Molecular Biology of the Cell,* 3rd edition. Garland Publishing, Inc., New York.

B. Lewin. 1994. *Genes* V. Oxford University Press, Oxford, England.

S.L. Wolfe. 1995. *An Introduction to Cell and Molecular Biology.* Wadsworth Publishing Company, Belmont, CA.

Additional Readings

J. Archambault and J.D. Friesen. 1993. Genetics of eukaryotic RNA polymerases I, II, and III. *Microbiol. Rev.* 57:703-724.

J.A. Atwater, R. Wisdom, and I.M. Verma. 1990. Regulated mRNA stability. *Annu. Rev. Genet.* 24:519-541.

D. Baltimore and A.A. Berg. 1995. DNA-binding proteins: A butterfly flutters by. *Nature* 373:287-288.

T. Beardsley. 1991. Smart genes. *Sci. Am.* 265(2):86-95.

E.M. DeRobertis, G. Oliver, and C.V.E. Wright. 1990. Homeobox genes and the vertebrate body plan. *Sci. Am.* 26:46-52.

R.E. Dickerson. 1992. DNA structure from A to Z. *Methods Enzymol.* 211:67-111.

J.F.X. Diffley. 1994. Eukaryotic DNA replication. *Curr. Opinion Cell Biol.* 6:368-372.

N. Dillon and F. Grosveld. 1993. Transcriptional regulation of multigene loci: multilevel control. *Trends Genet.* 9:134-137.

R. Drapkin, A. Merino, and D. Reinberg. 1993. Regulation of RNA polymerase II transcription. *Curr. Opin. Cell Biol.* 5:469-476.

G. Dreyfuss, M.J. Matunis, S. Piñol-Roma, and C.G. Burd. 1993. hnRNP proteins and the biogenesis of mRNA. *Annu. Rev. Biochem.* 62:289-321.

W.J. Gehring. 1992. The homeobox in perspective. *Trends Biochem. Sci.* 17:277-280.

S.C. Harrison. 1991. A structural taxonomy of DNA-binding domains. *Nature* 353:715-719.

J.D. Hellman and M.J. Chamberlin. 1988. Structure and function of bacterial sigma factors. *Annu. Rev. Biochem.* 57:839-872.

M.W. Hentze. 1991. Determinants and regulation of cytoplasmic mRNA stability in eukaryotic cells. *Biochim. Biophys. Acta.* 1090: 281-292.

J.W.B. Hershey. 1991. Translational control in mammalian cells. *Annu. Rev. Biochem.* 60:717-755.

P.F. Johnson and S.L. McKnight. 1989. Eukaryotic transcriptional regulatory proteins. *Annu. Rev. Biochem.* 58:799-839.

T.K. Kerppola and C.M. Kane. 1991. RNA polymerase: Regulation of transcript elongation and termination. *FASEB J.* 5:2833-2842.

M. Kozak. 1992. Regulation of translation in eukaryotic systems. *Annu. Rev. Cell Biol.* 8:197-225.

S. Kustu, A.K. North, and D.S. Weiss. 1991. Prokaryotic transcriptional enhancers and enhancer-binding proteins. *Trends Biochem. Sci.* 16:397-402.

K.J. Marians. 1992. Prokaryotic DNA replication. *Annu. Rev. Biochem.* 61:673-719.

M. McKeown. 1992. Alternative mRNA splicing. *Annu. Rev. Cell Biol.* 8:133-155.

W.C. Merrick. 1992. Mechanism and regulation of eukaryotic protein synthesis. *Microbiol. Rev.* 56:291-315.

H.F. Noller. 1984. Structure of ribosomal RNA. *Annu. Rev. Biochem.* 53:119-162.

H.F. Noller. 1991. Ribosomal RNA and translation. *Annu. Rev. Biochem.* 60:191-227.

R. Nussinov. 1990. Signal sequences in eukaryotic upstream regions. *Crit. Rev. Biochem. Molec. Biol.* 25:185-224.

C.O. Pabo and R.T. Sauer. 1992. Transcription factors: structural families and principles of DNA recognition. *Annu. Rev. Biochem.* 61:1053-1095.

L. Patthy. 1991. Exons-original building blocks of proteins? *Bioessays.* 13:187-192.

D. Rhodes and A. Klug. 1993. Zinc fingers. *Sci. Am.* 268(2):56-65.

P. Schimmel. 1991. Classes of aminoacyl-tRNA synthetases and the establishment of the genetic code. *Trends Biochem. Sci.* 16:1-3.

P.A. Sharp. 1994. Split genes and RNA splicing. *Cell* 77:805-815.

C.C. Thompson and S.I McKnight. 1992. Anatomy of an enhancer. *Trends Genet.* 8:232-236.

A.A. Travers. 1987. Structure and function of *E. coli* promoter DNA. *CRC Crit. Rev. Biochem.* 22(3):181-219.

T.S.F. Wang. 1991. Eukaryotic DNA polymerases. *Annu. Rev. Biochem.* 60:513-552.

The cloning experiments of Herbert Boyer, Stanley Cohen, Paul Berg, and their colleagues in the early 1970s ushered in the era of recombinant DNA technology (see Chapter 2, The DNA Revolution). A gene can now be separated from the rest of the chromosomal DNA and transferred to a well-studied foreign host cell for further study. Genetic material introduced into foreign cells is replicated and passed on to progeny cells. Today many methods are available for isolating and characterizing genes and proteins. Selected genes are being transferred to organisms such as plants, animals, bacteria, and fungi for a variety of reasons: Commercially desirable products can be efficiently produced in host cells, genes and their proteins can be studied in ways not possible before, and new medical biotechnologies are being explored.

At the heart of most molecular genetic technologies is the gene. A gene must be isolated and characterized before it can be used in genetic manipulations. One method of isolating and amplifying a DNA of interest is to clone the gene by inserting it into a DNA molecule that serves as a vehicle or vector. When these two DNAs of different origin are combined, the result is a recombinant DNA molecule. The molecule is moved into a host—*Escherichia coli*, for example—where it can be reproduced. When the cells divide, each cell, or **clone**, in the colony contains one or more identical copies of the recombinant DNA molecule. Thus, the DNA contained within the recombinant molecule is cloned. Gene cloning has many uses. To name just a few, DNA can be amplified in host cells to obtain many copies for further study, a gene can be **expressed** to obtain a valuable protein product, and a gene and its expression can be studied in a living cell. In this chapter the basic methods of obtaining and analyzing recombinant DNA are described, with emphasis on methods of isolating and characterizing DNA.

CUTTING AND JOINING DNA

Recombinant DNA molecules cannot be easily generated without two types of enzymes: Restriction endonucleases act as scissors to cut DNA at specific sites, and DNA ligase is the "glue" that joins two DNA molecules together in the test tube. Restriction endonucleases cut both strands of the DNA sugar-phosphate backbone. One class, Type II restriction endonucleases, recognizes specific sequences within the DNA molecule. The recognized sequences are usually four to six base pairs and are palindromic (Table 4.1): the sequence on each DNA strand is the same when read in the same direction—for example, in the 5' to 3' direction:

5'GCCAATTGGC3'
3'CGGTTAACCG5'

Each restriction enzyme recognizes a specific sequence and cuts at a particular place within that sequence. The enzyme cuts the double strand of DNA by breaking covalent bonds between the phosphate of one deoxyribonucleotide and the sugar of an adjacent deoxyribonucleotide.

Restriction endonucleases are found primarily in bacteria, where they cut, or fragment, foreign DNA of bacteriophage before the invading DNA can replicate within the host bacterial cell to produce new phage that would ultimately destroy the host. The bacterial cells are resistant because their DNA is chemically modified, primarily by the addition of methyl groups, to mask most of the restriction endonuclease recognition sites so that these will not be cut.

Restriction enzymes are named for the organisms from which they are isolated. For example, *Eco*RI is isolated from *Escherichia coli* RY13: *Eco* comes from the first letter of the genus name and the first two letters of the species name; R is for the strain type and I is for the first enzyme of that type. Thus, *Bam*HI is isolated from *Bacillus amyloliquefaciens* strain H, and *Sau*3A is isolated from *Staphylococcus aureus* strain 3A.

Restriction enzyme cleavage of a sugar-phosphate backbone can produce a double-stranded DNA fragment with either blunt or staggered ends. When both strands of the molecule are cut at the same position, the ends are flush, and no nucleotides are left unpaired; this is a **blunt end**. When each strand of the molecule is cut at a different position so that one strand (either the 5' or the 3') overhangs by several nucleotides, these single-stranded ends can spontaneously base pair with each other—that is, they are "sticky," or cohesive.

The enzyme DNA ligase can join DNA fragments that have complementary **sticky ends** or blunt ends. Ligase catalyzes the formation of covalent bonds between the sugar and the phosphate of the adjacent nucleotides, requiring only that one nucleotide have a free 5' phosphate and the adjacent one have a 3' hydroxyl group. Ligase does not discriminate between DNAs with different origins. Thus, two DNA fragments cut from the chromosomes of two different organisms by restriction enzymes are joined together by DNA ligase. The two fragments are now one DNA molecule. By this "cut and paste" technique a recombinant DNA molecule is produced. It is then transferred to a host cell where it is amplified (replicated) to be studied further.

SEPARATING RESTRICTION FRAGMENTS AND VISUALIZING DNA

The visualization of DNA from restriction enzyme digestions and other manipulations enables the results to be directly visualized. Agarose gel electrophoresis is a tech-

Table 4.1	Examples of Some Restriction Endonucleases and Their Properties			
			Characteristics	
Restriction Endonuclease	Source (Bacterial Species)	Target Site (Cuts at Arrow)	Recognizes (No. Base Pairs)	Product
Eco RI	*Escherichia coli* R13	↓ G-A-A-T-T-C C-T-T-A-A-G ↑	6	4-base-long sticky ends
Hha I	*Haemophilus haemolyticus*	↓ G-C-G-C C-G-C-G ↑	4	2-base-long sticky ends
Sma I	*Serratia marcescens*	↓ C-C-C-G-G-G G-G-G-C-C-C ↑	6	Blunt ends
Hae III	*Haemophilus aegyptius*	↓ G-G-C-C C-C-G-G ↑	4	Blunt ends

J. Ingraham and C. Ingraham, *Introduction to Microbiology*. Copyright © 1995 Wadsworth Publishing Co.

nique for separating DNA fragments by size and visualizing them after staining. To make the gelatinous agarose, a powder of purified agar (isolated from seaweed) is mixed with buffer, boiled, and poured into a mold where the agarose (generally 0.7 to 2.0%) gels and solidifies into a slab. A toothed comb forms "wells" in the molten agarose, and samples are loaded into the wells after the agarose solidifies and the well-forming comb is removed. The agarose slab is submerged in a buffer solution and an electric current is applied to electrodes at opposite ends of the slab to establish an electric field in the gel and the buffer. Since the sugar–phosphate backbone is negatively charged, the DNA fragments migrate toward the positive electrode.

Pores between the agarose molecules act like a sieve that separates molecules by size. Larger molecules move more slowly than smaller molecules. Increasing the percentage of agarose produces smaller pores; these increase the resolution of smaller fragments by impeding all but the smaller molecules. Lowering the percentage of agarose produces larger pores; these increase the resolution of larger fragments by allowing greater separation among them in the gel.

Since DNA by itself is not visible in the gel, ethidium bromide is added to make the DNA bands visible. Ethidium bromide molecules intercalate between the bases causing the DNA to fluoresce orange when the gel is illuminated with ultraviolet light. The lengths of the DNA fragments can be determined by comparing their position in the gel to reference DNAs of known lengths also in the gel. A DNA fragment migrates a distance that is inversely proportional to the logarithm of the fragment length in base pairs over a limited range in the gel.

Thus, using agarose gel electrophoresis allows the restriction fragment lengths to be determined.

CELL TRANSFORMATION

To introduce new DNAs into organisms or to use hosts for gene cloning, methods of gene transfer are required. A variety of methods are available for transforming bacteria, yeast, plants, and animals with foreign DNA. Some bacteria are naturally **competent**—that is, they have the ability to take up exogenous DNA. However, many bacteria must be treated chemically to become competent. Exposing bacteria to a salt solution such as calcium chloride and applying a heat shock in the presence of recombinant DNA is usually sufficient for transferring DNA into cells; that is, effecting **transformation**. However, eukaryotes require more complicated methods.

In organisms, like fungi, algae, and plants, that have cell walls, enzymes are used to degrade the walls, producing **protoplasts**, or cells without walls. Foreign DNA then can be introduced into protoplasts by **electroporation**: Protoplasts are exposed to a brief electrical pulse, which is thought to introduce transient openings in the cell membrane through which the DNA molecules enter. After transformation, cells are washed and cultured so that the cell wall re-forms and cell division begins. Transformed plants can be regenerated from cell culture (see Chapter 6, Plant Biotechnology).

Another method of transforming cells is microprojectile bombardment, or **biolistics** (Figure 4.1). Very small (4μm) microprojectiles made of gold or tungsten are coated with DNA and shot at high velocity from a

Figure 4.1 Particle gun used for transforming cells with DNA-coated microprojectiles.

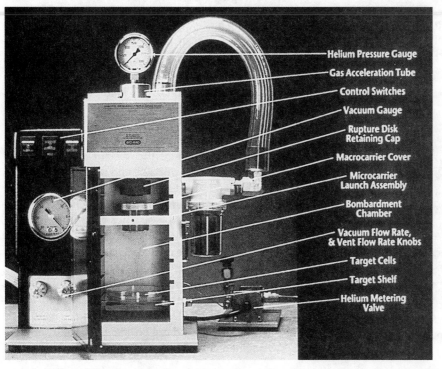

Helium Pressure Gauge
Gas Acceleration Tube
Control Switches
Vacuum Gauge
Rupture Disk Retaining Cap
Macrocarrier Cover
Microcarrier Launch Assembly
Bombardment Chamber
Vacuum Flow Rate, & Vent Flow Rate Knobs
Target Cells
Target Shelf
Helium Metering Valve

particle gun into cells or tissue. Since the projectiles penetrate the cell, the walls do not have to be removed. Biolistics shows much promise for use in gene therapy, whereby cells of organs in living animals can be transformed with a corrective gene.

Lacking walls, animal cells are easily transformed by electroporation, as well as by a method that precipitates DNA onto the cell surface with a calcium phosphate solution. Viruses often are used to transfer DNA into animal cells for the stable integration into the chromosome. (Viruses are also an important tool in the transformation of plant and bacterial cells.) However, the introduction of genes into all the cells of a multicellular animal requires a different method, called **microinjection**. Unlike plant cells, animal cells cannot regenerate into an entire organism. To express a gene in all the cells of an animal, thus generating a transgenic organism, fertilized eggs or very early embryos are transformed by microinjection (Figure 4.2). DNA is injected directly into the nucleus of animal cells with an extremely fine pipette. After DNA is transferred into the cell, it is integrated into the chromosome, and the transformed fertilized egg is implanted into an animal for the completion of development.

CLONING VECTORS

Often a foreign DNA is introduced into a host cell by insertion into a cloning vehicle or vector that can transport an inserted DNA of interest into the host cell. Many of these vectors replicate independently in a host cell. A cloning vehicle must (1) have an origin of replication so that the DNA can be replicated within a host cell; (2) be small enough to be isolated without undergoing degradation during purification; (3) have several unique restriction sites for cloning a DNA fragment so that the vector will be cut only once, and several restriction sites for insertion will be available; (4) have selectable mark-

ers for determining whether the cloning vehicle has been transferred into cells and to indicate whether the foreign DNA has been inserted into the vector.

Bacterial Vectors

The greatest variety of cloning vectors has been developed for *E. coli* because of the major role they have played in recombinant DNA experiments since the 1970s. Other cloning vectors are available for bacteria, such as *Bacillus subtilus*, as well as for yeast, fungi, animals, and plants.

Plasmids Bacteria harbor plasmids: circular double strands of DNAs that are **extrachromosomal**; that is, are not part of the bacterial chromosome. Often multiple copies of these plasmids are present in the cell. Plasmids have diverse functions. Some encode substances for antibiotic resistance and bacteriocins—agents that kill or inhibit similar bacterial strains or species. Others perform physiological functions, such as pigment production, degradation of compounds, and dinitrogen fixation. Toxin-producing and virulence plasmids encode endotoxins and hemolysins, while some plasmids confer resistance to metals such as mercury, cadmium, nickel, and zinc. Because plasmids are relatively small and easy to manipulate, they have been engineered as cloning vectors with unique restriction sites for insertion fragments of foreign DNA—up to approximately 10 kilobases in length (one kilobase or kb equals 1,000 nucleotides). When transported into bacterial cells, recombinant plasmid vectors are readily replicated, often in high copy number.

An early vector, pSC101, used by Cohen, Boyer, and their colleagues (discussed in Chapter 2) replicated to yield only one or two copies in each cell. Col E1, a high copy plasmid developed in 1974, produced several thousand copies in one cell. Col E1 was further modified so that several unique restriction sites were available for the insertion of DNA. One of these modified plasmids was pBR322, constructed in Boyer's laboratory (Figure 4.3). The name pBR322 is derived thus: p identifies the molecule as a plasmid; BR identifies the original constructors of the vector (Bolivar and Rodriquez); 322 is the identification number of the specific plasmid (other examples are pBR325, pBR327, and pBR328).

pBR322 is derived from three naturally occurring plasmids: the ampicillin-resistance gene from the plasmid R1, the tetracycline-resistance gene from pSC101, and the replication region from pMB1. Until recently, pBR322 was one of the most commonly used plasmids, for several reasons. First, the molecule is small, having only 4,363 base pairs, and can be isolated easily. Consequently, this vector can accommodate DNA of up to 5 to 10 kb. Second, pBR322 has several unique restriction sites where the plasmid can be linearized for inserting a DNA fragment. Finally, the genes encoding resistance to

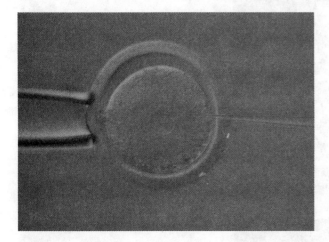

Figure 4.2 Microinjection of DNA into a pronucleus of a fertilized animal egg.

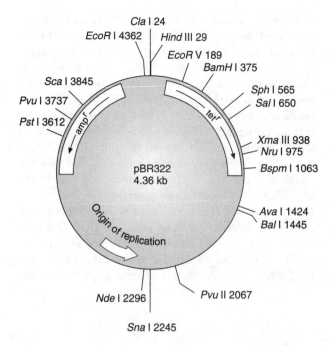

Figure 4.3 Restriction map of pBR322 showing unique restriction sites, the ampicillin and tetracycline resistance genes, and the origin of replication.

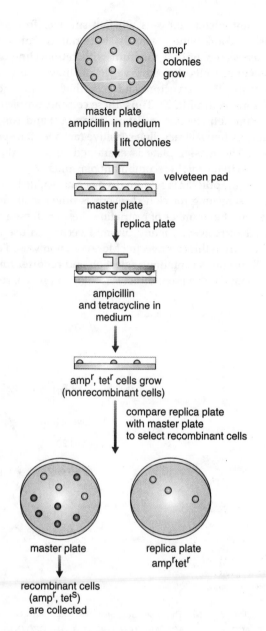

Figure 4.4 Selecting for recombinant cells after transformation.

ampicillin (amp) and tetracycline (tet) are used for plasmid and DNA insert selection. The DNA fragment of interest can be inserted into one of the antibiotic-resistance genes, inactivating that gene. The other antibiotic gene remains active and can be used to select for bacteria carrying the plasmid. Therefore, if a foreign DNA is inserted into the tetracycline gene, the bacteria containing such a plasmid would be ampicillin-resistant but tetracycline-sensitive ($amp^r tet^s$). Insertional inactivation is a powerful way to determine whether the vector contains a DNA insert. Ligation and transformation are inefficient processes; consequently, in a population of bacterial cells that have been transformed, not all cells have received a recombinant plasmid. In addition, plasmids without DNA inserts are present and may be transferred to cells. A screening process selectively kills cells by antibiotics to identify those bacterial cells that contain a recombinant plasmid: DNA is inserted to inactivate an antibiotic resistance gene so that the presence of recombinant molecules in bacterial cells can be detected (Figure 4.4).

To selectively kill cells with antibiotics, **colonies** must first be plated from the original, master plates to identical replica plates. Different antibiotics are added to the medium to identify (that is, select) cells with recombinant molecules. To maintain the original bacterial colonies so that recombinant colonies from the master plates can later be further analyzed, exact copies are made by **replica plating**. The technique is conducted as follows. After colonies of bacterial cells have grown in a medium such as agar, a sterile velveteen pad is pressed

against the master plate (the dish holding the colonies). Cells from the colonies adhere to the velveteen (other transfer media can be used, such as nylon filter paper). The location of the cells on the velveteen will present a mirror image of the locations of the original colonies on the master plate. The velveteen pad is then pressed against media in a second and third plate, transferring cells to them. The locations of these cells will now be identical to the original colonies on the master plate. Recombinant colonies are identified (and therefore selected) by the addition of different antibiotics to the replica plates.

When cells are placed onto ampicillin-containing agar medium, only cells with pBR322 will be able to survive; untransformed cells will not survive. To distinguish recombinant plasmids, bacterial colonies are replica-plated onto a medium with tetracycline and ampicillin. Cells that survive do not have the DNA insert in pBR322 vectors and cells that die have the DNA insert in pBR322. Thus, these cells are ampicillin resistant/tetracycline resistant (amp^{r}tet^r) and ampicillin resistant/tetracycline sensitive (amp^{r}tet^s), respectively. The master plate is examined to select those colonies that contain recombinant plasmids.

Today, pBR322 is seldom used because of its limitations—screening for cloned inserts is time-consuming, and a limited number of restriction sites for cloning are available. However, many plasmid vectors in use are derived from this early vector. More commonly used are small plasmids containing selection and **reporter** functions that allow a more direct screening approach, such

as the pUC plasmids developed in 1982. These small vectors (less than four kb in length) contain a polycloning site made up of multiple restriction sites, where foreign DNA can be inserted. A series of pUC vectors containing the ampicillin resistance gene for plasmid selection has been constructed (e.g. pUC8, pUC9, pUC18, pUC19). The multiple cloning sites differ in the type of restriction sites and their orientation (Figure 4.5). Selectable markers for insert selection are not limited to antibiotic resistance genes; the genes may encode enzymes that catalyze metabolic reactions. A chemical indicator aids in the identification of recombinant plasmids.

One type of selection, called α-**complementation**, allows the detection of DNA inserts in pUC plasmids. These vectors contain a portion of the *lacZ* gene (called *lacZ'*) that encodes the first 146 amino acids for β-galactosidase. The polycloning site resides in this coding

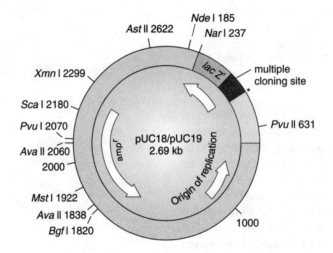

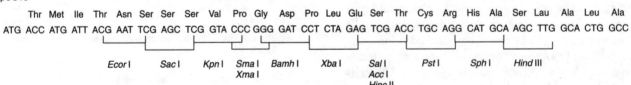

Multiple cloning sites
pUC18

| Thr | Met | Ile | Thr | Asn | Ser | Ser | Ser | Val | Pro | Gly | Asp | Pro | Leu | Glu | Ser | Thr | Cys | Arg | His | Ala | Ser | Lau | Ala | Leu | Ala |
| ATG | ACC | ATG | ATT | ACG | AAT | TCG | AGC | TCG | GTA | CCC | GGG | GAT | CCT | CTA | GAG | TCG | ACC | TGC | AGG | CAT | GCA | AGC | TTG | GCA | CTG | GCC |

Ecor I   *Sac* I   *Kpn* I   *Sma* I / *Xma* I   *Bamh* I   *Xba* I   *Sal* I / *Acc* I / *Hinc* II   *Pst* I   *Sph* I   *Hind* III

pUC19

| Thr | Met | Ile | Thr | Pro | Ser | Leu | His | Ala | Cys | Arg | Ser | Thr | Leu | Glu | Asp | Pro | Arg | Val | Pro | Ser | Ser | Asn | Ser | Leu | Ala |
| ATG | ACC | ATG | ATT | ACG | CCA | AGC | TTG | CAT | GCC | TGC | AGG | TCG | ACT | CTA | GAG | GAT | CCC | CGG | GTA | CCG | AGC | TCG | AAT | TCA | CTG | GCC |

Hind III   *Sph* I   *Pst* I   *Sal* I / *Acc* I / *Hinc* II   *Xba* I   *Bamh* I   *Sma* I / *Xma* I   *Kpn* I   *Sac* I   *EcoR* I

* In pUC18, *EcoR* I
  In pUC19, *Hind* III

Figure 4.5 Restriction map of pUC18/pUC19 showing the ampicillin resistance gene, the origin of replication, and the multiple cloning site within the *lacZ* gene. The orientation is reversed in pUC18 and pUC19 with the *EcoR*I site and the *Hind*III site of the multiple cloning site immediately downstream from the lac promoter in pUC18 and pUC19, respectively.

region. If the *lacZ* region is not interrupted by inserted DNA, the amino terminal portion of the β-galactosidase polypeptide (lacZ') is synthesized. An *E. coli* deletion mutant called *lacZ*ΔM15 harbors a chromosomal mutant of *lacZ* that encodes only the carboxyl end of the β-galactosidase. Both the plasmid and chromosomal *lacZ* fragments encode nonfunctional proteins. However, by α-complementation the two partial proteins can associate and form a functional β-galactosidase. In cells, β-galactosidase normally hydrolyzes the sugar lactose into glucose and galactose. The α-complementation is indicated by the presence of blue colonies on special media. The blue color indicates that the plasmid-encoded lacZ' has combined with the partial lacZ from the **complementary** fragment of the chromosomal *lacZ* gene residing on the chromosome of *E. coli*, generating an enzymatically active β-galactosidase that turns a chromogenic agent blue. The chromogenic lactose analog, X-gal or 5-bromo-4-chloro-3-indoyl-β-D-galactopyranoside, when added to the medium, is broken down by β-galactosidase and produces a blue color. When the plasmid lacZ' gene fragment on the plasmid is interrupted by the insertion of a foreign DNA fragment, colonies appear white on media.

Bacteriophage A virus that infects a bacterium is called a bacteriophage. Viral DNA can be engineered for use as a cloning vector; the first to be used was from a lambda bacteriophage, in 1974. Today many variations of lambda exist. Lambda phage vectors are derived from the 50-kb wild-type double-stranded genome that has single-stranded complementary ends of 12 nucleotides (**cohesive termini** or *cos*) that can base pair. The *cos* ends (important for the **lytic** pathways) base pair forming a circular DNA molecule once the phage DNA is inside the host cell. DNA replication then occurs from the circular molecules, producing linear lambda DNA made up of several 50-kb phage DNA end on end. In the lytic pathway (cycle), the host cell lyses after phage reproduction, releasing progeny virus. Lytic phages are used to clone and amplify a DNA of interest. The resulting **plaques**—cleared areas on the medium where host cells have lysed—contain millions of recombinant phage particles that can be isolated. In the **lysogenic** pathway, the bacteriophage genome is integrated into the host cell, no cell lysis occurs, and the phage genome replicates along with the host genome. For example, in gene therapy, animal viruses are used to integrate a cloned therapeutic gene into the genome.

All phage vectors used as cloning vectors have been disarmed for safety purposes and can function only in special laboratory conditions. Recombinant DNA containing the viral DNA and the DNA of interest are packaged into viral particles in the test tube. Host bacterial cells are infected with recombinant phage DNA, the DNA replicates within the host cells, and progeny phage are produced when the host cell undergoes lysis. Plaques then become visible on plates where cells in the colony have lysed.

The DNA of lambda-type phage can accommodate only an additional 3 kb or 5% of its genome, for a total size of 52 kb. DNA that is too large or too small cannot be packaged into the head of the viral particle. Fortunately, a significant portion of the viral genome can be deleted without adversely affecting packaging and infection of *E. coli* host cells. Removal of one-third of nonessential DNA from the central portion of the phage allows a DNA insert up to about 20 kb to be ligated into the phage DNA. The deletion contains most of the DNA necessary for integration and excision from the *E. coli* genome and is not necessary for use as a cloning vector. Vectors that have a segment of nonessential DNA removed, making room for foreign DNA, are called replacement vectors. Another type of phage vector, called an insertion vector, has one restriction site for the insertion of DNA between five to ten kb in size.

Cosmids A significant disadvantage of using plasmids and bacteriophage as vectors is that only relatively small DNA fragments can be inserted. Larger DNA fragments can be cloned using engineered hybrids of phage DNA and plasmids called **cosmids**. These are plasmids with a small portion of lambda DNA, the *cos* sites. A cosmid vector not only is composed of *cos* sites for packaging into phage particles, but also contains a plasmid replication origin for replication in bacterial hosts and genes for plasmid selection (for example, antibiotic resistance genes to confer resistance to anitbiotics such as ampicillin or kanamycin). The cosmid vector is packaged in a protein coat *in vitro* as with a bacteriophage vector; however, after the packaged DNA infects *E. coli* host cells, the DNA replicates as a plasmid rather than as bacteriophage DNA, and the cells are not lysed. Bacterial colonies rather than plaques are formed on petri plates. Cosmid vectors are small (some are only 2.5 kb), and since they are packaged for infection of host cells if *cos* sites are separated by 37 to 52 kb, they accommodate large inserts of foreign DNA. Typically 35 to 45 kb can be cloned into cosmid vectors.

Vectors for Other Organisms

Although many cloning experiments are carried out using *E. coli* as the host cells, other organisms also are used that require different types of cloning vectors. When the goal is to obtain a protein product such as insulin or growth hormone, or to modify the properties of a specific organism, such as to introduce pest resistance into soybean, the cloning vector must be compatible with the organism used. Cloning vectors are available for yeast and other fungi, plants, insects, fish, and mammals.

Yeast Artificial Chromosomes (YACs) Yeast artificial chromosomes (YACs) are useful for eukaryotic molecular studies. The yeast chromosome has the following necessary components: (1) a **centromere**, which distributes the chromosome to the daughter cells during cell division; (2) a **telomere** at the end of the yeast chromosome to ensure that the end is correctly replicated and to protect against degradation; and (3) an **autonomously replicating sequence** (ARS), which consists of specific DNA sequences that enable the molecule to replicate. YACs also have a gene that provides a way to detect an inserted DNA fragment. These components are joined together to make a yeast artificial chromosome that can replicate in yeast host cells.

YACs are especially useful for cloning large DNA fragments. Many animal genes being studied can be 200 kb or more in length, requiring a cloning vector that accommodates very large fragments. As noted earlier, many vectors accommodate only rather small DNA inserts. YAC vectors accept fragments between 200 to 1,500 kb, allowing complete genes as well as gene clusters to be cloned for study. YAC libraries are available, and are very useful for large genomes, such as human genomes. They enable the researcher to isolate and sequence specific regions of the genome (see Chapter 9, The Human Genome Project).

Plant Cloning Vectors DNA is being cloned into plants for several purposes: to generate resistance to disease, pests, and herbicides; to improve crop yields, quality, and nutritional value; to develop new ornamental plant characteristics; and to increase the shelf life of many common fruits and vegetables. Several cloning vectors have been constructed that allow for efficient transfer of DNA to plant cells. The most commonly used are plant viruses, such as tobacco mosaic virus, or TMV, and the **Ti plasmid** of the soil bacterium, *Agrobacterium tumefaciens*.

A. tumefaciens is a soil microorganism that induces crown gall formation in many species of dicotyledonous plants. This bacterium infects tissue wounds in plants (such as in the stem) and induces plant cells to proliferate, resulting in a cancerous tissue mass, or crown gall, near the infection site. The bacterium's large Ti plasmid—greater than 200 kb—is what induces crown gall formation: Ti plasmid genes are involved in infection and they induce plant cell division that leads to the tumorlike growth. A special region on the plasmid, the **T-DNA** (tumor-inducing DNA), containing approximately eight genes that encode the disease characteristics, is incorporated into the plant's genome. Some of these genes direct the synthesis of unusual compounds, called opines, that the bacterial cells use as nutrients for growth. Thus, *A. tumefaciens* redirects transcription and translation within the plant host cell for its own use.

The properties of the Ti plasmid make it an efficient cloning vector in certain plants. The T-DNA, which integrates into the plant chromosomes, can be used to transfer foreign genes into plants (Figure 4.6). For use in biotechnology, the engineered Ti plasmid lacks some of the genes that contribute to the cancerous properties.

One of the challenges facing scientists is the development of efficient methods of transferring foreign DNA into all the cells of a plant. One way is to integrate the DNA of interest into a few cells, which, once transformed, divide and give rise to the whole plant (see plant cell culture in Chapter 6, Plant Biotechnology). Another way is to transfer DNA to cells in the embryo by soaking seeds in a solution containing recombinant *A. tumefaciens* bacteria.

A selectable marker gene included in the cloning vector enables scientists to identify plants that harbor T-DNA with the foreign gene or **cDNA**. A reporter gene, connected to the DNA of interest and under the same control, is used to indicate whether the foreign gene or cDNA is being expressed in the plant. Many reporter genes are available that encode enzymes with readily assayed activities. One important reporter used today is the luciferase gene taken from either the firefly or bacteria. When the

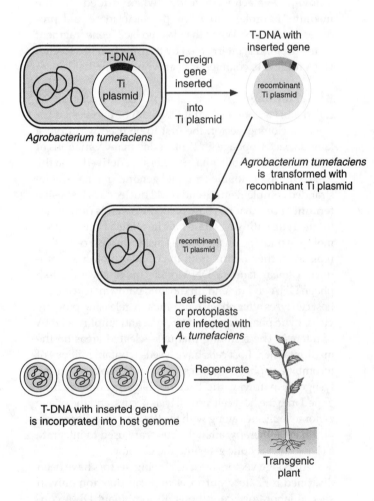

Figure 4.6 The use of *Agrobacterium tumefaciens* to transfer DNA into plants. A gene of interest is cloned into the T-DNA of the *A. tumefaciens* plasmid (along with a selectable marker gene), and *A. tumefaciens* is transformed with the recombinant Ti plasmid. Transformed *A. tumefaciensis* is used to infect plant cells in culture; the T-DNA, including the foreign gene, is transferred into plant chromosomal DNA. Plants regenerated from cultured cells are tested for the presence of the foreign gene.

firefly luciferase gene is expressed in the presence of the substrate luciferin, the plant gives off detectable **bioluminescence** (Figure 4.7). Reporter genes enable the investigator to quantify the level of expression and to identify the tissue in which the gene is being expressed.

Mammalian Cell Vectors The synthesis of complex eukaryotic gene products requires eukaryotic hosts (such as human mouse cells). Often, eukaryotic gene products must be modified in ways not possible with bacterial cells, such as by post-translational cleavage, glycosylation, and amino acid modifications such as phosphorylation, acetylation, sulfation, and acylation. Although cDNAs can be used with prokaryotic host cells, if a gene encoding a protein product contains introns or if the protein requires processing after synthesis, mammalian host cells often are used. Many animal proteins can be produced only in eukaryotic hosts, since only these cells have the machinery to process RNA and protein. Many important pharmaceutical compounds for therapeutic use are produced in this way; among them are Factor VIII used by hemophiliacs to aid in blood clot formation, β-interferon, growth hormone, and erythropoietin. Vectors specific for animal hosts have been developed for a variety of uses (see Chapter 10, Medical Biotechnology).

The first eukaryote-infecting virus to be used for cloning, and one of the most-studied viruses, was Simian virus 40, or SV40, a small, circular, double-stranded DNA tumor virus. Although the virus can infect the cells of several mammalian species, including monkey cells, its host range is limited. Moreover, only a limited amount of foreign DNA can be inserted because the DNA must be packaged into a viral coat or capsid.

To provide space for inserted foreign DNA, both nonessential and essential genes have been deleted in vectors derived from SV40. Thus, a helper virus containing the necessary genes is required to propagate and package recombinant SV40 virus. Finally, only transient expression of an inserted gene has been achieved.

Retroviruses are single-stranded RNA viruses that show much promise for use as vectors in a wide variety of animal cells, including human. The viral genome contains two single-stranded RNA molecules that are held together by hydrogen bonds at the 5' ends. To replicate, the virus uses the enzyme **reverse transcriptase** to make a double-stranded DNA molecule from the RNA template. This DNA integrates stably into the chromosomes of dividing host cells, where transcription and translation of the provirus (i.e., the integrated DNA) occur. Some of the synthesized RNAs are packaged into virus particles.

Retroviruses are very useful cloning vectors for several reasons: (1) Most of the genes in the provirus can be replaced with foreign DNA if a helper virus is used during infection of host cells, allowing the insertion of large fragments of DNA; (2) retroviruses can infect cells and transfer their DNA with high efficiency; (3) they have a broad host range and so can transform a wide variety of mammalian species and cell types; (4) the double-stranded proviral DNA that integrates into the chromosome is very stable; and (5) the virus does not kill host cells, since the DNA integrates into the chromosome.

The adenovirus is a double-stranded DNA virus into which fragments greater than 30 kb can be inserted. Like the retrovirus, it can infect cells with high efficiency and has a broad host range. Since the pathogenicity in humans is low, the adenovirus makes a desirable vector for gene therapy. Moreover, unlike the retrovirus, adenovirus does not have to infect dividing cells.

CONSTRUCTING AND SCREENING A DNA LIBRARY

Vectors are used to compile a "library" of DNA fragments that have been isolated from the genomes of a variety of organisms. This collection of fragments is used to isolate specific genes and other DNAs. The relationship of a single DNA fragment to a **DNA library** is analogous to that of one book to a library containing thousands of books. Searching for one specific piece of DNA is similar to looking for one book in a library.

DNA fragments are generated by cutting the DNA with a specific restriction enzyme. These fragments are ligated into vector molecules and the collection of recombinant molecules is transferred into host cells, one molecule in each cell. This library is searched, that is, screened, with a molecular **probe** that specifically identifies the target DNA.

To allow scientists to learn more about the organization and regulation of genes, libraries for a variety of

Figure 4.7 A tobacco plant transformed with the firefly luciferase gene. The bioluminescence indicates the gene was incorporated into the plant DNA and is being expressed.

organisms have been constructed to **map** and sequence their genomes (see Chapter 9, The Human Genome Project). For example, libraries are available for the bacteria *Bacillus subtilis* and *E. coli,* the yeast *Saccharomyces cerevisiae,* the nematode *Caenorhabditis elegans,* the fruit fly *Drosophila,* and the plant *Arabidopsis.* To achieve such ambitious goals, researchers must construct ordered, overlapping genomic clones representing the entire genome; that is, all the DNA in each chromosome.

Two main types of libraries can be used to isolate specific DNAs: genomic and cDNA. A genomic library contains DNA fragments that represent the entire genome of an organism. A **cDNA library** is derived from mRNA of a tissue and consequently contains only sequences that are expressed at a given time in a specific tissue.

Genomic Library

Soybeans will be used as an example to demonstrate how a genomic library is constructed. First, total nuclear DNA is isolated from soybean cells and cut with a specific restriction enzyme. At the same time, a cloning vector (which can be a plasmid, cosmid, or bacteriophage) is cut with the same restriction enzyme so that the vector is linearized and the ends are complementary to those of the genomic DNA fragments. The two DNAs—genomic fragments and vector—are mixed together in a test tube and DNA ligase is added to form recombinant molecules.

The recombinant DNAs are introduced into host cells, usually *E. coli,* if plasmids, cosmids, or bacteriophage vectors are used. Transformed bacterial cells, each containing recombinant plasmids or cosmids, multiply when plated onto antibiotic-containing medium. Each colony contains a soybean DNA fragment cloned into the vector. Plaques, cleared areas where *E. coli* cells have lysed, contain recombinant virus with soybean genome inserts.

The collection of plaques or colonies that together contain all the DNA fragments of a genome constitute the library. For example, the human genome comprises more than 100,000 cosmid clones, or approximately 10,000 YAC clones. The soybean genomic library usually includes at least one complete soybean genome cut into fragments and housed within *E. coli* cells. It is best to have clones representing four or five genomes so that all of the genetic information is represented at least once. From the average size of the DNA inserted into the vector and from the genome size of the organism, a simple calculation tells researchers how many library clones are required to represent the entire genome.

cDNA Library

Some organisms have very large genomes, and a genomic library may yield too many different clones to manipulate easily, even when a host–vector system is used that can accommodate large DNA inserts. When the library must be screened for a particular gene, especially in a plant or animal, the number of clones to be screened may be unmanageable. Thus, a library that includes only expressed genes from a certain type of cell may be more feasible (for example, leaf-specific cDNAs from soybean). This library, called a cDNA library, dramatically reduces the total amount of DNA to be cloned since only a fraction of genomic DNA is expressed by any given cell type. Genes are expressed differentially; consequently, in a complex multicellular organism, specialized cells produce cell-specific proteins. Although a cell has a full complement of genes, only a specific set of genes (in addition to "general housekeeping" genes) will be expressed, and all the others will be silent.

To make, for example, a soybean cDNA library, RNA from leaf cells is isolated and used as a template to make DNA by a method called complementary DNA (cDNA) synthesis. Reverse transcriptase catalyzes the reverse synthesis of complementary DNA from the mRNA template. The mRNA then is degraded with a **ribonuclease** or an alkaline solution, and DNA polymerase is used in the synthesis of the second DNA strand. Double-stranded **DNA linkers** with ends that are complementary to the cloning vector are added to the double-stranded DNA molecule before ligation into a cloning vector. Recombinant clones are introduced into bacteria. Thus, the library is composed of cDNAs from the expressed mRNAs in soybean leaf cells.

The differences between genomic and cDNA libraries are significant, and they influence the choice of library for scientific research. First, the noncoding introns present within most eukaryotic genes are not included in the cDNA, since the mRNA has already undergone post-transcriptional modification before isolation for cDNA synthesis. In addition, a cDNA library does not contain regulatory elements associated with genes, such as promoters and enhancers; nor does it contain the portion of the genome that does not code for RNA—that is, noncoding DNA. Fewer clones compose the library, making the screening process much simpler and less time-consuming. If noncoding DNA (such as promoter sequences) is to be isolated from a library, a genomic library must be used.

Screening Libraries

Once a library, whether genomic or cDNA, has been constructed, the recombinant clones that contain a specific DNA insert can be detected by the screening process called nucleic acid **hybridization**. For screening, a specific DNA sequence is used as a probe to identify the clones or plaques containing the appropriate target sequence. The library of bacterial colonies or plaques is transferred from master plates (the petri dishes) to membranes made of nylon or nitrocellulose (a cellulose treated with nitrates and pressed into paper) in a manner similar to replica plating (Figure 4.8). Membranes are placed over the colonies or plaques, pressed against them to transfer the colonies or plaques to the membranes, and lifted. More than one membrane can be used for each master plate to produce multiple membranes for screening the library in duplicate or triplicate.

Within the library are perhaps one to several colonies or plaques that contain the DNA of interest. A DNA probe is designed to complement the target DNA and base pair (i.e., hybridize) with it. The result is a DNA hybrid: one target DNA strand hydrogen bonded to one probe strand. The membranes first are treated so cells in colonies are lysed and cell debris is removed. The remaining DNA is denatured to obtain single-stranded molecules. Single-stranded DNA molecules bind tightly to the membrane by the sugar-phosphate backbones, and the unpaired nucleotides are free to base pair with a complementary DNA probe. Hybridization of single-stranded probes to these membranes is used to identify specific recombinant DNA molecules from the bacterial colonies or bacteriophage plaques transferred to the membranes. Although the probe and target DNA may not be completely complementary, enough base pairing must be established to produce a stable hybrid.

To detect whether the DNA probe and the target DNA bound to the membrane have hybridized, the probe is first made radioactive or is labeled by a nonradioactive method that results in fluorescence or staining after hybridization. After hybridization, the excess probe is washed from the membranes, which are then exposed to photographic film. Where hybridization has occurred, radiolabeled DNA makes an autoradiographic image.

Probes must be sufficiently complementary to the target DNA for hybridization to occur. A complementary DNA probe can be a piece of DNA isolated from another organism or cDNA. For example, a **heterologous probe** (that is, the same DNA fragment from another species) from spinach might be used to hybridize to soybean DNA. For the nucleotide sequence of the probe to base pair with the target sequence, the spinach DNA must have enough sequence conservation; that is, the majority of the nucleotide sequence of the spinach DNA must be the same as the soybean DNA. Stringency can be controlled during hybridization by adjusting temperature and salt concentration. Under conditions of highest stringency, hybridization occurs only when the bases between probe and template base pair almost perfectly. Under conditions of lowered stringency, base pairing between partially complementary sequences can occur and **mismatches** are tolerated. Thus, closely matching, but nonidentical DNAs, can be used as probes.

The probe also can be made by a DNA synthesizer. If a portion of the amino acid sequence of the protein is known, the possible nucleic acid sequences can be deduced. A short DNA fragment called an **oligonucleotide** can be synthesized, labeled radioactively, and used as a probe in hybridizations. The genetic code is degenerate (see Chapter 3, Gene Expression), and the specific codons for amino acids sometimes preferentially used by an organism (codon usage) usually are unknown. Consequently, a degenerate oligonucleotide probe must be made to account for all the different codons that may encode an amino acid. For example, to make an oligonucleotide that encodes the amino acids isoleucine-aspartic acid-methionine-tryptophan-glutamic acid-glutamine, the degenerate oligonucleotide would look like this:

$$\text{AUAGACAUGUGGGAACAA}$$
$$\text{C} \quad \text{U} \qquad\qquad \text{G} \quad \text{G}$$
$$\text{U}$$

Expression Libraries

Expression libraries are made with a cloning vector that contains the required regulatory elements for gene expression, such as the promoter region. In an *E. coli* expression vector, an *E. coli* promoter is placed next to a unique restriction site where DNA can be inserted. When a foreign gene or cDNA is cloned into an expression vec-

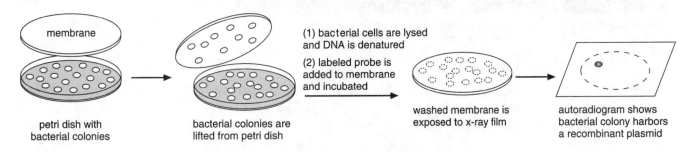

Figure 4.8 Colony hybridization is used to identify bacterial cells that harbor a specific recombinant plasmid. Bacterial cells transferred to membranes are disrupted and the DNA denatured by an alkaline solution. The DNA bound to the membrane is hybridized to a radioactively labeled DNA probe. X-ray film is exposed to detect the colonies containing the DNA of interest on the membranes.

tor in the proper reading frame (see Chapter 3, Gene Expression), the gene is transcribed and translated in the *E. coli* host cell. When a library is made using an expression vector, only DNAs—that is, genes and cDNAs, inserted in-frame within the vector are expressed. Expression libraries are useful for identifying a clone containing the gene or cDNA of interest when an **antibody** to the protein encoded by that gene or cDNA is available.

Antibodies are made by eukaryotic cells in response to proteins and other molecules that are recognized as foreign. Antibodies can be used instead of a nucleic acid as a probe to isolate a clone containing a sequence of interest that is transcribed and translated. A radioactively labeled antibody can be used to identify a specific protein made by one of the clones of the expression library. Antibody binding, a technique similar to nucleic acid hybridization, identifies the clone containing the gene expressing the specific protein.

Expression vectors and a variety of host cells are used routinely to produce large quantities of a specific protein and have important commercial applications. Table 4.2 lists a few host cells and their important recombinant products.

SOUTHERN BLOT HYBRIDIZATION

In recombinant DNA technology, the location of specific regions within a cloned gene or DNA insert often must be known before the fragment can be sequenced or studied further. For example, the exact location of a 2-kb gene must be identified within a 15-kb insert so that it can be isolated for further analysis or sequencing. Restriction endonuclease digestion reduces the large DNA insert to DNA fragments, which then are separated by agarose gel electrophoresis. To identify the fragment that contains the gene of interest, a specific DNA probe, such as a small region of the gene of interest, can be used to hybridize to the DNA fragments.

In the mid 1970s, Edward Southern developed a simple technique, called **Southern blotting**, by which DNA fragments are transferred from gels to a special membrane that then is used for hybridization. During Southern blotting, these DNA fragments are denatured by an alkaline (high pH) buffer and the single strands are transferred to a nylon or nitrocellulose membrane cut to the size of the agarose gel. A "sandwich" is made: The membrane is placed on the gel and paper towels are stacked on top of the membrane (Figure 4.9). Buffer in a trough moves by capillary action through a wick placed under the agarose slab, facilitating the transfer of DNA from the gel to the membrane. The membrane is a replica of the agarose gel and is used in hybridizations as described earlier for colony or plaque hybridizations. If a probe hybridizes to fragments on the membrane, photographic film placed next to the membrane will be exposed where the probe has hybridized to a specific DNA band or bands. Southern hybridization is often used to identify a specific gene fragment from the often many DNA bands on a gel.

Southern blot hybridizations are used to detect differences in banding patterns between organisms or within DNA regions. Differences in restriction fragment lengths of DNA, called **restriction fragment length polymorphisms**, or RFLPs, are used to generate individual DNA "fingerprints." DNA **fingerprinting** is a powerful tool in forensic analysis (see Chapter 11, Forensics and DNA Profiling). Individuals can be identified by their DNA fingerprints, and since RFLPs are inherited, such fingerprinting can identify maternally and paternally inherited DNAs to settle paternity disputes. Another important application is in medical **diagnostics**, where differences in RFLPs are used to identify a genetically inherited disease. Some defective genes differ from their normal counterparts by only one nucleotide; however, if the mutation is in a restriction enzyme recognition site, the enzyme may not be able to cut the DNA, thereby generating a different DNA banding pattern after hybridization with a probe specific to the DNA

Table 4.2 Examples of Host Cells Used to Produce Recombinant Proteins		
Mammalian cells	Saccharomyces cerevisiae (yeast)	Insect baculovirus system
Growth hormone	Insulin	Adenosine deaminase
Erythropoietin	Epidermal growth factor	Erythropoietin
Interferon	α1 antitrypsin	Interferon
Interleukin-2	Granulocyte-macrophage colony stimulating factor	Poliovirus proteins
Monoclonal antibodies		Influenza virus hemagglutinin
Tissue plasminogen activator	Hepatitis B virus surface antigen	Tissue plasminogen activator
Blood-clotting factors	Hepatitis C virus protein	Rabies glycoprotein

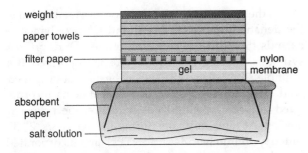

Figure 4.9 Southern blotting. DNA is transferred from an agarose gel to a membrane for hybridization by making a sandwich of the gel, membrane, filter paper, and paper towels. A salt solution that moves by capillary action facilitates the transfer of DNA from the gel to the membrane.

region. A comparison of the hybridization patterns enables medical clinicians to detect normal and variant genes in an individual.

A similar blotting and hybridization method, called Northern blotting, can be used to probe RNA molecules separated by denaturing agarose gel electrophoresis. Instead of DNA, RNA is transferred from the gel to a membrane in a similar way to that described for DNA transfer. Northern blotting is useful for identifying isolated RNAs and for studying the expression of specific genes.

POLYMERASE CHAIN REACTION

Traditional DNA isolation methods rely on the construction and screening of genomic or cDNA libraries, a process that takes many steps. A method developed in 1985 enables a researcher to rapidly isolate a specific DNA without building and screening a library. The method, called the **polymerase chain reaction** (PCR), uses the components of DNA replication to "replicate" a specific DNA in the test tube.

PCR enables the scientist to selectively amplify specific DNA sequences without using hybridization. Any DNA sequence can be isolated from the total DNA of an organism. Generally, however, the sequence of the region that flanks the DNA to be amplified must be known so that **primers** used in amplification can be synthesized.

Two short oligonucleotide primers are used that flank the DNA region to be amplified. The primers anneal or hybridize to the target sequence, one on each strand of the double-stranded DNA molecule. The oligonucleotides define the limit of the region to be amplified, and the DNA polymerase replicates the DNA between the primers using all four of the deoxyribonucleotides (dGTP,dATP,dCTP,dTTP) provided in the test tube (Figure 4.10). In an amplification (i.e., replication) cycle, the template is denatured by high temperature,

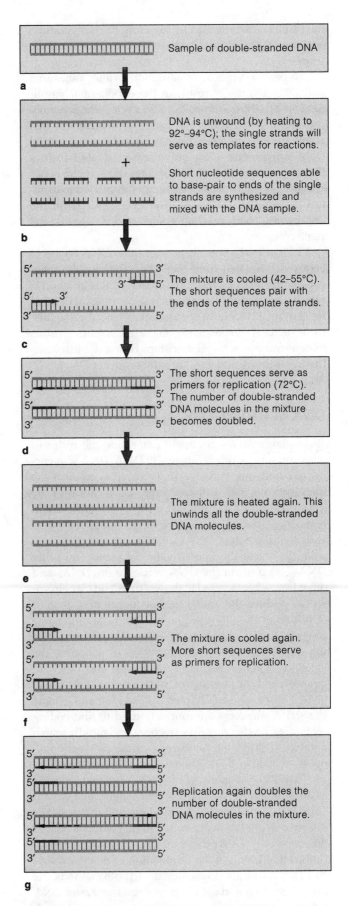

Figure 4.10 The polymerase chain reaction (PCR) is an *in vitro* DNA replication method.

the primers are annealed by lowering the temperature, and the DNA polymerase extends the DNA from the primers.

Repeated cycles of denaturation, primer annealing, and DNA synthesis result in the exponential amplification of DNA. About 25 to 40 cycles are generally conducted using a thermal cycler, an instrument that automatically controls temperatures and times. A special DNA polymerase—Taq polymerase, isolated from a thermophilic bacterium, *Thermus aquaticus*, living in hot springs—is stable at the high temperature used to denature the template DNA. The product generated from the polymerase chain reaction is analyzed by agarose gel electrophoresis.

PCR allows a specific gene or other DNA region to be rapidly isolated from total DNA without the time-consuming task of screening a library. PCR is used to: (1) rapidly isolate specific sequences for further analysis or for cloning, (2) identify specific genetic loci for diagnostic or medical purposes, (3) generate DNA fingerprints to determine genetic relationships or to establish identity in forensics, and to (4) rapidly sequence DNA.

DNA SEQUENCING

Today, laboratories routinely determine the order of deoxyribonucleotide sequences in DNA. DNA sequencing methods are valuable tools; they (1) confirm the identity of genes isolated from DNA libraries by hybridization or amplified by the polymerase chain reaction, (2) determine the DNA sequence of promoters and other regulatory DNA elements that control expression, (3) reveal the fine structure of genes and other DNAs, (4) confirm the DNA sequence of cDNAs and other DNAs synthesized in the test tube, and (5) help scientists deduce the amino acid sequence of a gene or cDNA from the DNA sequence. Today, large genome projects, such as the Human Genome Project, are yielding information about the evolution of genomes; the location of genes, regulatory elements, and other sequences; and the presence of mutations that give rise to genetic diseases (see Chapter 10, The Human Genome Project). Computers are routinely used to find coding sequences, predict protein sequences, and rapidly search databases (EMBL and Genbank, for example) for homologous protein and DNA sequences of known function.

Two DNA sequencing methods were first reported in 1977. Allan Maxam and Walter Gilbert at Harvard University developed a chemical method that uses the selective degradation of bases, and (2) Frederick Sanger developed a DNA synthesis method (i.e., test tube replication) that involves the termination of newly synthesized DNA strands at specific sites. In both methods, the DNA is labeled and gel electrophoresis separates DNA fragments. Today, most investigators use the Sanger sequencing method.

In the Sanger method, a DNA primer is annealed to the denatured template DNA, and DNA polymerase extends the sequence from the primer. During DNA synthesis, labeled nucleotides are incorporated (for example, ^{32}P-labeled dATP) when the nucleotide is specified. Alternatively, instead of the label being incorporated into the growing DNA sequence, the primer is labeled. At random points during the synthesis, 2′, 3′-**dideoxyribonucleotides** (ddNTPs) are incorporated into the growing strands which then terminate DNA synthesis (Figure 4.11). Four different ddNTPs are used: dideoxyadenine, dideoxycytosine, dideoxyguanine, and dideoxythymine triphosphates—one for each of the nucleotide termination reactions. DNA synthesis cannot proceed when, for example, the dideoxyribonucleotide, ddATP, is incorporated into the DNA strand instead of dATP, because the absence of a 3′-OH group prevents the addition of a new deoxyribonucleotide.

The DNA to be sequenced is aliquoted into four separate test tubes, one for each of the four nucleotide bases; the tubes contain all four nucleotides (dATP, dTTP, dGTP, and dCTP, one of which is radioactively labeled), DNA polymerase, and a small amount of one of the four ddNTPs. For example, in the "A" tube (the other tubes being "C," "T," and "G"), all the normal nucleotides are added, but the only ddNTP is ddATP, which terminates DNA synthesis at adenine (Figure 4.12). The "A" tube contains many identical DNA molecules that are being sequenced; however, termination occurs only at "A" sites. The resulting strands are of different lengths, since termination is a random process. If all the strands are pooled together, a termination will have occurred at each "A." In other words, all possible "A" sites in the sequence will have an incorporated ddGTP. The same process occurs in the "C," "T," and "G" tubes with their respective dideoxynucleotides. After termination, the DNAs in all four tubes are denatured by heating and the DNA strands are then separated by differences in length through electrophoresis on a denaturing polyacrylamide gel (a chemical polymer that keeps the single-stranded DNAs from hydrogen bonding to one another). The band pattern is detected by **autoradiography**. X-ray film is placed against the sequencing gel and is exposed and developed. The dark bands on the film correspond to the DNA fragments that

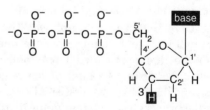

Figure 4.11 The structure of a dideoxynucleotide shows the substitution of an H instead of an OH on the 3′ carbon of the deoxyribose (shaded area).

DNA template to be sequenced:

3' GGGCCTAAGCCTTAAACTGAAGGTTATGCCCCCTTAGCC 5'

The sequencing primer is annealed to the denatured template to initiate DNA sequencing:

3' GGGCCTAAGCCTTAAACTGAAGGTTATGCCCCCTTAGCC 5'
   | | | | | | | | | | |
5' CCCGGATTCGG3'

Deoxyribonucleotides used:

dCTP, dATP, dTTP, dGTP

Dideoyribonucleotide used to teminate DNA sequencing in the "A" tube:

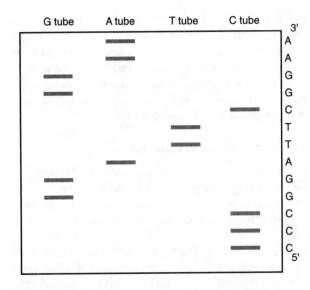

Sequencing Reactions in the tube (A* indicates sequencing was terminated by a dideoxy ATP):

3' GGGCCTAAGCCTTAAACTGAAGGTTATGCCCCCTTAGCC 5'
5' CCCGGATTCGGA*

3' GGGCCTAAGCCTTAAACTGAAGGTTATGCCCCCTTAGCC 5'
5' CCCGGATTCGGAA*

3' GGGCCTAAGCCTTAAACTGAAGGTTATGCCCCCTTAGCC 5'
5' CCCGGATTCGGAATTTGA*

3' GGGCCTAAGCCTTAAACTGAAGGTTATGCCCCCTTAGCC 5'
5' CCCGGATTCGGAATTTGACTTCCA*

3' GGGCCTAAGCCTTAAACTGAAGGTTATGCCCCCTTAGCC 5'
5' CCCGGATTCGGAATTTGACTTCCAA*

3' GGGCCTAAGCCTTAAACTGAAGGTTATGCCCCCTTAGCC 5'
5' CCCGGATTCGGAATTTGACTTCCAA*

3' GGGCCTAAGCCTTAAACTGAAGGTTATGCCCCCTTAGCC 5'
5' CCCGGATTCGGAATTTGACTTCCAATA*

3' GGGCCTAAGCCTTAAACTGAAGGTTATGCCCCCTTAGCC 5'
5' CCCGGATTCGGAATTTGACTTCCAATACGGGGA*

3' GGGCCTAAGCCTTAAACTGAAGGTTATGCCCCCTTAGCC 5'
5' CCCGGATTCGGAATTTGACTTCCAATACGGGGAA*

Figure 4.12 DNA sequencing: the "A" tube reaction. A hypothetical DNA sequence is used to illustrate how dideoxy sequencing is conducted. Either the primer or one of the deoxyribonucleotides is radiolabeled to allow detection of the banding pattern on x-ray film. Three other reaction tubes for "G", "C", and "T" would be used to obtain the DNA sequence.

have migrated to a certain point according to their length. The base sequence is read directly from the film, beginning from the bottom of the gel (Figure 4.13).

The Maxam and Gilbert method of sequencing uses double-stranded DNA molecules that are radioactively labeled at the 5' end. The DNA is denatured and added to four tubes, one for each of the four nucleotides. To each tube is added a certain chemical that modifies then removes a specific base from the DNA strand. For example, a chemical is added to the "C" tube that removes cytosine. After the chemical reaction, the DNA backbone is cut where the base is missing. This process generates

Figure 4.13 Autoradiogram of a sequencing gel showing the bands that are read to obtain the sequence. Sequence is read from bottom to top.

many radioactively labeled fragments of varying lengths, each ending where a specific base was eliminated. As with the Sanger method, DNA is separated by size on a denaturing gel, a researcher detects the radiolabeled DNA fragments on the gel by autoradiography, and the sequence is read.

DNA sequencing has been automated to rapidly sequence thousands of base pairs—a requirement in laboratories that daily sequence long stretches of DNA (such as from the human genome). A variety of methods are available; some protocols rely on labeling the terminal dideoxynucleotides (ddATP, ddGTP, ddTTP, ddCTP) with fluorescent dye, each emitting a different wavelength (color) upon fluorescence. Unlike conventional sequencing and gel electrophoresis, in this automated method all four reactions (G,A,C,T) occur in one tube and the fragments are separated in one lane of the gel. A laser beam scans the lane to detect the different wavelengths emitted by the different nucleotides. This method is called chemiluminescent DNA sequencing.

PCR also is used in a variation of Sanger dideoxy DNA sequencing, a method called cycle sequencing. Annealing of the sequencing primer, primer extension (i.e., DNA synthesis), and chain termination all can be conducted in the thermal cycler, with one tube for each dideoxynucleotide. A radioactively labeled primer is used at the 5' end.

PROTEIN METHODS

The study of proteins is an important component of recombinant DNA technology. Proteins are the molecules scientists ultimately seek to target for modification of quantity and sequence. The structural and functional characteristics of a protein must be elucidated before the protein's structure is altered. The important amino acids must be identified and the ultimate effect of the modification on, for example, enzyme activity or protein stability must be determined. A few of the methods used to study proteins are discussed below.

Protein Gel Electrophoresis

Electrophoresis is used to separate individual proteins by size and charge. The proteins are separated by either one- or two-dimensional polyacrylamide gel electrophoresis; polyacrylamide is used instead of agarose because it gives better resolution. Usually the one-dimensional gel is a vertical slab, unlike the horizontal slab for separating nucleic acids. Proteins are visualized as bands (Figure 4.14) either by a dye that binds to the proteins in the gel, or by radiolabeling and autoradiography.

Proteins are visible as spots rather than bands in a two-dimensional separation (Figure 4.15). First, the proteins are separated by charge in a slender tube gel (a pH gradient is established in the gel), in a process called isoelectric focusing. The tube is then placed horizontally along the top of a vertical slab gel, and proteins are electrophoresed out of the tube and into the gel, which separates them by size. A detergent called sodium dodecyl sulfate (SDS) that is used in the second dimension binds to the hydrophobic regions of proteins, giving the proteins an overall negative charge. Thus, proteins are

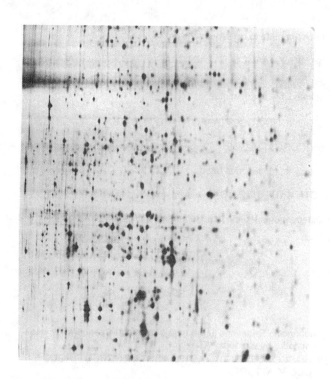

Figure 4.15 Human HeLa cell proteins separated by two-dimensional gel electrophoresis are visualized as spots.

resolved by differences in size rather than charge. Negatively charged proteins move toward the positive end of the gel. Separating proteins in two dimensions achieves greater resolution of the molecules.

Protein Engineering

Protein engineering is an exciting area that enables researchers to improve specific protein characteristics. By modifying the protein's gene sequence, they can change protein structure to enhance protein function in specific ways. Changes can include increased stability, such as resistance to degradation, pH change, temperature, oxidation, and contamination; enhanced enzyme activity; a change in substrate specificity or enzyme activity in extreme or less-than-optimum conditions, and increased nutritional value. The researcher must know the amino acid sequence (often deduced from the DNA sequence), the chemical properties of the protein, the predicted three-dimensional structure, which is usually generated by a computer (sometimes from analysis of the purified protein), and in some cases the specific amino acids that contribute to the catalytic function of the enzyme. The properties of proteins are modified by changing certain amino acids.

Several methods are available for changing the amino acids of proteins by mutating the corresponding nucleotides of the DNA. Frequently only one nucleotide must be mutated to produce the desired amino acid change. One method, oligonucleotide-directed mutagenesis, uses a short, single-stranded DNA, an oligonucleotide

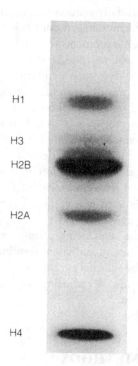

Figure 4.14 The five histone proteins that associate with eukaryotic DNA are shown. Proteins are separated into discrete bands by gel electrophoresis and stained or radiolabeled for visualization.

15 to 20 bases in length, that is complementary to the gene region to be mutated except for the individual bases to be changed. The first step in the mutagenesis procedure is to hybridize the oligonucleotide to the cloned DNA to be mutated. A viral vector, a bacteriophage called M13, is used in the cloning process. M13 has a single-stranded DNA stage that is packaged within a viral coat and a double-stranded phase or **replicative form** (RF) that is synthesized after cell infection. The RF, which is similar in structure to a plasmid, is used for ligating the DNA. From these recombinant molecules, single-stranded molecules are then generated. The oligonucleotide containing a specific change in sequence is then annealed to the single-stranded circular DNA molecule, except at the specific mismatch site that cannot base pair with it (Figure 4.16). The oligonucleotide serves as a primer for replication by DNA polymerase, and DNA ligase seals the ends of the newly synthesized circular DNA. The resulting molecule is a double-stranded, circular DNA with a mismatch region where the oligonucleotide mutation is located. The DNA is transferred to *E. coli*, where it is replicated. One strand of the original DNA molecule has the mutation from the oligonucleotide and the other does not. Thus, some replicated RF double-stranded molecules will have the mutation in their DNA sequence and others will not;

the latter are called wild-type. Progeny phage are produced within the bacterial host cells and are released as a single-stranded DNA packaged in a viral protein coat. Mutant plaques are determined by hybridizing the oligonucleotide (used to generate the mutation) to the resulting plaques containing either wild-type or mutant DNA. Since this mutagenesis procedure is inefficient and generates far fewer mutant molecules than expected, methods for enriching mutated DNA are used. Once obtained, the double-stranded mutant DNA (gene or cDNA) is isolated, placed in an expression vector, transferred to appropriate host cells such as bacteria or yeast, and expressed as a mutant protein for further study.

Another way to mutate proteins is to substitute a DNA fragment, or "cassette," containing the selected nucleotide mutations for the same DNA fragment in the organism. The DNA region containing the nucleotides to be changed is removed with a restriction enzyme and replaced with the cassette. This replacement is conducted *in vitro* in a vector that is then transferred to *E. coli* for replication.

A variety of other mutagenesis methods are used, some technically complex and using PCR, such as PCR-amplified oligonucleotide-directed mutagenesis. The method used depends on the researcher's goal: to obtain random, multiple-point mutations by producing single nucleotide changes at different intervals in the DNA; to introduce deletions or insertions of nucleotides; or, if the exact nucleotides to be changed are unknown, to generate a series of DNAs with different mutations in a specific area for further study.

Monoclonal Antibodies

Antibodies or **immunoglobulins** (Igs) are used extensively in modern biotechnology to detect genetic and acquired diseases, as therapeutic agents in medicine, and as tools in basic research (see Appendix C). Immunoglobulins are protein molecules produced by a variety of cells from advanced vertebrates like mammals; they protect the organism by binding to and eliminating foreign molecules, called **antigens**. Five classes of immunoglobulins are synthesized, and serve a variety of functions. The basic immunoglobulin structure is conserved and consists of two different polypeptides, one light chain (L) and one heavy chain (H), that associate as H_2L_2. The antigen binding site varies highly among the immunoglobulins, and is determined by the amino terminus, called the hypervariable region, which is derived from the amino acid sequences of both the light and heavy chains.

Immunoglobulins are synthesized by **B lymphocyte cells**, which are produced by cells of bone marrow and circulate in the blood and lymph. Antigen specificity is determined by the particular immunoglobulin that a B lymphocyte synthesizes. Normally B lymphocytes are inactive and do not divide; however, in response to a cir-

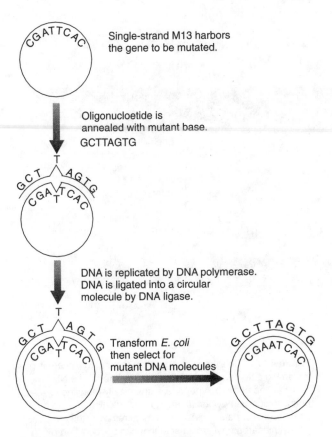

Single-strand M13 harbors the gene to be mutated.

Oligonucloetide is annealed with mutant base.
GCTTAGTG

DNA is replicated by DNA polymerase. DNA is ligated into a circular molecule by DNA ligase.

Transform *E. coli* then select for mutant DNA molecules

Figure 4.16 Oligonucleotide-directed mutagenesis involves the introduction of a desired base change into the target DNA during replication of the DNA.

culating antigen to which it binds, the B lymphocyte, with the help of a **T lymphocyte**, begins to divide and synthesizes and releases antibody against the antigen. All B lymphocyte progeny secrete the same specific antibody, thus increasing the level of that antibody in the system and destroying the antigen. Such an antibody is called monoclonal.

Antibodies can be produced in the laboratory against a variety of selected antigens. When an antigen such as a protein is injected into an animal (usually mice, rabbits, or goats), a mixture of antibodies—that is, polyclonal antibodies—having different antigen specificities are produced and isolated. Only a fraction of these antibodies will be specific to the foreign protein.

By contrast, **monoclonal antibodies** (MABs) provide researchers with an unlimited amount of very specific antibodies. However, B lymphocytes do not divide in culture; cells must be fused with ones that are able to divide indefinitely in culture. To immortalize an antibody-producing cell, spleen cells usually obtained from mice are fused with immortal cancer cells from the same lineage—for example, a cancerous mouse cell, often a myeloma cell that is of B cell origin—that can divide indefinitely. The fusion product, called a **hybridoma**, secretes an antibody against a particular antigen. The cells derived from the hybridoma are all clones and thus produce the same specific monoclonal antibody.

Monoclonal antibodies have many important applications; a few are briefly reviewed. Sometimes, after electrophoresis, a method called **Western blotting** (similar to Southern blotting) is used to identify a specific protein in a gel: Proteins are blotted onto a membrane (the Western blot) and a labeled monoclonal antibody is used as a probe to identify its target protein. Monoclonal antibodies also are used to identify a specific clone from an expression library. Their specificity makes these antibodies important tools in medical diagnostics; they increase the sensitivity of **immunoassays** for detecting a variety of hormones (such as concentrations of progesterone and estrogen) and antigens (blood group antigens, for example). Other sensitive diagnostics use monoclonal antibodies to detect a variety of diseases. Monoclonal antibodies are being studied to evaluate their therapeutic potential to help fight disease, boost the immune system, and prevent organ rejection. In this last case, antibodies synthesized by the recipient's T lymphocytes against the newly transplanted organ are removed from the body by monoclonal antibodies. Monoclonal antibodies also can aid in the prevention of graft-versus-host disease (GVH), which occurs when donor T-cells from a bone marrow transplant attack the host. The donor T-cells are removed by monoclonal antibody binding.

Protein Sequencing

Experiments sometimes require that the amino acid sequence be determined for a protein. For example, if a protein has been purified and a DNA probe is unavailable for hybridization and gene isolation, the protein can be partially sequenced. Oligonucleotide probes, generated from the amino acid sequence, can be used for PCR or hybridization to isolate the gene.

The linear order of amino acids in a protein can be elucidated by a process called Edman degradation. In this method, the amino acids that make up the amino terminal end (the NH$_2$–terminus) of the polypeptide strand are determined one at a time by chemical cleavage (Figure 4.17). The first amino acid is chemically modified to a phenylthiocarbamoyl amino acid and removed by acid treatment for identification. The next amino acid is now at the beginning of the chain (NH$_2$ terminus), and the process is repeated until all the amino acids of the protein chain are identified. This method has been automated and often is conducted by an amino acid sequenator.

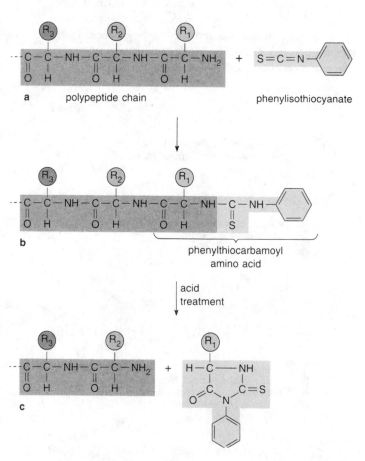

Figure 4.17 The Edman degradation method for determining the amino acid sequence of a protein. (a and b) The NH$_2$-terminal end of the polypeptide is chemically modified to phenylthiocarbamoyl amino acid by the addition of phenylisothicyanate. (c) Acid treatment removes the phenylthiocarbamoyl amino acid that can be identified by analytical methods. The process, repeated for each amino acid at the NH$_2$-terminus, continues until the polypeptide is sequenced.

Table 4.3 Some Applications of Recombinant DNA Technology

Field	Application	Importance
Basic Biology	DNA sequencing	Answers questions about gene structure, gene function, and relatedness of genes and organisms
	Directed mutagenesis	Answers questions about gene function
Medicine	Therapeutic proteins	Makes human proteins for treating diseases such as diabetes, pituitary dwarfism, hemophilia
	Gene therapy	Treats genetic diseases such as cystic fibrosis
	Improved vaccines	Produces more effective vaccines with fewer side effects
	Diagnosis	Allows rapid, accurate diagnosis of infections and other diseases
	Veterinary medicine	Better diagnosis, prevention, and treatment of disease
Industry	Altering microorganisms	Improved production of antibiotics, amino acids, vitamins, and enzymes; also improved disposal of waste, including persistent toxic chemicals
Agriculture	Altering plants	More rapid breeding of disease-resistant and improved plants (e.g., tomatoes that stay fresh-tasting longer)
	Altering farm animals	More rapid development of superior breeds
Criminal investigation	DNA fingerprinting	Can determine if a biological sample such as blood, semen, or tissue is from a particular person

J. Ingraham and C. Ingraham, *Introduction to Microbiology*, Table 7.2. Copyright © 1995 Wadsworth Publishing Co.

APPLICATIONS OF RECOMBINANT DNA TECHNOLOGY

Recombinant DNA technology has revolutionized methods of conducting research in a variety of fields, not only from molecular and cell biology to biochemistry, but also in fields such as ecology and anthropology. Because of the complexity of eukaryotic genomes, genes and their regulation can be studied only when a gene or several genes have been isolated from the genome and are *in vitro* or in host cells. Recombinant DNA technology also plays a major role in modern biotechnology. Advances in agricultural, medical, and environmental biotechnologies rely on recombinant DNA technologies, as Table 4.3 indicates. For example, the manufacture of important products (therapeutic human proteins, for example) from cloned genes; the analysis and repair of genetic disorders; and the production of transgenic plants, animals, and bacteria can be achieved only with recombinant DNA methods. Exciting new applications have the potential to provide us with new medical treatments, to augment current methods of removing toxic wastes and pollutants from the environment, and to increase agricultural productivity.

General Readings

F.M. Ausubel, R. Brent, and R.E. Kingston, eds. 1993. *Current Protocols in Molecular Biology*. Wiley, New York.

M.A. Innis, D.H. Gelfand, J.J. Sninsky, and T.J. White. 1990. *PCR Protocols: A Guide to Methods and Applications*. Academic Press. San Diego, California.

K.B. Mullis, F. Ferre, and R.A. Gibbs, eds. 1994. *The Polymerase Chain Reaction*. Birkhauser Verlag, Basel, Switzerland.

P. Peters. 1993. *Biotechnology: A Guide to Genetic Engineering*. Wm. C. Brown, Dubuque, Iowa.

J. Sambrook, E.F. Fritsch, and T. Maniatis. 1989. *Molecular Cloning: A Laboratory Manual*. Vol. 1-3, 2nd edition, Cold Spring Harbor Laboratory Press, New York.

Additional Readings

J. Alam and J.L. Cook. 1990. Reporter genes: application to the study of mammalian gene transcription. *Anal. Biochem.* 188:245-254.

P. Balbas and F. Bolivar. 1990. Design and construction of expression plasmid vectors in *E. coli. Methods Enzymol.* 185:14-37.

F. Bolivar, R.L. Rodriguez, P.J Greene, M.C. Betlach, H.L. Hyneker, H.W. Boyer, J.H. Crosa, and S. Falkow. 1977. Construction and characterization of new cloning vehicles. II. A multipurpose cloning system. *Gene* 2:95-113.

D.T. Burke, G.F. Carle, and M.V. Olson. 1987. Cloning of large segments of exogenous DNA into yeast by means of yeast artificial chromosome vectors. *Science* 236:806-812.

M.R. Davey, E.L. Rech, and B.J. Mulligan. 1989. Direct DNA transfer to plant cells. *Plant Mol. Biol.* 13:273-285.

J. Geisselsoder. F. Whitney, and P. Yuckenberg. 1987. Efficient site-directed *in vitro* mutagenesis. *BioTechniques* 5:786-791.

H.G. Griffin and A.M. Griffin. 1993. DNA sequencing: Recent innovations and future trends. *Appl. Biochem. Biotechnol.* 38:147-159.

M. Grunstein and D.S. Hogness. 1975. Colony hybridization: A method for the isolation of cloned DNAs that contain a specific gene. *Proc. Natl. Acad. Sci. USA* 72:3961-3965.

G. Kohler and C. Milstein. 1975. Continuous cultures of fused cells secreting antibody of predefined specificity. *Nature* 256:495-497.

A.M. Maxam and W. Gilbert. 1977. A new method for sequencing DNA. *Proc. Natl. Acad. Sci. USA* 74:560-564.

Y. Nosoh and T. Sekiguchi. 1990. Protein engineering for thermostability. *Trends Biotechnol.* 8:16-20.

N. Rosenthal. 1994. Stalking the gene—DNA libraries. *New Engl. J. Med.* 331:599-600.

F. Sanger, S. Nicklen, and A.R. Coulson. 1977. DNA sequencing with chain-terminating inhibitors. *Proc. Natl. Acad. Sci. USA* 74:5463-5467.

E.M. Southern. 1975. Detection of specific sequences among DNA fragments separated by gel electrophoresis. *J. Molec. Biol.* 98:503-517.

W.R. Tolbert. 1990. Manufacture of biopharmaceutical proteins by mammalian cell culture systems. *Biotechnol. Adv.* 8:729-739.

R. Walden and J. Schell. 1990. Techniques in plant molecular biology—progress and problems. *Eur. J. Biochem.* 192:563-576.

T.A. Waldmann. 1991. Monoclonal antibodies in diagnosis and therapy. *Science* 252:1657-1662.

J.G. Wetmur. 1991. DNA probes: applications of the principles of nucleic acid hybridization. *Crit. Rev. in Biochem. Molec. Biol.* 26:227-259.

T.J. White. 1996. The future of PCR technology: diversification of technologies and applications. *Trends Biotechnol.* 14:478-83.

G. Winter and C. Milstein. 1991. Man-made antibodies. *Nature* 349:293-299.

J. H. Yang, J. H. Ye, and D.C. Wallace. 1984. Computer selection of oligonucleotide probes from amino acid sequences for use in gene library screening. *Nucleic Acids Res.* 12:837-843.

R.A. Young and R.W. Davis. 1983. Efficient isolation of genes using antibody probes. *Proc. Natl. Acad. Sci. USA* 80:1194-1198.

M.J. Zoller and M. Smith. 1982. Oligonucleotide-directed mutagenesis using M13-derived vectors: an efficient and general procedure for the production of point mutations in any fragment of DNA. *Nucleic Acids Res.* 10:6487-6500.

Commercial Production of Microorganisms

Industrial Fermenters

Single-Cell Protein

Bioconversions

The Use of Immobilized Cells

Microorganisms and Agriculture

Ice-Minus Bacteria

Microbial Pesticides

Bacillus thuringiensis

Baculoviruses

Products from Microorganisms

Metabolites

Enzymes

Antibiotics

Fuels

Plastics

Bioremediation

Oil Spills

Wastewater Treatment

Chemical Degradation

Heavy Metals

Oil and Mineral Recovery

Oil Recovery

Metal Extraction

Microorganisms and the Future

Microorganisms are used in industrial settings to produce many important chemicals, antibiotics, organic compounds, and pharmaceuticals. Using living organisms as chemical synthesis factories reduces many of the risks and complexities of industrial syntheses, allowing costly and polluting raw materials to be replaced by less expensive processes. The by-products of biosynthetic reactions are usually less toxic and hazardous than those of industrial chemical reactions.

Bacteria also are playing a more prominent role in environmental cleanup of industrial spills and accidents, chemical leaching, and crude oil spills. Unfortunately, radiation leaks, toxic herbicides and pesticide residues, acid rain, chemical effluents and accumulations of lead and mercury are the result of industrial practices.

Microorganisms have been exploited for hundreds of years. Consequently, we know far more about their biochemical properties than we know about plants and animals. Since the extent of their commercial and industrial exploitation far exceeds the scope of a single chapter, we examine only a few of the most important uses of microorganisms.

COMMERCIAL PRODUCTION OF MICROORGANISMS

Biochemists define fermentation as an anaerobic process that generates energy by the breakdown of organic compounds; the end products can be microbial metabolites such as lactic acid, enzymes, the alcohols ethanol and butanol, and acetone. Industrial users of fermentation have broadened the definition to include (1) any aerobic process that produces microorganisms (**biomass**) as the end product and (2) biotransformation (tranformation by microbial cells of a compound added to the fermentation medium of a commercially valuable compound). In large-scale fermentations, useful metabolites, proteins, carbohydrates, and lipids are provided to enable the propagation of microbial cells, or biomass. Aerobic or anaerobic microorganisms are cultured under controlled conditions in large chambers or fermenters sometimes called bioreactors. Most fermentations, whether anaerobic or aerobic, require a number of steps: (1) sterilization of the fermentation vessel and associated equipment; (2) preparation and sterilization of the culture medium; (3) preparation of a pure cell culture for inoculation of the medium in the fermentation vessel; (4) cell growth and synthesis of the desired product under a specific set of conditions; (5) product extraction and purification or cell collection; and (6) disposal of expended medium and cells, and the cleaning of the bioreactor and equipment.

Industrial Fermenters

A goal that drives development in fermentation technology is to improve the productive performance of microorganisms by optimizing their growth conditions. Most of the microorganisms used in industrial fermentations today are aerobic. Besides requiring oxygen for rapid growth, they need a consistent pH in the growth medium, temperature control, a supply of nutrients for rapid growth, and an antifoaming agent to alleviate excess foaming during aeration of the culture. (The culture is aerated either by bubbling air into the medium or by agitating it.)

Fermenters have ports for adding the gases, nutrients, culture inoculum, and antifoaming agents. A probe continuously monitors pH, and a separate port allows acid or alkali to be added to adjust the pH. Temperature is regulated by a water jacket that circulates cold water around the outside of the culture vessel. An exit valve allows microbial cells, medium, or microbial products to be collected. Sterility is essential; the sterilized vessel and culture medium are kept free of contamination by the use of filters, such as for admitting sterilized gases.

Different applications dictate the use of different types of fermenters. Figure 5.1 shows the most common type, the **stirred tank reactor**, which relies on an agitator to circulate oxygen. Another common type is the **airlift fermenter**, which supplies oxygen to the culture through an intake valve in the bottom of the culture vessel. Air flow into the fermenter creates high pressure that vigorously circulates the culture medium.

Products and cells are collected from fermenters by either of two methods, **continuous fermentation** or **batch culturing**. In continuous fermentation, nutrients are fed into the fermenter while an equal volume of products, cells, and medium is collected. (Sometimes the collected cells are added back to the fermenter.) Thus, continuous fermentation allows continuous culture growth to be maintained for long periods of time. This method is most suitable for obtaining compounds that are produced proportionally to cell growth, such as primary metabolites (vitamins and enzymes, for example). Continuous fermentation is also very useful for treating wastewater and for degrading organic industrial wastes. The waste compounds are carbon and nutrient sources for microbes.

In batch culturing, the cells, products, and medium are collected after the fermentation process has termi-

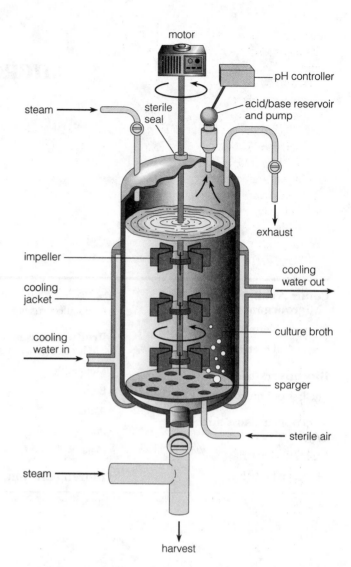

Figure 5.1 Industrial fermenters produce very large volumes of cell cultures. The fermenter shown is a stirred tank reactor.

nated, and the fermenter is set up for the next fermentation. This method is preferred when synthesis of the products does not depend on the amount of cell biomass; antibiotics are an example. The cells and liquid are separated by sedimentation or filtration, and the metabolites and enzymes are extracted from the collected liquid. Cells also can be concentrated for later use, or, if cellular products are to be obtained, cells must be lysed and the product purified and concentrated. In fed-batch fermentation, a variation of batch culturing, fresh medium is added periodically during fermentation.

Single-Cell Protein

Microbes have been used as food and food supplements for several thousand years. For example, yeast cells have been used as a supplement in soups, sausages, and animal feed to contribute additional protein, minerals, and vitamins. A monoculture of algal, bacterial, or fungal cells has a protein content that is 70 to 80% of its dry weight. When such a monoculture is grown in large volumes for use as human or livestock feed supplements, it is called single-cell protein, or SCP. SCP is high in such nutrients as minerals, vitamins, carbohydrates, and lipids, as well as essential amino acids, such as lysine and methionine, that are often lacking in plant protein.

SCP is produced by using inexpensive substrates to supply nitrogen and carbon—substrates such as methanol from natural gas, carbohydrate-containing raw materials, and wastes from cheese production and pulp mills. An early example of single-cell protein is baker's yeast from *Saccharomyces cerevisiae*. The cyanobacterium *Spirulina* also has been used as SCP in Israel, Mexico, and Taiwan. Photosynthetic microorganisms require only water, a nitrogen source such as ammonia or nitrate, sunlight, and minerals (often present in the water supply), and can be grown in outdoor ponds, which eliminate the need for costly culturing facilities. Photosynthetic eukaryotic algae also are commercially produced in dried or tablet form.

In the future, SCP may be produced using wastes or by-products of industrial processes or waste treatment facilities. Although a few SCPs are well-established for human and livestock consumption, others that have potential are ruled out for the present because they contain toxic compounds. Nucleic acids are potentially toxic but can be degraded by nucleases. However, other extremely toxic compounds may also be concentrated in SCP, such as hepatotoxins produced in cyanobacteria, or heavy metals absorbed from the substrate. Cost-effective methods to remove toxins from SCP must be developed if it is to be widely distributed. In a world in which most people receive inadequate nutrition, SCP may eventually provide a low-cost solution.

BIOCONVERSIONS

Bioconversions occur when microorganisms modify a given compound to a structurally related compound. Microorganisms can use substrates that are not usually present in their environment. Synthesis of these unusual substrates can result in products that are more useful than those resulting from synthesis of the naturally occurring substrate. Enzymes convert the substrate into a metabolite that is not normally a constituent of the cell. The compound may be of no value to the cell for cell function or growth, but the cell metabolizes it nevertheless and yields a product that has commercial value. These bioconversions are exploited commercially to produce a vast array of chemicals and pharmaceuticals.

Bioconversions are useful when a multistep chemical synthesis is more expensive or inefficient—or indeed is impossible to achieve—in the test tube. Industrial conversions often employ a mixture of both chemical synthesis and cellular reactions to generate a given end product.

Biotechnology research is producing bacteria that are more efficient bioconverters (for example, they might transform a substrate in fewer steps). New enzymatic pathways can be transferred into bacteria, or existing pathways can be modified to enable microorganisms to synthesize new, commercially desirable end products.

Important examples of bioconversions include the syntheses of steroids. To obtain the end product prednisone—an anti-inflammatory drug used for allergic reactions to poison ivy and for some **autoimmune diseases** such as arthritis—chemical synthesis requires 37 steps, beginning with a product extracted from specific plants. Substituting microbial bioconversions at specific points in the synthesis reduces the process to 11 steps. Bioconversions significantly reduce the cost and shorten the time required to synthesize this important steroid. Other examples of microbial bioconversions are the synthesis of the amino acid L-phenylalanine from phenylpyruvic acid and the conversion of D-sorbitol to L-sorbose—an important intermediate in the synthesis of vitamin C—by *Acetobacter suboxydans* and its enzyme sorbitol dehyrdrogenase. In the synthesis of ascorbic acid from glucose, only one bacterial bioconversion step is currently available; all other steps must be conducted chemically. If other microorganisms can provide additional enzymatic reactions, the chemical steps will be further reduced. In the near future, microorganisms may supply most if not all of the enzymatic steps. Recombinant DNA technology may allow us to give microorganisms all the enzymatic steps needed for completely converting glucose to ascorbic acid. Recently, a gene from *Corynebacterium* sp. encoding one of the enzymes in vitamin C synthesis has been transferred into *Erwinia herbicola*, which supplies another step in vitamin C synthesis, thus reducing the number of chemical steps in the synthesis.

The Use of Immobilized Cells

In some cases, cell immobilization is a preferred alternative to fermentation for synthesizing commercially important compounds. Cells are immobilized by chemical cross-linking or by a polymer matrix such as natural agar or carrageenan, which comes from algae. (Cells also can be microencapsulated.) Immobilization allows cells to be concentrated, thereby increasing cell densities that can result in an increase in product for a given reaction volume. Cells that are localized (that is, bound to a substrate) also may increase the efficiency of biotransformation and other synthetic reactions since the substrate and enzymes will be in close contact.

The method presents a few disadvantages that sometimes preclude its use. Since the cell contains many enzymes, undesirable biochemical reactions may occur that alter the desired product. This problem can sometimes be solved by immobilizing only the purified enzyme. The immobilizing substrate also may interfere with cell processes and reduce catalytic activities. Thus, for certain reactions, traditional fermentation remains the more desirable method.

MICROORGANISMS AND AGRICULTURE

Microorganisms are playing an increasingly important role in agriculture as pesticides and fertilizers to increase crop yields. In addition, their genes are being transferred to plants to confer resistance to insect pests and herbicides (see Chapter 6, Plant Biotechnology).

Ice-Minus Bacteria

In frost-sensitive plants such as deciduous fruit trees, corn, tomatoes, beans and various subtropical plant species, frost injury results when ice crystals form within cells or between cells. Crystals enter plant cells and damage membranes by mechanical disruption. Upon thawing, plants become soggy and limp because cells have ruptured. The loss of important crops such as oranges and strawberries to frost each year is extremely costly to the grower and to the consumer, who pays dramatically higher prices for fruit and fruit juices.

Bacteria such as *Pseudomonas viridiflava*, *Erwinia herbicola*, and *Xanthomonas campestris* are active in ice nucleation at temperatures above –5°C. *Pseudomonas syringae*, a microbial colonizer of plant surfaces, produces a protein that promotes the formation of ice crystals between 0 and 2°C. If these ice-nucleating bacteria are not present, frost damage does not occur until the temperature

is between –6 to –8°C. The gene for the ice-nucleating protein was deleted from the chromosome of *P. syringae* by Steven Lindow and his colleagues at the University of California at Berkeley. In 1983, safety concerns voiced by the public prevented field testing of the genetically modified *P. syringae*. In 1987, these "ice-minus" bacteria were tested on strawberry and potato plants. When sprayed onto plants, these bacteria readily colonize plant surfaces before the wild variety can do so, thus providing frost resistance (Figure 5.2). Concerns over the use of recombinant *P. syringae* prompted scientific investigators to explore the potential of naturally occurring ice-minus bacteria. In 1992, Frost Technologies Corporation registered with the Environmental Protection Agency (EPA) a mixture of three naturally occurring bacterial strains (two *P. syringae* strains and one *P. fluorescens*) for use in frost control. The trade name is Frostban B. The EPA has specified that because bacteria used for biological control, including those for frost control, decrease the presence of the wild-type bacteria, they are pesticides and must be registered as such. At this time, the cost of registration, efficacy testing, and environmental impact studies makes biological pesticides expensive to produce.

Microbial Pesticides

As plants have evolved, they have acquired a variety of insect defense mechanisms. However, when vast agricultural areas are devoted to monocultures, these resistance mechanisms do not always provide enough protection—insects evolve to become resistant to protective mechanisms. Each year, to prevent insect infestations that would devastate important crops, chemical pesticides must be sprayed several times during a growing season because crops are so susceptible to insect larvae.

Chemical pesticides have been widely used since the early 1940s. One of the first was DDT, or dichlorodiphenyltrichloroethane, a chlorinated hydrocarbon.

Figure 5.2 The potato plant on the right has been sprayed with ice-minus *Pseudomonas syringae* and supercooled but is not damaged by frost, while the plant on the left has been frozen because of the presence of ice nucleation-active bacteria.

DDT was an extremely effective, and very widely used, pesticide that attacked insect nervous systems and muscle tissue. Other common chlorinated hydrocarbon pesticides are chlordane, lindane, dieldrin, and toxophene. All contain arsenic, lead, or mercury, and are called persistent pesticides because they are not degraded by bacterial action or sunlight. By 1960, over 100 million acres of agricultural fields in the United States were treated with chemical pesticides. Winds picked up dust from contaminated soils and spread it around the world. In the 1960s, at the height of DDT use, rain distributed 40 tons of DDT each year on England alone.

Unfortunately, insects eventually become resistant to any pesticide, requiring that more and more of the chemical be applied. Moreover, beneficial insects—those that prey on pests—are often more susceptible than the target pests to the chemicals, and their numbers decline. As organisms consumed other organisms that had absorbed DDT in their diet, the pesticide accumulated in the food chain, becoming more concentrated at successively higher trophic levels. DDT ended up in the fatty tissues of animals and birds at the top of the food chain. The populations of many birds of prey such as bald eagles, peregrine falcons, sparrow hawks, and brown pelicans declined dramatically because the egg shells became too thin and breakable to protect the embryos. The number of young produced plummeted during the 1960s; the eastern peregrine falcon became extinct, and many other birds were threatened or endangered. Although DDT was banned in 1972, once its effect on birds was realized, the half-life (the time required for one-half of the chemical to degrade into other compounds) may be as long as 20 years. Even today, 25-odd years after DDT was banned, seals, penguins, and fish in the Antarctic show traces of DDT in their fat. The long-term effects on human and animal health will never be known completely.

Microorganisms and insect viruses may provide a source of insecticides that are **biodegradable** and insect-specific. Although over 100 fungal, viral, and bacterial strains that infect insects have been identified, only a few of these have been exploited commercially. Bacterial pesticides degrade rapidly in the environment because the active components are labile molecules. Unlike chemical pesticides, these organic pesticides are proteins that are produced by microorganisms such as *Bacillus thuringiensis*. Once exposed to sun, rain, and other environmental elements, the proteins break down rapidly. Recombinant DNA technology can be used to increase the potency and stability of the biological insecticide, to move genes encoding toxins into other hosts such as plants, and to increase the range of specificity. Many crop plants such as corn, potatoes, cotton, tobacco, and tomatoes are being engineered to have increased insect resistance; they produce a compound that is toxic to specific insects but does not affect humans and wildlife.

Recent developments include the engineering of insect-killing viruses to increase their killing potency. However, these types of viral insecticides have caused concern among scientists and laypeople alike, who ask whether these "superviruses" might kill beneficial insects or transfer their genes to related viruses. The use of *Bacillus thuringiensis* (Bt) and **baculovirus** as biopesticides is the focus of intense reasearch and development. The insertion of Bt toxin genes into plants in particular shows much promise.

Bacillus thuringiensis The soil bacterium *Bacillus thuringiensis* has been used as a biopesticide for many years. Bt is a Gram-positive bacterium that forms **endospores** within a **sporangium** during adverse environmental conditions or when nutrient availability is low. Sporulating cells produce parasporal crystalline inclusions composed of aggregated precursor proteins; the crystalized proteins are lethal to insect larvae once ingested (Figure 5.3). The crystal proteins, referred to as δ-endotoxins or insecticidal crystal proteins (ICPs), are toxic to specific insect pests.

The Bt toxin protects crops from insect pests without affecting mammals, fish, and birds. It has been used by farmers and home gardeners for 30 years without negative side effects. Dormant spores are dusted on crops and later become active and cover plants with Bt cells. Now, through recombinant DNA technology, the Bt toxin genes have been inserted into *Pseudomonas* bacteria, which are Gram-negative, nonpathogenic leaf-colonizers (Figure 5.4). These bacteria produce ICPs and are heat-killed prior to their use on crop plants. This method of delivery by encapsulation overcomes the problem of protein instability that results from exposure to environmental conditions in the field. Bt toxin genes also have been inserted into crops so that transgenic plants can produce their own ICPs (Figure 5.5).

More than 50 subspecies (as classified by flagellar serotypes) are recognized within the *B. thuringiensis*

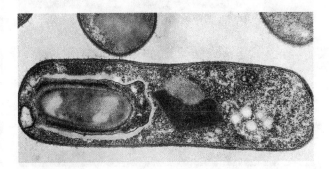

Figure 5.3 HD-1 strain of *Bacillus thuringiensis*. The bipyramidal crystal protein (dark triangular structure) has insecticidal activity and is adjacent to a **spore** (oblong structure) surrounded by a sporangium. The crystal is composed of three closely related Cry 1A proteins, each approximately 130,000 daltons.

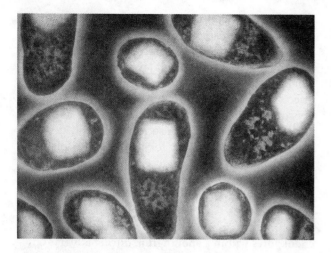

Figure 5.4 Transformed *Pseudomonas fluorescens*, a harmless leaf saprophyte, harbors a Bt crystal and expresses insecticidal activity. Production of Bt toxin is very high in *P. fluorescens* cells.

subspecies of *B. thuringiensis* is effective against all insect pests.

When crystals are ingested by insect larvae, protoxins are solubilized (that is, dissociated from the crystal) and digested in the carboxy-terminal region, where exposure to the alkaline midgut and digestive proteases activates them. These smaller toxic proteins bind to specific receptors on the midgut epithelial cell membranes. Within minutes, the gut is paralyzed and the insect ceases to feed. Death occurs within three to five days after a generalized paralysis. The toxins are thought to disrupt ion channels in cell membranes and may increase the flux of important ions such as potassium, thereby altering the osmotic balance. The specific mode of interaction of the ICPs with the cell membrane may depend on the type of toxin produced. Since parasporal crystals are sensitive to ultraviolet light, sunlight inactivates the ICPs.

Although toxin instability is undesirable, it may be that the transient presence of pesiticide is what prevents the development of insect resistance. (Such resistance may occur through a decrease in the binding affinity of the toxin for its receptor, or perhaps through a decrease in specific receptors for a particular toxin.) In the future, we are likely to see less labile sprays, more stable toxins, increased host ranges, effective methods of killing insect borers (which are rarely affected by Bt toxin applications), and less-expensive biopesticides.

The *B. thuringiensis* protoxin proteins and their plasmid-encoded genes have been studied extensively. The central portion of the protein determines insect specificity. Recombinant DNA technology may allow scientists to one day create hybrid or fusion proteins by

species, each with different pesticidal activity, established by host-range testing. Examples include the subspecies *B.t. tenebrionis*, which produces a toxin against beetles (e.g., potato beetles, boll weevils); *B.t. kurstaki*, which kills larvae of butterflies, moths (e.g., gypsy moths, tent caterpillars, cabbage worms, tobacco hornworms), spruce budworm, and cabbage worm; and the *B.t. israelensis* toxin, which is effective against dipteran larvae such as mosquitoes and black flies. Unlike chemical pesticides, these toxins are effective against the larvae of specific insects, and only if the ICPs are ingested. Because of toxin selectivity, no one Bt protoxin from any

a

b

Figure 5.5 The Bt toxin gene has been transferred to crop plants to confer insect resistance. (a) Transgenic plants are resistant to insect larvae damage (left), while plants without the Bt gene are susceptible to damage by larvae. (b) Transgenic cotton (left) after challenge with lepidopteran larvae are resistant to attack, unlike the cotton from a plant that does not contain the Bt gene (right).

introducing gene fragments encoding the central portion of different toxin proteins into recipient Bt toxin genes. A single subspecies could then synthesize several types of Bt toxins. In experiments, subspecies that have acquired additional toxin genes affected additional types of insect larvae. Fusion proteins were effective against insects that normally are not targets of either individual toxin, and the combined toxins were sometimes more potent against established target insects than either single toxin, thus demonstrating synergism. As additional toxin genes and fragments encoding core protein regions are exchanged and mutated, new and more potent toxin-producing *B. thuringiensis* strains are likely to be produced.

Also on the horizon are insect-resistant plants that may enable farmers to use less pesticide on their crops. Special care will be required so that no insecticide-resistant insects are selected. As they do with chemical insecticides, insects can develop resistance rapidly because their high reproduction rates allow them to rapidly accumulate mutations in response to selection pressure. To ensure longer-lasting resistance, plants could be transformed with several types of resistance genes. Alternatively, expression of the Bt gene could be induced by a nontoxic chemical applied to the field so that the protoxin would be present only during a specific time period. Another strategy for reducing the number of resistant insects is to supply reservoirs of insect-sensitive plants in fields; by allowing wild-type, nonresistant insects to breed, these plants reduce selection for resistant insects.

Baculoviruses Baculoviruses are invertebrate-specific DNA viral pathogens. They infect mostly larval stages of various insect orders such as Lepidoptera, Diptera, Hymenoptera, Coleoptera, and Homoptera. Since the larval forms of insects are the most damaging to crop plants, baculoviruses are potentially useful as biopesticides. These viruses do not harm nontarget organisms, since they have specific host ranges. However, unless the host range can be extended through biotechnology, their usefulness as pesticides will be rather limited.

Insects are infected by ingesting plant material contaminated with the virus. The viral particles are released into the gut lumen, where digestive enzymes break down the crystalline protein polyhedron that protects the particles. Eventually the viral DNA is transported to the nuclei of insect cells, where the virus replicates. Viral particles are produced, cells lyse, and the virus infects other cells and tissues (Figure 5.6). The progressive infection of cells results in mass destruction of tissues. By the time the insect dies, there is practically only virus within the insect cuticle. The virus is released onto plant material, ready to be consumed by other insect larvae. The virus can spread by wind and rain, as well as by birds and other animals that feed on dead insects. Animal

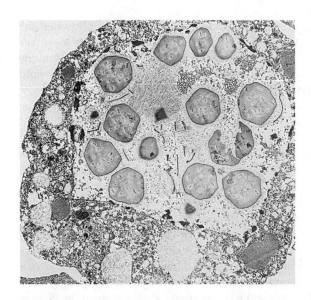

Figure 5.6 Transmission electron micrograph of a cultured *Spodoptera frugiperda* (a moth whose larva is called fall armyworm) cell nucleus infected with baculovirus particles (Baculovirus *Autographa californica* MNPV). An occlusion body is in the process of enlarging and occluding virions that are observed near the nuclear envelope. The electron dense chromatin-like material is the virogenic matrix where nucleocapsid assembly is thought to occur.

digestive enzymes do not break down the protein coat of the virus, which is deposited with the feces in locations that may be far from the original location.

Novel engineered baculoviruses may provide an alternative to the use of chemical pesticides. The main disadvantage of using baculovirus insecticide is the length of time that the infected insect takes to die: After exposure to the virus, the insect continues to feed and damage crop plants for several days. The insertion of foreign genes may enhance the virulence of these viruses or disrupt the host insect's metabolism. For example, a useful insertion would be a gene encoding an insect hormone or enzyme that affects the life cycle and eliminates feeding. Biotechnology offers the opportunity to increase the effectiveness of baculovirus insecticides and to propagate engineered baculovirus in insect culture for commercial use.

Juvenile hormone is responsible for larval growth and maintenance of the larval stage. Juvenile hormone esterase inactivates juvenile hormone so that larvae metamorphose into pupae and cease feeding. If the insect juvenile hormone esterase gene is transferred to baculoviruses and expressed in the host insect upon infection, feeding and growth should be significantly reduced. This effect was observed in experiments in which the tobacco budworm (*Heliothis virescens*) esterase gene was transferred to baculovirus and esterase was expressed in the host upon viral infection. One drawback, however, is that larvae are sensitive to juvenile hormone only during specific stages of development: In the experiments, the effect was observed only when the recombinant virus infected first instar larvae—other stages were insensitive to changes in juvenile hormone levels. To be commercially useful, the esterase would have to be effective at other larval stages. Alternatively, transferring insect toxin genes to baculoviruses may be an effective way to control insects. Although toxins tend to be insect-specific, their transfer to baculovirus may extend their range; for example, the neurotoxin gene from the North African scorpion *Androctonus australis Hector* was transferred to baculovirus and has shown extended specificities. A scorpion toxin gene has been inserted into the *Autographa californica* multicapsid nuclear polyhedrosis virus (AcMNPV), a baculovirus that kills the larvae of butterflies and moths by paralyzing them. Research has demonstrated that the AcMNPV becomes much more lethal after the scorpion gene is added. Field trials are in progress in which the virus is sprayed on crops infested with lepidoptera larvae. (Other field tests have involved the use of an AcMNPV from which a gene has been deleted. The gene prevents infected larvae from molting and pupating; the deletion of this gene decreases the amount of leaf material the larvae consume, since they do not need to feed for metamorphosis.)

Studies on the use and safety of recombinant viruses in the field are ongoing. Primary concerns focus on the question of whether recombination with related viruses promotes the spread of foreign genes and will negatively affect nontarget insects. The issues of viral dispersal and persistence can be addressed with additional research. The use of disabled virus mutants will help ensure safety in the field and restrict the spread of recombinant viruses.

PRODUCTS FROM MICROORGANISMS

As we saw in Chapter 1, the products of microbial fermentation have been used for thousands of years. Food and beverages were the first products of microbial fermentation—beer, wine, bread, and cheeses are only a few examples. Much more recently, the production of antibiotics through fermentation was a biotechnological breakthrough that revolutionized medicine after World War II. Today, fermentation using microorganisms yields a still wider variety of commercially important compounds. Microbial fermentation provides flavorings, nutrients, and colorings for a wide variety of foods. For example, among the many fungal varieties used in cheese manufacturing, the fungus *Penicillium roquefortii* gives blue cheese its characteristic flavor. Amino acids are used as flavor enhancers (e.g., alanine, aspartic acid, and glutamic acid) and nutritional supplements (e.g, threonine, methionine, and lysine), as well as in indus-

trial biochemical syntheses. Microbial fermentation produces pharmaceutically active compounds such as anti-inflammatory agents, antidepressants, anticoagulants, and coronary vasodilators. The ability to genetically alter microorganisms through recombinant DNA technology has expanded the list of commercially important compounds obtained by fermentation. Microbial cells can now synthesize compounds with important therapeutic value, such as interferons and interleukins for the treatment of a variety of cancers; factor VIII for hemophilia; erythropoietin for anemia; human insulin for diabetes, α_1-antitrypsin for the treatment of emphysema; nerve growth factor to promote the repair of nerve damage; and human growth hormone for the treatment of pituitary dwarfism.

In the following sections, a few of the important applications of commercial fermentation are examined. These include the production of important metabolites, enzymes, antibiotics, fuels, and plastics.

Metabolites

Two types of metabolites are synthesized by microbial cells. Primary metabolites are produced during the organism's growth phase; these compounds are essential to an organism's metabolism and can be intermediary metabolites or end products. Secondary metabolites are not essential to cell growth or function and are characteristically produced quite late in the growth cycle; they are the end products of metabolism and are usually derived from primary metabolites or the intermediates of primary metabolites. Although the specific functions of secondary metabolites often are unknown, these compounds probably give the organism a selective advantage against its competitors in its natural environment. Table 5.1 shows examples of commercially important primary and secondary metabolites.

Table 5.1 Examples of Primary and Secondary Metabolites Produced by Fermentation

Primary Metabolites	Secondary Metabolites
Amino acids	Antibiotics
Vitamins	Pigments
Nucleotides	Toxins
Polysaccharides	Alkaloids
Ethanol	Many active pharmacological compounds (e.g., the immunosuppressor cyclosporin, hypotensive compound dopastin)
Acetone	
Butanol	
Lactic acid	

Enzymes

Enzymes isolated from microorganisms have applications in the production of foods, cosmetics, and pharmaceuticals and in the synthesis of industrial chemicals and detergents (Table 5.2). Industrially, enzymes are used to convert substrates, often from plant material, into commercially valuable products. An example is the production of high-fructose corn syrup from corn starch, a highly branched polymer composed of covalently bonded glucose monomers. The synthesis uses three enzymes—bacterial α-amylase, glucoamylase, and glucose isomerase. Corn is steamed under high pressure to make it more susceptible to enzymatic action. After it has cooled, α-amylase is added to hydrolyze the glucose polymer into short-chain polysaccharides. Glucoamylase completely hydrolyzes the polysaccharides to glucose by breaking bonds of the branching cross-links. By a transformation reaction, the enzyme glucose isomerase then converts glucose into fructose. Although the product is commonly called high-fructose corn syrup, it is a mixture of glucose and fructose. Because this process yields a lower cost per unit of production than the process for producing sucrose, fructose has replaced sucrose as the major sweetener in the United States. This industrially important bioconversion uses the bacteria *Bacillus amyloliquefaciens* for α-amylase, *Aspergillus niger* for glucoamylase, and *Arthrobacter sp.* for glucose isomerase.

Recombinant DNA technology provides a way to mass produce a commercially important enzyme. The gene encoding a newly identified enzyme can be moved into commonly used cells such as *E. coli* for mass production through fermentation or for genetic manipulation to change, for example, enzyme activity or substrate specificity (see Protein Engineering, in Chapter 4).

Antibiotics

Antibiotics are small metabolites with antimicrobial activity that are produced by Gram-positive and Gram-negative bacteria as well as fungi. Their role in nature is probably to enable the antibiotic producer to effectively compete for resources in the environment by killing or inhibiting the growth of competitor microorganisms. The first antibiotic used was penicillin, discovered in 1929 by Alexander Fleming. Since the widespread use of penicillin during World War II, antibiotics have been produced commercially by microbial fermentation. Table 1.2 identifies microorganisms and the important antibiotics they produce.

A wide variety of antibiotics, many of them from the Gram-positive soil bacterium *Streptomyces*, are available today to treat many bacterial infections. These antimicrobial drugs act in different ways by (1) disrupting the plasma membrane of microbial cells, (2) inhibiting cell wall synthesis, and (3) inhibiting synthesis of important metabolites such as proteins, nucleic acid, and folic acid.

Today many bacteria are becoming resistant to the present arsenal of antibiotics used to fight infections. Scientists and clinicians are searching for new antibiotics to combat antibiotic resistance. One way to identify new antibiotics is to screen purified secondary metabolites for antimicrobial activity. This is an arduous task that may not yield new antibiotics. A second way is to "feed" unusual substrates and substrate analogs to microorganisms that can be used in antibiotic biosynthetic pathways. Chemical variants of existing antimicrobial compounds may be produced (an example is the synthesis of novel β-lactams, a family of antibiotics that includes penicillin). Other methods include chemically modifying preexisting antibiotics or synthesizing in the laboratory chemicals with antimicrobial activity.

Recombinant DNA technology is also playing an important role in the development of powerful new antibiotics. The manipulation of genes for antibiotic synthesis enables scientists to produce novel antibiotics that differ in structure from the original antibiotic. Methods include combining antibiotic biosynthetic pathways and the targeted genetic modification of pathways. The genes for two related but independent antibiotic biosynthetic pathways can be placed in the same microorganism. The enzymes of one pathway are able to act on the intermediates produced in the other pathway. That is, the compounds produced in one antibiotic biosynthetic pathway can be used as substrates in the other pathway. In this way, new structural variants are

Table 5.2 Examples of Microbial Enzymes and Their Uses	
Enzyme	**Uses**
Lipase	Enhances flavor in cheese making
Lactase	Breaks down lactose to glucose and galactose; lactose-free milk products
Protease	Detergent additive; hydrolyzes suspended protease proteins in beer that form during brewing for a less cloudy chilled beer
α-amylase	Used in production of high fructose corn syrup
Pectinase	Degrades pectin to soluble components, reduces cloudiness in chilled wine, fruit juice
Tissue plasminogen activator (TPA)	Dissolves blood clots

generated that differ from the intermediates produced in both pathways. These new compounds can have different antimicrobial activities. Novel antibiotics in the isochromanequinone family are produced by moving cloned genes for a specific isochromanequinone antibiotic biosynthetic pathway on plasmids into host organisms that already synthesize a particular antibiotic in the same antibiotic family (antibiotics include actinorhodine, granaticin, and medermycin). Variants have been made by cloning the genes for actinorhodin from *Streptomyces coelicolor* and placing them into other species of *Streptomyces* that synthesize different antibiotics.

Another way to produce new antibiotics is to feed unusual substrates to a microorganism that contains a pathway from another organism. For example, one biotechnology company has produced intermediates of cephalosporin (closely related to penicillin) by feeding adipic acid to *Penicillium chrysogenum* that harbors the cephalosporin biosynthetic pathway genes from *Streptomyces* or *Cephalosporium*.

Fuels

The industrialized world consumes enormous quantities of fuels that are nonrenewable resources, and the nations of the developing world are burning increasing amounts as well. The earth harbors only finite amounts of coal, oil, and natural gas; eventually supplies will run out unless alternatives are found. All produce pollution when burned, although in different amounts. We must find cleaner, renewable alternatives. One such alternative is methane, a natural gas produced by anaerobic bacteria in swamps and landfills. These bacteria may serve as a renewable resource in the future. They are inexpensive to culture, since they use domestic and agricultural sewage and industrial effluents as nutrients for growth. Landfill gas already is harnessed for energy. ("Scrubbing" removes the undesirable gaseous compounds and the separated methane is burned, producing CO_2 and H_2O.)

One potential fuel that is extremely attractive with respect to its environmental impact is hydrogen, since combustion produces only energy and water. There have been two main obstacles to adopting hydrogen as a fuel of choice. First, pure hydrogen is extremely combustible (much more so than gasoline), so there are great concerns about the practicality of storing it safely in conditions of normal use, such as in the fuel tank of an automobile. Second, the high cost of extraction precludes its commercial use. Here again bacterial fermentation shows great promise. The bacterial enzyme hydrogenase readily produces hydrogen gas from two hydrogen ions. Research using both *Clostridium* bacteria and the eukaryotic algae *Chlorella* looks promising, since both organisms can produce hydrogen gas for pro-

longed periods. Though the production of fuel gas from natural organisms is still in the experimental phase, there is much promise for the future.

Plastics

A tremendous amount of plastic is used worldwide. Most of the plastic in the United States is used for products that must be durable and resist deterioration. The rest is used for packaging (wraps, lids, caps, sacks, bottles, etc.) and for coatings, sheeting, and photographic and video film. Polyethylene, polypropylene, polystyrene, polyvinyl chloride, or polyethylene terephthalate are the types most commonly used. Most of these synthetic polymers are produced from the naphtha fraction of petroleum or from natural gas, and are heavy polluters because they are not biodegradable.

Most plastics end up in landfills, on our shores, or in the ocean. Several hundred thousand tons are discarded into the marine environment each year. If such disposal stopped today, plastic litter would continue to wash up on our shores for a century or longer. The heavy toll on marine mammals, turtles, and birds has been well-documented. More than a million marine animals die each year by choking, suffocating, drowning, or starving after ingesting or becoming entangled in discarded nondegradable plastics. Plastic is now found in the gut of almost all seabirds examined. Plastic particles mistaken for food and ingested reduce the capacity for feeding; malnutrition and starvation follow.

The so-called biodegradable or photodegradable plastics require ultraviolet light, oxygen, and other elements in the environment to slowly break them down or partially degrade them. Although some plastics contain additives that are truly biodegradable, there is not a plastic on the market that is completely biodegradable because their usefulness as products (e.g., packaging) requires that their refractory hydrocarbon-based polymers resist degradation.

Burning plastic wastes has not been a solution, since combustion releases airborne pollutants; some plastics produce hydrogen cyanide, and burning polyvinyl chloride (PVC) produces hydrogen chloride. These harmful chemicals must be removed from the combustion products before they are released into the air.

At least a partial solution to this devastating environmental problem is the development and use of bioplastics that are readily broken down by natural elements and the action of microorganisms. Biotechnology may one day provide the tools to synthesize such natural plastics, and do so using methods of production that do not pollute the environment. Current research looks promising, and advances have been made with both bacteria and plants (see Chapter 6, Plant Biotechnology). Some of the chemicals that are required for plastic production may be produced by microorganisms. For example, alkene oxides are used extensively in plastics and polyurethane foams. The conventional method of producing these oxides generates pollutants, but a bacterium, *Methylococcus capsulatus*, found in the hot springs in Bath, England, produces alkene oxides as a gas at 81°C. Use of this bacterium may simplify the collection of alkene oxides and eliminate pollution.

Living organisms are capable of synthesizing many complex polymers and catalyzing reactions at various temperatures. A group of very diverse microorganisms (including examples from the genera *Alcaligenes, Azospirillum, Acinetobacter, Clostridium, Halobacterium, Microcystis, Pseudomonas, Rhizobium, Spirulina, Streptomyces,* and *Vibria*) produces polymers in the poly(β-hydroxyalkanoate), or PHA, family, which are used for energy and as a storage form of cellular carbon. These microorganisms can accumulate from 30 to 80% of their dry weight in PHA (Figure 5.7). PHA compounds are produced and stored in intracellular inclusions during unbalanced growth conditions—conditions in which there is excess carbon, such as glucose, but too little of another nutrient, such as ammonia, or mineral, such as magnesium, sulfate, or phosphate. All PHA compounds are linear polymers or polyesters of β-hydroxyalkanoic acids containing different alkyl groups at the β-position.

The most extensively characterized PHA is poly(β-hydroxybutyrate), or PHB. Polyesters with different chemical and physical properties, for a variety of uses, are produced by controlling the chain lengths of repeating units. The desired polymers are produced by supplying the microorganism defined nutrients in correct ratios. The metabolic flexibility of many microorganisms allows them to synthesize new varieties of bioplastics to replace petrochemical-based plastics in use today. Growth is initially maximized to provide a large amount of biomass; the medium is then changed to create an imbalance by depleting one nutrient, and growth slows.

Figure 5.7 Transmission electron micrograph of the bacterium *Pseudomonas olevorans*, showing the accumulation of electronlucent poly (β-hydroxyalkanoates) (PHA) inclusions.

A polymer-forming substrate is supplied that the cells use to concert the compound into the desired polymer. Homopolymers of PHB have been produced that have chemical and physical properties almost identical to the chemically synthesized, nonbiodegradable polypropylene. Recombinant DNA technology provides a means by which genes can be modified to produce new, useful polymers that can bypass costly and sometimes hazardous chemical processes.

Being 100% biodegradable microbial bioplastics, PHA polymers have many important potential industrial and medical applications; for example, as biodegradable carriers for fertilizers, insecticides, herbicides, and fungicides. They could be used in the medical field for such surgical paraphernalia as sutures, pins, and staples, artificial blood vessels, bone replacements, capsules for pharmaceuticals, and containers or bags for intravenous solutions, medications, and nutrients. Medical use of biopolymers would dramatically reduce the amount of nonbiodegradable waste. Bioplastics also could be used in packaging as containers, bottles, bags, and the like. PHA compounds could be used in disposable diapers, a product that is not biodegradable and the disposal of which has become a great concern as the nation's landfills are reaching their maximum capacity.

BIOREMEDIATION

Hydrocarbons are abundant in the environment and come from both human sources (industry, tanker accidents, transportation of petroleum products, atmospheric fallout, fires, mining) and natural sources (phytoplankton, natural seeps). **Biodegradation** is the natural process whereby bacteria and fungi are able to break down hydrocarbons and produce carbon dioxide, water, and partially oxidized biologically inert molecules as by-products of metabolism. The hydrocarbons are oxidized and provide energy for the microorganisms. Hundreds of species of bacteria, yeasts, microalgae, cyanobacteria, and filamentous fungi have been identified that can metabolize hydrocarbons. Each species is able to break down a few specific types of hydrocarbons. The types of molecules that can be degraded depend on the metabolic pathway in each species. Polycyclic aromatic hydrocarbons (PAHs) are an example of a group of complex hydrocarbons. **Bioremediation** is the process of reclaiming or cleaning up contaminated sites using microorganisms to remove or degrade toxic wastes and other pollutants from the environment (a variant process using plants is called phytoremediation). The natural breakdown of hydrocarbons is accelerated by adding fertilizer (and sometimes trace metals and other micronutrients as well as microorganisms), thus providing a source of nitrogen and phosphorus that may be limiting in the natural environment.

Toxic wastes such as crude oil, refined petroleum products, and heavy metals accumulate in the environment through spills and other accidents, dumping, and leaching. Since these toxic compounds do not degrade naturally, or do so extremely slowly, they persist in the environment and may poison wildlife and cause cancer in humans. Such compounds are called **xenobiotics** (*xeno* means "foreign") because they are foreign to a living organism and do not occur naturally in the environment. Examples are such volatile organic compounds as benzene, vinyl chloride, trichloroethylene, polychlorinated biphenyls, and polycyclic aromatic hydrocarbons, and such semivolatile compounds as phenol, naphthalene, benzopyrene, and benzofluorene. Chlorophenols have been widely used to preserve wood, hides, textiles, paints, and glues, and as a disinfectant to keep industrial cooling waters free of organisms such as algae. Pentachlorophenols (PCPs) are used to preserve wood. Wood treatment facilities and sawmills have been significant sources of contamination; their wastes are now treated as hazardous. Polychlorinated biphenyls (PCBs) have been widely used as cooling fluids in electrical equipment and some industrial cooling systems. They are no longer manufactured in the United States and are being removed from most uses. Polybrominated biphenyls (PBBs) are widely used as flame retardants in textiles, carpets, and plastics. PVC is a common plastic that is often abundant in municipal waste.

Many industrial processes and other activities may contribute to the production of a toxic compound. For example, polycyclic aromatic hydrocarbons, or PAHs, come from sources as diverse as incineration of refuse and waste, forest fires, charcoal broiling, automobile exhaust, refinery and oil storage wastes, and spills from oil tankers, to name just a few.

The disposal of toxic wastes presents a formidable challenge worldwide. Incineration and chemical treatment of wastes are costly methods, are unavailable in many areas, and may themselves create additional environmental problems.

Oil Spills

Oil is composed of a variety of hydrocarbons; its specific composition depends on whether it is unrefined—the crude oil that makes up most ocean spills—or has been distilled to separate its components by molecular weight into products such as very volatile gasolines, diesel fuel, jet fuels, and heating oils.

The world consumption of petroleum is enormous. In 1989, one trillion gallons per day were consumed. During transport some oil is spilled, polluting oceans and shorelines in many areas of the world. Tanker accidents, which receive much media attention, release approximately 100 million gallons of crude oil into the environment every year, yet they account for only a small portion of the oil that is spilled: The largest tanker spill occurred in 1978, when the Amoco Cadiz lost 67 million gallons. Although that spill was six times the size of the 1989 Exxon Valdez spill, it nevertheless accounted for only 8% of the oil lost at sea in 1978. Every year, some 900 trillion gallons are released into the sea.

Oil spills from tanker accidents and tank ruptures have serious environmental consequences and affect vast numbers of wildlife. However, people are more directly affected by small releases into the environment, such as petroleum leaking from underground storage tanks and contaminating soil and groundwater. Microorganisms are responsible primarily for the biodegradation of evaporated and dissolved molecules of crude oil and refined products in the environment. Molecules that have not evaporated or dissolved in the sea will degrade over time—although the process may require centuries.

Bioremediation of petroleum requires microorganisms with hydrocarbon-oxidizing enzymes and the ability to bind to hydrocarbons. Also required are nutrients for microbial growth; the major limiting factor in the bioremediation of petroleum-contaminated water and soil is the availability of phosphorus and nitrogen. These can be added in the form of fertilizers (Figure 5.8). Currently, this is the only enhancement used to clean up oil

Figure 5.8 Bioremediation of an oil-contaminated beach after the Exxon Valdez tanker accident off the coast of Alaska. An untreated area of beach is covered with oil (left), while a section treated with fertilizer (Inipol) showed a dramatic reduction in oil on cobble sediment shortly after treatment (right).

wastes and spills. In the future, biotechnology may produce microbes with enhanced oil degrading abilities.

Wastewater Treatment

The best-known application of bioremediation technology to date is microbial aerobic oxidation of wastewater. To treat domestic sewage and industrial aqueous wastes, bacteria are introduced into wastewater in a carefully controlled environment containing nutrients for microbial growth. In most wastewater treatment plants, microbes (often called biomass) are free floating, but sometimes they are immobilized on permeable plastic films and the wastewater flows over them. The treatment of soils and sludges requires the use of water, nutrients, and microbes, which are sometimes attached to a substrate, often the soil or sludge particles. Contaminated soils can be treated by a variety of *ex situ* methods (that is, removed for treatment off-site) or *in situ* methods (that is, on-site). Two common *in situ* methods are pumping contaminated water to the surface before adding nutrients to encourage bacterial action, or percolating water and nutrients into the contaminated soil.

Artificially constructed wetlands also are effective for treating urban runoff, industrial effluents, landfill leachate, coal pile seepage, agricultural wastes, acid mine drainage, and municipal and domestic wastewater. In these systems, soils and aquatic plants with attached microbes adsorb and degrade organic and inorganic wastes. Wastewater also can be treated *in situ* in waste treatment ponds and lagoons, as well as in oil-field production pits. Often such lagoons or ponds contain a sludge composed of hydrocarbons, which must be broken down. First, aerators agitate the sediment and liberate the hydrocarbons; then nutrients, often in the form of commercial fertilizers, are added to promote the growth of microorganisms. If the indigenous population is not effective, the lagoon or pond can be inoculated with additional microorganisms.

Often, however, indigenous microorganisms, even after inoculation, cannot efficiently degrade toxic chemicals (and the chemicals may themselves inhibit microbial growth). For example, resins and aromatics, such as penta-, tetra-, and naphtheno-aromatics, are very resistant to biodegradation. Most aromatic hydrocarbons containing more than five rings degrade slowly and some not at all. A comparison of half-lives is telling: benzopyrene, a five-ring compound, 200 to 300 weeks; pyrene, a four-ring compound, 34 to 90 weeks; naphthalene, a two-ring compound, only 2.4 to 4.4 weeks. Solvents of highly chlorinated hydrocarbons are not readily degraded by natural microbial populations. In fact, the number of halogen atoms directly affects the rate of degradation; the more halogens on the molecule, the slower the degradation. Effective treatment requires

bacteria that are genetically engineered to treat these specific types of pollutants. In the future more of these microbial products of biotechnology will be available for bioremediation, offering a cost-effective and environmentally friendly means of removing toxic compounds.

Chemical Degradation

In the mid 1960s, several microorganisms were discovered to have the ability to degrade pesticides, herbicides, and many organic chemicals. Today many species of bacteria and fungi are known to oxidize a wide range of compounds. Strains of *Pseudomonas*, the most common soil bacteria, have degraded more than 100 organic compounds. The bacteria use the chemicals as a carbon source and metabolize the compound using enzymes of biodegradative pathways. The genes encoding the enzymes of the metabolic pathway may reside on the chromosome or plasmids or both. The plasmids encoding these enzymes are usually large, from 50 to 200 kb, and currently are being isolated and studied in the laboratory.

The most abundant components of herbicides and pesticides, such as DDT, are halogenated aromatic chemicals. Chemicals that contain the halogen elements astatine, bromine, chlorine, fluorine, or iodine are hazardous pollutants found at many toxic waste sites. Many halogens that are important industrial chemicals—for example, the dry-cleaning solvent carbon tetrachloride, and the insulation in electrical equipment (polychlorinated biphenyls)—are carcinogenic and toxic to fish and wildlife. Halogenated compounds also occur naturally in the environment; most contain chlorine. More than 200 halogenated compounds are known to be produced by algae, bacteria, and sponges.

The nonhalogenated aromatic chemicals are called polycyclic aromatic hydrocarbons. These are hazardous chemicals composed of two or more benzene rings in linear, clustered, or angular arrangements. Examples are naphthalene, anthracene, pyrene, and fluoranthene. These compounds occur as by-products of industrial processing, cooking, and fossil fuel combustion. They enter the environment in chemical and sewage effluents and though leakage, aerial fallout, industrial accidents, the use and disposal of petroleum products, and natural oil seeps. Their toxicity, environmental persistence, and lipid solubility increase as the number of benzene rings increases.

Dehalogenation, the process of removing halogens, converts most halogenated aromatic chemicals to chemicals that are not toxic. Dehalogenation occurs by an enzymatic reaction, using dioxygenase, that replaces the halogen on the benzene ring with a hydroxyl group.

The same enzymes that transform halogenated aromatic compounds also convert the polycyclic aromatic hydrocarbons to other chemicals such as catechol or protocatechuate. Nonhalogenated and halogenated compounds are broken down to either catechol or protocatechuate depending on the starting material and bacterial pathway used (Figures 5.9 and 5.10). Catechol and protocatechuate are further broken down to either acetyl-CoA and succinate or pyruvate and acetaldehyde, naturally occurring molecules that are metabolized by most aerobic organisms. The end products are determined by the type of chemical modification that the benzene rings undergo (Figure 5.11). Various degradative routes often are available, since different microorganisms may have specific catabolic pathways for chemical degradation.

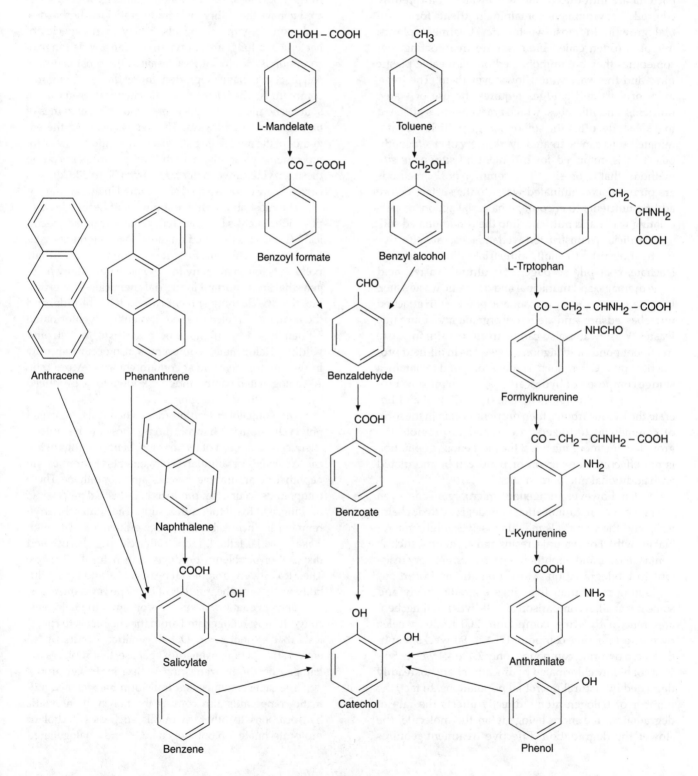

Figure 5.9 The bioconversion of aromatic compounds to catechol by bacteria.

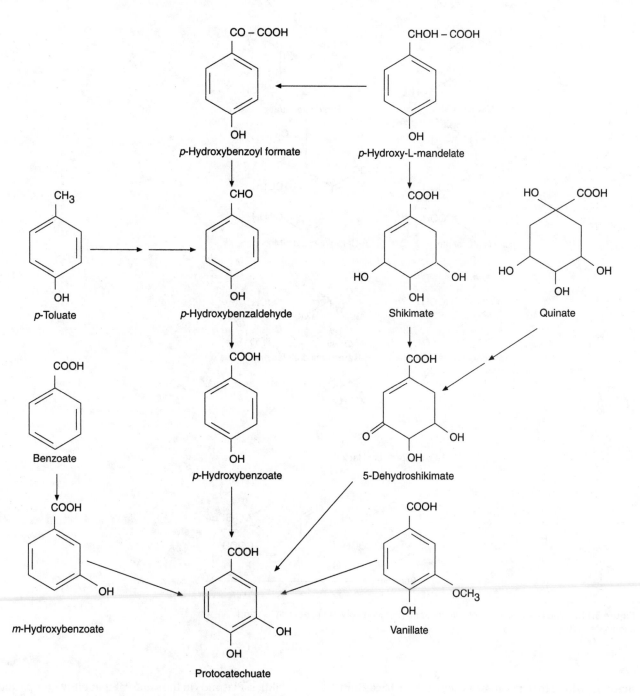

Figure 5.10 The bioconversion of aromatic compounds to protocatechuate by bacteria.

A wide variety of bacteria, algae, and fungi can metabolize polycyclic aromatic hydrocarbons. For example, cyanobacteria in the genera *Nostoc, Anabaena, Aphanocapsa,* and *Oscillatoria* oxidize naphthalene. Bacterial degradative pathways may be engineered to expand the range of chemicals that can be degraded. Since wastes generally are mixtures of toxic chemicals— oil spills and those containing heavy metal wastes are examples—bacterial strains that can degrade a wide array of chemicals are highly desirable. As mentioned earlier, degradative pathways often are plasmid-encoded; for example, the TOL plasmid degrades

toluene and xylene and the pJP3 plasmid degrades the herbicide 2,4-D (2,4-dichlorophenoxyacetic acid). By transferring plasmids encoding enzymes for a specific pathway from a donor strain to a recipient, we can generate bacteria that biodegrade a variety of chemicals. Usually plasmids are transferred by conjugation: Two strains are mated, and the plasmid DNA from the donor is transferred to a recipient strain. If an endogenous plasmid and the transferred plasmid have homologous regions of DNA, homologous recombination occurs to generate a fusion plasmid carrying the enzymes for more than one degradative pathway. Subsequent

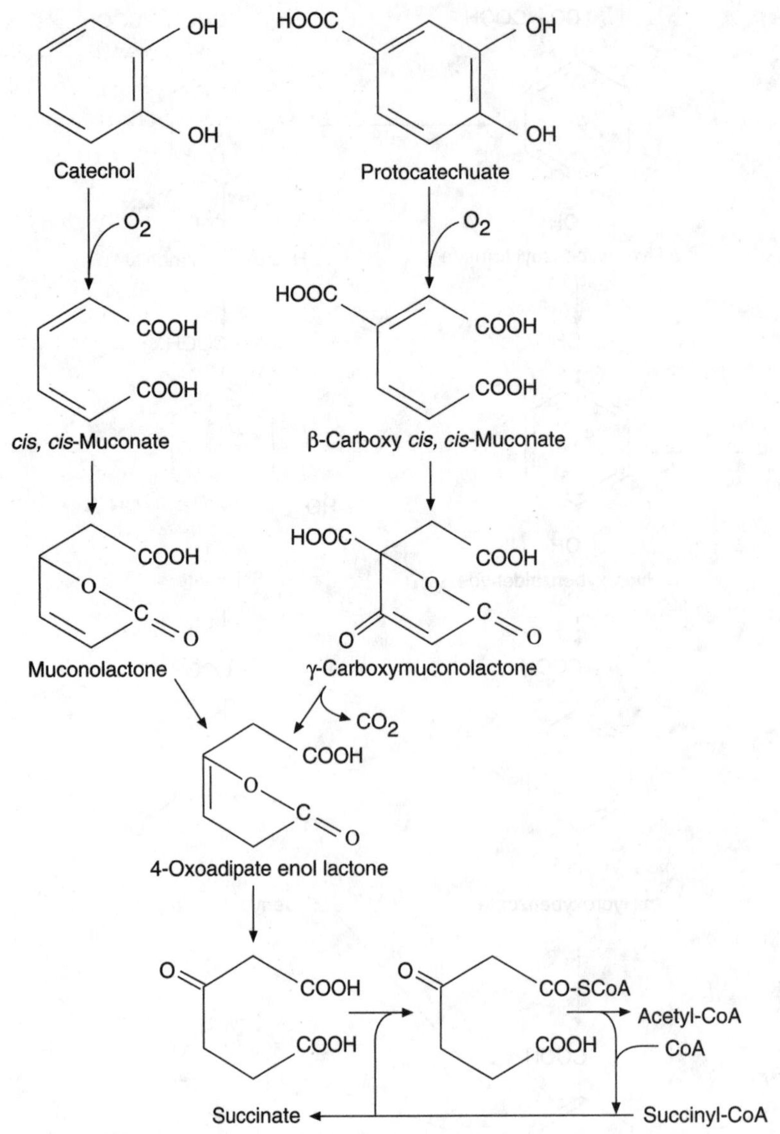

Figure 5.11a The bioconversion of catechol and protochatechuate to acetyl-CoA and succinate.

matings with other strains can generate a bacterium that can degrade a wide variety of chemicals. Such bacteria have been dubbed "superbugs."

The first engineered microorganisms with advanced degradative properties were generated by Ananda Chakrabarty and his colleagues in the 1970s. Plasmids were transferred into a bacterial strain that could degrade several compounds in petroleum (Figure 5.12). Essentially, the metabolic capabilities were expanded when new degradative pathways were transferred to the organism. Chakrabarty transferred a camphor-degrading plasmid (CAM plasmid) into a bacterium carrying a plasmid with the genes for degrading octane (OCT plasmid). Plasmids can be classified according to compati-

bilities. Plasmids in the same compatibility group cannot coexist in the same cell; that is, replicate and be stably maintained as separate molecules. As a result of their incompatabilities and homologous regions, the CAM and OCT plasmids recombined to form a fusion plasmid that encoded enzymes for both pathways. Thus, one bacterial strain could degrade both camphor and octane. Another bacterial strain with a plasmid for degrading xylene (XYL plasmid) received a plasmid encoding enzymes for degrading naphthalene (NAH plasmid). Because of their compatibility, the two plasmids could coexist in the same bacterial strain. The transfer of the CAM-OCT plasmid to the bacterial strain carrying the XYL and NAH plasmids resulted in one bacterial strain

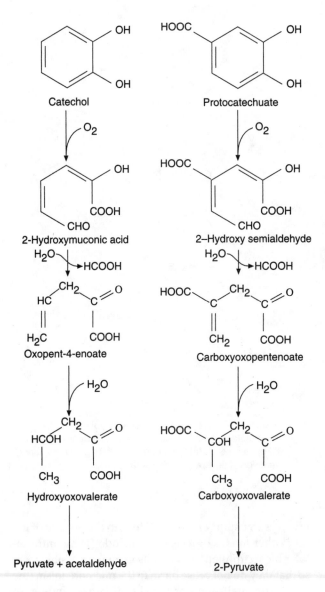

Figure 5.11b The bioconversion of catechol and protochatechuate to pyruvate and acetaldehyde.

harboring four degradative pathways encoded by plasmids. This strain grew very well on crude oil, and used camphor, octane, xylene, and naphthalene as sources of carbon.

Chakrabarty obtained the first U.S. **patent** for a genetically engineered microorganism. Although the bacterial strain never was used to clean up oil spills, the development of an "oil-eating" microbe was a significant achievement. Chakrabarty's work revolutionized industrial biotechnology in the sense that biotechnology companies now could use patents to protect their living "inventions" exactly as other companies protect, for example, pharmaceuticals and organically synthesized chemicals.

Chakrabarty also isolated a *Pseudomonas cepacia* strain that degrades the herbicide, 2,4,5-T (2,4,5-trichloroacetic acid) by mixing laboratory bacteria with bacterial strains from soils containing the herbicide. The bacteria degrade 2,4,5-T in soils by removing the chlorine, and degrade other halogenated compounds containing chlorine, fluorine, and bromine. An important step in degradation is the removal of the halogens. Cellular enzymes can degrade the dehalogenated compounds. Unfortunately, the products of bacterial degradation are sometimes more toxic than the initial compound; DDT, for example, is converted to the extremely potent and stable insecticide DDD (dichlorodiphenyldichloroethane). Nevertheless, strain modification shows much potential, and research is underway to develop bacterial strains through plasmid transfer that can grow under a variety of environmental conditions (temperatures, salinities, etc.) and degrade a variety of chemicals. These engineered organisms may one day be invaluable in cleaning up polluted soils, rivers, lakes, streams, and oceans.

Genetic engineering allows us also to produce new toxin-degrading bacterial strains by manipulating the genes within an organism's biodegradative pathways: to enhance degradative capabilities by increasing the levels at which enzymes are transcribed; to create mutants with altered enzyme activities that further degrade products of existing pathways into still less toxic compounds; to combine gene transfer and mutation to increase the number of compounds that are degraded; or to exploit new hosts that enhance degradation of specific chemicals. For example, after the addition of *Pseudomonas putida* genes encoding toluene dioxygenase, *E. coli* can oxidize carcinogenic chemicals such as trichloroethylene, benzene, toluene, xylene, naphthalene, and phenol. Transformed *E. coli* cells seem less susceptible than *Pseudomonas* to damage by some chemicals and can maintain degradation rates longer. Thus, bacterial species, such as *E. coli*, with an increased tolerance to toxic compounds may serve as efficient hosts for the biodegradative enzymes from other bacteria.

Future developments will provide new solutions for degrading and detoxifying wastes. Research on cell-free enzyme digestion of aqueous toxic wastes—particularly pesticides—in soils and streams seems very promising. The use of these enzymes *in situ* or in high-volume fermenters would facilitate large-scale cleanup. Microbes and their compounds can be used to chelate (bind) insoluble metals and produce soluble complexes for their removal (see next section). Bacteria release compounds that bind metals in a soluble form and transport them into the cell for metabolism. This process may be exploited in the future to concentrate absorbed or insoluble metals in polluted soils, sludges, groundwater, and aquifers.

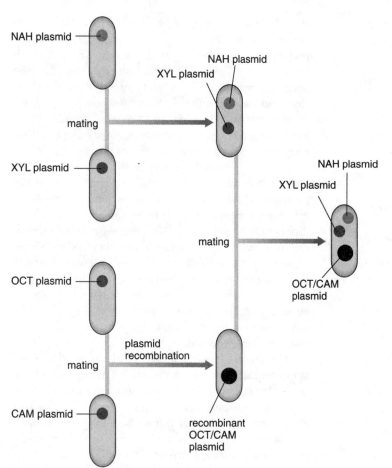

Figure 5.12 The production of a camphor (CAM)- , naphthalene (NAH)- , and octane (OCT)-degrading strain of bacteria through the recombination of plasmids during matings of several bacterial strains.

Heavy Metals

Metals, important elements in the earth's crust, are essential to living organisms. Several are required in trace amounts (examples are magnesium, manganese, copper, selenium); they are necessary for the activities of many enzymes in metabolic reactions (i.e., prosthetic groups) and for maintaining the integrity of membranes and other cellular structures. Many metals are toxic only above naturally occurring concentrations. By contrast, heavy metals, such as arsenic, cadmium, lead, and mercury, do not play a role in metabolic processes and are highly toxic even in low concentrations (for example, they can disrupt metabolic reactions, bind to DNA, and increase mutations). The main sources of heavy metals in the environment are smelters, power plants, waste incinerators, and traffic. These metals remain in the atmosphere for long periods of time (up to 10 days) and accumulate in the food web. Methods of removal rely on adsorption of metals to bacterial cells or to nonliving substrates, precipitation of contaminating metals, and transformation and degradation of toxic metal wastes by microorganisms.

Many bacteria have evolved effective resistance mechanisms, most of them plasmid-encoded, ensuring survival in the presence of high metal concentrations. The main mechanism for heavy metal resistance involves transport of the metal ions to the outside of the cell. Other modes of resistance include (1) the intracellular bioaccumulation of metals in an inaccessible form so that they do not cause harm to the microorganism (the compartmentalization or the chelation, i.e. binding, of metal ions) and (2) the chemical transformation of a toxic compound to a less toxic one. The latter process shows much promise for bioremediation.

The complete microbe-mediated transformation of hazardous metals to such nontoxic compounds as CO_2 and H_2O is not possible. However, processes that contribute to the conversion of toxic to less toxic substances or the sequestering of toxic metals by bacteria are being explored. Bacterial genes are being identified that function in bacterial resistance to mercury, cadmium, cobalt, zinc, chromate, copper, lead, tin, arsenic, and tellurium. Microorganisms that have been genetically modified to constitutively express operons for heavy-metal resistance are being studied to determine their effectiveness in detoxifying metal-containing compounds.

OIL AND MINERAL RECOVERY

The potential exists for the enhanced recovery of oil and minerals in the future. Oil and minerals are nonrenewable resources that must be extracted at great expense and often by compromising the environment. As oil reserves are reduced, prices increase and the reliance on foreign oil has serious international consequences. According to a 1985 estimate by the Department of Energy, the known U.S. oil reserve is estimated to have originally totalled 488 billion barrels. Of that, 133 billion barrels have been produced through 1983. Existing technologies can remove approximately 28 billion additional barrels. However, approximately 327 billion barrels cannot be recovered by conventional methods. Only 5% or less of the total U.S. production has been recovered using alternative technologies, such as microbial enhanced oil recovery (MEOR). By the year 2010, one third of the total oil production will most likely be by new technologies such as MEOR.

As the worldwide demand for minerals increases, new methods of extraction must be developed. Traditional methods of mineral extraction include chemical treatment and smelting of metal-rich ores. These methods are costly and generate airborne pollutants that become acid rain. Biotechnology offers opportunities to develop economical, energy-efficient, and environmentally sound methods of extracting metals such as copper, antimony, gold, silver, tin, cobalt, molybdenum, nickel, zinc, and lead.

Oil Recovery

Microbe-enhanced oil recovery (MEOR) may become an economically attractive alternative way of recovering oil. Bacteria can be used for oil recovery in several ways. Bacteria are injected into the oil reservoir along with nutrients for growth and colonization. Alternatively, the growth and metabolism of indigenous microbes are stimulated by the injection of appropriate nutrients. Microbial products such as extracellular polysaccharides and biosurfactants from fermenters are injected into the oil reservoir to enhance oil recovery. A polysaccharide called xanthan gum from bacteria is pumped with water into the ground near oil wells to increase oil viscosity. The oil that adheres to rock is loosened and flushed underground to the well as a dense paste with the water-gum mixture. Bacterial cells that grow and produce extracellular polysaccharides in pores, crevices, and highly permeable areas divert the flow of liquid to less permeable areas of the oil reservoir. These bacteria and their products improve oil recovery by bringing the recovery fluid (the water-gum mixture) into contact with a larger quantity of oil within the reservoir.

Anaerobic bacteria that produce a variety of solvents, such as acetone, and a variety of alcohols reduce interfacial and surface tension forces, thereby increasing oil mobility and releasing oil from rock surfaces and small cracks or pores. The solubilization of gases produced by microbes results in the precipitation of certain compounds that allows the oil to flow better. Because reservoirs lack oxygen, anaerobic bacteria generally prevail in these environments and are the microorganisms of choice for MEOR.

Technical difficulties must be solved before MEOR can be used on a large scale. Some of the most serious problems include the natural biodegradation of crude oil and microbial corrosion and "souring" (an increase in hydrogen sulfide content caused by sulfate-reducing bacteria) of the oil. Aerobic microorganisms are undesirable in oil recovery. If oxygen is introduced, hydrocarbon-degrading aerobes can grow and metabolize the lighter and more easily recovered oil fractions, including the low molecular weight aromatic and heterocyclic components of oil. Anaerobes will grow on the by-products of aerobic bacteria and produce undesirable reduced sulfur compounds. These processes reduce the quantity and quality of recovered oil.

Metal Extraction

Microorganisms using oxidation, convert insoluble metal compounds in ores to soluble compounds. Bacteria and fungi can be used in the recovery of metals from solution because their negatively charged cell surfaces of polysaccharides trap and concentrate positively charged metal ions. Thus, valuable metals can be recovered or hazardous industrial wastes containing toxic heavy metals can be removed by adsorption.

Four main groups of microorganisms are associated with ores and are involved in mineral extraction: (1) heterotrophic organisms that use organic compounds as carbon sources; (2) mesophiles from the genera *Thiobacillus* and *Leptospirillum* that have a 20 to 40°C optimum range (3) moderate thermophiles of the genus *Sulfobacillus* that have a 40 to 55°C optimum range; and (4) extreme thermophiles of the genera *Sulfolobus*, *Acidanus*, *Sulfurococcus*, and *Metallosphaera* that live at temperatures greater than 55°C. Many commercial extractions contain a mixture of bacterial species or strains, which act synergistically to increase the biochemical oxidation reactions.

Chemoautotrophic bacteria (a broad group of prokaryotes that include sulfur, iron, hydrogen, and nitrifying bacteria) obtain energy by the oxidation of inorganic materials such as minerals and require only carbon dioxide as a carbon source; they use, for example, substrates containing iron, such as ferrous ions (fe^{2+}), and reduced sulfur compounds such as sulfides (e.g., hydrogen sulfide—H_2S), elemental sulfur (S^0), and thiosulfate ($S_2O_3^{2-}$). The oxidation of minerals by chemoautotrophs can be exploited for the extraction of metals. In addition to a supply of oxygen, the oxidation of a sulfidic mineral requires a nutrient solution containing ammonium, magnesium, potassium, sulfate, nitrate, calcium, chloride, and phosphate ions. Also required are the trace elements zinc and manganese, which usually are present in the minerals being extracted. Thus, to catalyze mineral oxidation reactions under natural conditions, the bacteria must have access to the above nutrients in addition to oxygen.

Thiobacillus ferrooxidans seems to be the dominant oxidizing organism of most acidic environments at temperatures below 40°C. This microorganism was isolated in 1947 from ore deposits in a coal mine in West Virginia and now is used routinely to leach copper and uranium from rock. *T. ferrooxidans* also may be useful for the mining of cobalt, nickel, zinc, and lead. It oxidizes iron sulfide to iron sulfate for energy; one of the by-products is sulfuric acid, which, in conjunction with sulfate, leaches out metallic minerals while eroding away rock. *T. ferrooxidans* also can concentrate metals intracellularly; among these metals are cobalt, zinc, copper, nickel, arsenic, antimony, cadmium, and even strontium and rubidium.

In direct leaching of metals from ores, bacteria act on metal sulfides where the metal can be zinc, copper, nickel, cobalt, or lead. For example, the copper is leached by the oxidation of copper sulfide to copper sulfate:

$$CuS + 2O_2 \longrightarrow CuSO_4$$

Indirect leaching of metals involves the oxidation of FeS_2 (pyrite) to ferric sulfate and sulfuric acid. So, for example, if copper is to be extracted, the reaction would be:

$$4FeS_2 + 15O_2 + 2H_2O \longrightarrow 2Fe_2(SO_4)_3 + 2H_2SO_4$$

Ferric sulfate then oxidizes the copper sulfide to copper sulfate. The reaction is:

$$CuS + Fe_2(SO_4)_3 \longrightarrow CuSO_4 + {}_2FeSO_4 + S$$

For both direct and indirect leaching, the copper sulfate solution is collected and the copper is precipitated.

The final products of bacterial oxidation of an ore or mineral concentrate can be the treated solid material and a liquid. The valuable metal may be found in the solid or the liquid or both. For highly resistant (i.e., refractory) minerals containing gold or silver, the metal remains in the solid material. After oxidations of silver, antimony, lead, and tin minerals, the metal becomes an insoluble sulfate. Copper, zinc, cobalt, uranium, nickel, and molybdenum end up in the liquid phase.

Each type of oxidation process requires the use of different metal recovery methods for liberated gold in solids, for solubilized metals, and for the separation of metals in the form of insoluble sulfates from solid residues. The by-products in effluents of all extractions contain sulfuric acid, iron, and sometimes arsenic, which is toxic to the environment if not removed.

Modern biotechnology may be used to enhance the efficiency of metal extraction by engineering microorganisms with increased resistance to heavy metal ions such as cadmium and mercury, to acidic conditions, and to high temperatures. Bacterial cell adsorption of metals may become a common method of recovery in the future. Bacterial cell surfaces, including the many extracellular polysaccharide polymers that make up capsules and slime, are able to bind metals in the environment because of their anionic properties. Bacterial adhesion to mineral surfaces seems to occur in two phases. Electrostatic interactions (due to the electrostatic charge on the bacterial cell surface) probably play a role in initial adsorption, and a second, stronger bonding occurs through the formation of chemical bonds (e.g., sulfur-sulfur bonds) between the cell and the mineral surface. Different bacterial surface properties appear to bind different metals. Thus, specific bacteria could be used to obtain various metals or minerals. Metal selectivity might even be enhanced if the anionic properties of the surface could be modified for selective metal binding. Localized high concentrations of metals on surfaces and subsequent mineralization or scavenging by bacterial exopolymers may one day have important industrial applications and be used to concentrate and retrieve valuable metals.

MICROORGANISMS AND THE FUTURE

Microorganisms are fascinating, flexible, living machines that are capable of producing a vast array of useful compounds. New strains are being isolated and studied for eventual exploitation. We can expect to see microorganisms play a more prominent role in bioremediation and in the production of newly identified metabolites, novel compounds from bioconversions, and the expression of foreign proteins. The future is bright for microbial biotechnology.

General Readings

K. H. Baker. 1994. *Bioremediation*. McGraw-Hill, New York, NY.

J. Barrett, M.N. Hughes, G.I. Karavaiko, and P.A. Spencer. 1993. *Metal Extraction by Bacterial Oxidation of Minerals*. Ellis Horwood, London.

Y. Murooka and T. Imanaka, eds. 1994. *Recombinant Microbes for Industrial and Agricultural Applications*. Marcel Dekker, Inc., New York, NY.

P.F. Stanbury and A. Whitaker. 1984. *Principles of Fermentation Technology*. Pergamon Press Ltd., Oxford, England.

Additional Readings

R.M. Atlas. 1995. Bioremediation of petroleum pollutants. *Bio-Science* 45:332-338.

C.A. Batt and A.J. Sinskey. 1984. Use of biotechnology in the production of single cell protein. *Food Technol.* 38:108-111.

J.W. Blackburn and W.R. Hafker. 1993. The impact of biochemistry, bioavailability and bioactivity on the selection of bioremediation techniques. *Trends Biotechnol.* 11:328–333.

J.R. Bragg, R.C. Prince, E.J. Harner, and R.M. Atlas. 1994. Effectiveness of bioremediation for the Exxon Valdez oil spill. *Nature* 268:413-418.

B.C. Carlton and C. Gawron-Burke. 1993. Genetic improvement of *Bacillus thuringiensis* for bioinsecticide development. In L. Kim, ed. *Advanced Engineered Pesticides*. Marcel Dekker, Inc., New York, pp 43-61.

A.M. Chakrabarty. March 1981. Microorganisms having multiple compatible degradative energy-generating plasmids and preparation thereof. U.S. patent no. 4,259,444.

M. Charles. 1985. Fermentation scale-up: Problems and possibilities. *Trends Biotechnol.* 3: 134-139.

D.J. Cork and J.P. Krueger. 1991. Microbial transformations of herbicides and pesticides. *Adv. Appl. Microbiol.* 36:1-66.

T.C. Currier and C. Gawron-Burke. 1990. Commercial development of *Bacillus thuringiensis* bioinsecticide products. In J.P. Nakas and C. Hagedorn, eds., *Biotechnology of Plant-Microbe Interactions*. McGraw-Hill, New York, NY, pp 111-143.

R.J. Frederick and M. Egan. 1994. Environmentally compatible applications of biotechnology. *BioScience* 44:529-535.

J.R. Fuxa. 1991. Insect control with baculoviruses. *Biotechnol. Adv.* 9:425-442.

G.M. Gadd and C. White. 1993. Microbial treatment of metal pollution—a working biotechnology? *Trends Biotechnol.* 11:353-359.

K.D. Gagnon, R.W. Lenz, R.J. Farris, and R.C. Fuller. 1992. The mechanical properties of a thermoplastic elastomer produced by the bacterium *Pseudomonas oleovorans*. *Rubber Chem. Technol.* 65:761-777.

D. Ghosal, I.S. You, D.K. Chatterjee, and A.M. Chakrabarty. 1985. Microbial degradation of halogenated compounds. *Science* 228:135-142.

R.J. Giorgio and J.J. Wu. 1986. Design of large scale containment facilities for recombinant DNA fermentations. *Trends Biotechnol.* 4:60-65.

A. Kiessling. 1993. Nutritive value of two bacterial strains of single-cell protein for rainbow trout (*Oncorhynchus mykiss*). *Aquaculture* 109:119-130.

R. Lewis. 1989. Baculovirus for biocontrol and biotech. *BioScience* 39:431-434.

J.G. Leahy and R.R. Colwell. 1990. Microbial degradation of hydrocarbons in the environment. *Microbiol. Rev.* 54:305-315.

S.E. Lindow. 1993. Biological control of plant frost injury: The ice-story. In L. Kim, ed. *Advanced Engineered Pesticides*. Marcel Dekker, Inc., New York, NY, pp 113-128.

S.E. Lindow. 1994. Control of epiphytic ice nucleation active bacteria for management of plant frost injury. In L. Gusta, G. Warren, and R. Lee, eds. *Biological Ice Nucleation and its Applications*. Amer. Phytopathol. Soc. Press, St. Paul, Minnesota, pp 239-256.

R. Madoery and C. Gattone. 1995. Bioconversion of phospholipids by immobilized phospholipase A2. *J. Biotechnol.* 40:145-153.

J.E. McGhee. 1984. Continuous bioconversion of starch to ethanol by calcium-alginate immobilized enzymes and yeasts. *Cereal Chem.* 61:446-449.

P.R. Norris. 1990. Acidophilic bacteria and their activity in mineral sulfide oxidation. In H.L. Ehrlich and C.L. Brierley, eds., *Microbial Mineral Recovery*. McGraw-Hill Publishing Company, New York, pp 3-28.

Y. Poirier, C. Nawrath, and C. Somerville. 1995. Production of polyhydroxyalkanoates, a family of biodegradable plastics and elastomers, in bacteria and plants. *Bio/Technology* 13:142-150.

A. Pollio and G. Pinto. 1994. Progesterone bioconversion by microalgal cultures. *Phytochem.* 37:1269-1272.

R.C. Prince. 1993. Petroleum spill bioremediation in marine environments. *Crit. Rev. Microbiol.* 19:217-242.

J.L. Ramos, A. Wasserfallen, K. Rose, and K.N. Timmis. 1987. Redesigning metabolic routes: Manipulation of TOL plasmid pathway for catabolism of alkylbenzoates. *Science* 235:593-596.

M.C. Ronchel. 1995. Construction and behavior of biologically contained bacteria for environmental applications in bioremediation. *Appl. Envir. Microbiol.* 61:2990-2994.

L. Standberg, K. Köhler, and S. O. Enfors. 1991. Large-scale fermentation and purification of a recombinant protein from *Escherichia coli*. *Process Biochem.* 26:225-234.

R.P.J. Swannell and I.M. Head. 1994. Bioremediation comes of age. *Nature* 368:396-397.

J. Van Rie, W.H. McGaughey, D.E. Johnson, B.D. Barnett, and H. Van Mellaert. 1990. Mechanism of insect resistance to the microbial insecticide *Bacillus thuringiensis*. *Science* 247:72-74.

M.D. White, B.R. Glick, and C.W. Robinson. 1994. Bacterial, yeast and fungal cultures: effect of microorganism type and culture characteristics on bioreactor design and operation. In J.A. Asenjo and J. Merchuk, eds., *Bioreactor System Design*. Marcel Dekker, Inc., New York, NY, pp 1-34.

H.A. Wood and R.R. Granados. 1991. Genetically engineered baculoviruses as agents for pest control. *Annu. Rev. Microbiol.* 45:69-87.

T. Yamamoto and G.K. Powell. 1993. *Bacillus thuringiensis* crystal proteins: Recent advances in understanding its insecticidal activity. In L. Kim, ed. *Advanced Engineered Pesticides*. Marcel Dekker, Inc., New York, NY, pp 3-42.

M. Yoshida and H. Minato. 1987. Assessment of the pathogenicity of bacteria used in the production of single cell protein. *Agri. and Biol. Chem.* 51:241-242.

G.J. Zylstra. 1994. Molecular analysis of aromatic hydrocarbon degradation. In S. J. Garte, ed., *Molecular Environmental Biology*. CRC Press, Inc., Boca Raton, Florida, pp 83-115.

Biotechnological developments and genetic engineering are revolutionizing agriculture. Some major benefits of this revolution will be crops that are resistant to pests and disease, livestock that are resistant to viruses, and microbes to increase resistance to frost damage. These benefits will increase food production.

The focus of this chapter is the genetic manipulation of plants using the tools of biotechnology. Transgenic plants are created by integrating new DNA into a plant's DNA to become a permanent part of the plant's genome. Seeds, tissues, or cells are transformed and then regenerated into a whole plant. Transgenic plants can be used to study how a plant controls the transcription and translation of its genes. Although these plants are important to basic research, the knowledge acquired often leads to new applications. Natural characteristics are being altered or new characteristics are being added to make plants more desirable. Many plants are being genetically manipulated: tomato, soybean, carrot, canola, potato, corn, rice, sunflower, rye, beet, alfalfa, pear, apple, and cabbage are just a few. Sophisticated plant breeding methods are also generating many new varieties of agronomically important plants.

According to the Biotechnology Industry Organization, in 1996, over six million acres of land worldwide were to be planted with genetically engineered plants for commercial use. More than 15 genetically engineered agricultural products already are on the market, and many more can be expected in the next three to six years.

Modifying plants to increase their agricultural or horticultural value is only one use of genetic engineering. Plants can also be modified for use as living bioreactors or factories to synthesize rare or valuable products.

PLANT TISSUE CULTURE AND APPLICATIONS

The ability to regenerate whole plants from cells or tissues has been invaluable to plant biotechnology. This ability—called **totipotency**—is unique to plants. Totipotent cells are **somatic cells** that have retained the ability to divide and differentiate into a mature plant if placed in the appropriate environment. Organ culture—the propagation of plant organs such as the **ovary** or **anther**—was achieved in the 1920s, and **tissue culture**—the regeneration of a complete plant—in the 1930s.

Plant tissue culture offers many practical benefits. For one, plants propagated by seed can be cultured *in vitro* to yield thousands of identical plants; a number of agronomically important plants, such as the oil palm in the tropics, have been successfully propagated by large-scale cloning. Specific characteristics such as disease and herbicide resistance also can be selected for while plants are in culture. Regenerating flowering plants in culture offers an additional advantage: There are no viruses in cells of meristematic regions—regions of plant tissue made up of cells that divide to give rise to the tissues and organs of flowering plants. Thus, plants regenerated from **meristematic tissue** will be virus-free. Each growing season, commercial potato and strawberry producers start with virus-free plants from tissue culture.

Plant tissue culture is invaluable when traditional plant breeding cannot generate plants with desired traits. Often, micropropagation shows faster results than traditional plant breeding (Table 6.1).

Micropropagation

Researchers and horticulturists have exploited plant regeneration to propagate large numbers of clones—identical plants originating from a single parent tissue. To clone plants, they induce cultures of plant cells or tissues to divide and differentiate into a new plant. Meristematic regions of plants such as a shoot apex, composed of a meristem and embryonic leaves, can be cultured *in vitro*. Cultured plant tissue must be supplied with various ratios and concentrations of mineral nutrients, certain vitamins, sucrose, and plant growth hormones such as **auxin** and **cytokinin**. Adding very small amounts of auxin and cytokinin to the culture induces multiple shoots to appear from a single-shoot apex within several weeks of culture. These shoots can be separated and each grown into a complete plant through clonal micropropagation (sometimes called *in vitro* propagation)—a means of producing many genetically identical plants (**clones**) with desirable traits from a single individual in a small space and relatively quickly (Figure 6.1).

Clonal micropropagation is practiced in both laboratories and plant nurseries. The economic value of propagation using tissue culture is obvious, yielding different species and varieties of ornamental plants, such as those sold in supermarkets; agricultural crops such as strawberry, banana, potato, and tomato; and a variety of medicinal plants. Micropropagation forms the basis of a multimillion dollar industry.

Today's high-yielding oil palms on plantations in places such as Malaysia were generated beginning in the 1960s by tissue culture methods. These palms not only produce more oil—approximately 30% more than normally cultured palms—but are significantly shorter, making harvesting less expensive. High-yielding micropropagated palms are replacing old plantation trees, since palm oil is used extensively in foods.

From Callus to Plant

Cloning plants by micropropagation uses actively dividing cells, but regeneration can be performed using cells from **explants**—mature, differentiated tissue removed

Table 6.1 Comparison of Micropropagation and Traditional Culture Methods for Tubers over Two Years		
	Traditional	Micropropagation
Year 1	100 g tuber	100 g tuber
	↓	↓
	One mature plant	One preconditioned plant
		↓
		10 nodal cultures
		↓
	1600 g tubers	Shoot multiplication
	↓	↓
Year 2	----------------------------	650,000 minitubers Dormant
	16 mature plants	617,500 plants --------- Season
	↓	↓
	16 × 600 g tubers	617,500 × 500 tubers
	= 25.6 kg	= 308,750 kg

Note: The reduced number of plants from minitubers between year one and two reflects an assumed loss of 5% on establishment in soil.

S. H. Mantell et al., *Principles of Plant Biotechnology*, Table 6.5. Blackwell Scientific Publications, London, 1985. Used by permission of Blackwell Science Ltd.

Figure 6.1 Plant tissue culture room where plants are regenerated from cells and tissues.

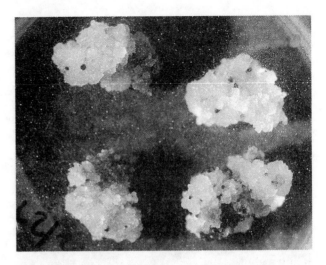

Figure 6.2 Callus is an undifferentiated mass of cells from which plants can be regenerated.

Somatic Embryogenesis

One method of plant tissue culture produces embryolike structures, called embryoids, from cultured somatic plant tissues. Somatic embryogenesis induces cultured somatic, or nonsex, cells (gametes are the sex cells) to form embryoids from vegetative tissues. The source of vegetative tissue can be mature plants, reproductive tissues excluding the zygote, or embryonic tissues, and **cotyledons** and **hypocotyls** from young plants. Embryoids can be made to form directly from tissue by using hormones to disrupt normal tissue development; auxins and reduced nitrogen are usually used to initiate **embryo** formation. Embryoids also can be formed from callus tissue; a change in hormones induces these cells to form embryoids instead of shoots and roots. During callus formation, numerous embryoids form on the surface, and each will develop into a normal plant. Thousands of plantlets can be produced by this method.

Somatic embryoids also can be produced in a liquid suspension culture. The source material can be single cells or tissue such as leaf or callus. If tissue is used, the cells must be separated; the cell walls are digested with pectinase and cellulase to yield protoplasts. In the liquid medium, cell walls regenerate around the protoplasts, cell division begins, and the new cells aggregate to form small clumps. These cell aggregates can be plated onto solid culture medium to induce the formation of shoots and roots. Important crop plants have been produced using protoplast regeneration and include rice, sweet potato, potato, tomato, flax, sunflower, and cabbage.

Liquid propagation is the preferred method for large-scale commercial production. Propagation in liquid cultures using large-volume bioreactors yields millions of embryoids. Food crops such as asparagus, celery, tomato, potato microtubers, and numerous ornamentals such as begonias, African violet, Boston fern,

from a plant. Cells from explants are induced to divide and eventually form a complete plant. The three stages of culturing are preparation of the explant, multiplication, and pretransplant. To prepare the explant, a plant part such as a piece of stem, leaf, or root is disinfected with a solution such as sodium hypochlorite and cultured aseptically in a specific culture medium. During the multiplication phase, the number of cells increases for later propagation. Cells cultured in the presence of low concentrations of auxin and cytokinin form a **callus**: an undifferentiated mass of loosely arranged cells that arise from the parent tissue (Figure 6.2). Calli are either placed in liquid medium and grown in suspension or are placed on a solid substrate to be cultured. Once callus cultures are established, they can readily be subdivided and recultured—the primary reason why clonal propagation is so valuable. When exposed to combinations of specific growth regulators—growth hormones—in the medium, plants form shoots and roots (Figure 6.3). The third stage, the pretransplant stage, allows for the formation of roots, for further shoot development and maturation, and for the development of photosynthesis. Cytokinins are reduced or eliminated and auxins are added to promote root development.

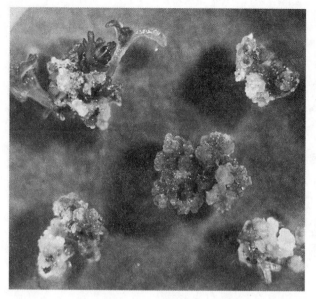

a

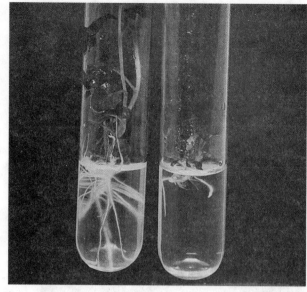

b

Figure 6.3 Shoots (a) and roots (b) develop from callus tissue in the presence of specific plant growth hormones.

and lilies have been successfully propagated. Automation allows 8,000 or more plantlets to be transferred to sterile soil per hour.

Somatic embryoids may eventually be packaged as functional seeds if they can be encapsulated for distribution. Each embryoid will be enveloped in a protective, hydrated gel containing nutrients, growth regulators, fungicides, and other supplements, allowing one to clonally propagate somatic embryoids.

Protoplasts can be transformed with foreign genes (to introduce new characteristics) using such methods as microinjection and electroporation (described in Chapter 4). The genetic characterisitcs of cells from two unrelated plant species or genera, or from plants of the same species with different characteristics, can be combined by an additional method: In protoplast fusion, two cells are fused using chemical fusigenic agents or electroporation,

their genetic material is mixed, and the fusion produces a somatic hybrid with new characteristics (Figure 6.4). Protoplast fusion is particularly useful if species are sexually incompatible or cross with difficulty. Mature, genetically modified plants are regenerated from these protoplasts (Figure 6.5). Hybrids are screened for desirable traits.

Somaclonal Variation

Mutant plantlets are sometimes obtained when somatic embryos derived from single cells are grown into mature plants. Plant breeders have exploited this phenomenon to find new types of genetic variation that improve crop plants' disease and herbicide resistance. Regenerated variants can then be used for breeding or vegetative propagation. The resulting plants differ from the origi-

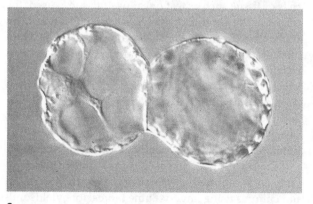

a

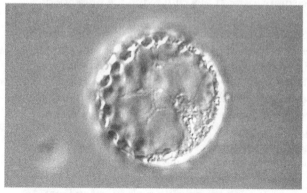

b

Figure 6.4 Two tobacco plant protoplasts, obtained by digesting away the cell wall, are fused (a and b) to produce a cell that acquires some of the characteristics of both genetic backgrounds and can be regenerated into a plant with some traits from both parental plants.

a   b

Figure 6.5 (a) Left: a wild diploid non-tuber bearing species (*Solanum brevidens*) that is resistant to potato leaf roll virus (PLRV) and frost. Right: a diploid tuber-bearing potato clone from *Solanum tuberosum*. Center: protoplasts fused from the two parental lines of potato generated a tetraploid somatic hybrid that was resistant to PLRV and taller. (b) Flower and calyx (leaves not shown) of *S. brevidens* (left) and *S. tuberosum* (right) are smaller than the fusion hybrid (center).

nal parent plant if, during growth and development, genetic material is rearranged when chromosomes break and rejoin or when DNA fragments change position in the plant genome. These mutations are produced randomly so that the regenerated mature plant is not genetically identical. Although the process of **somaclonal variation** is not entirely understood, it is one way that new, potentially desirable traits, such as disease resistance, are obtained if they can be stably maintained and passed on to progeny plants.

Valuable Germplasm

Environmental degradation, urbanization, and changes in farming practices are among the major reasons that valuable germplasm is becoming extinct. Endangered plant species must be conserved, and plant tissue culture is important to the conservation effort. Researchers are identifying new ways to store plant tissue for future use; cryopreservation is an example.

The germplasm of many wild relatives and ancient varieties of important crop plants can be stored or cultured for later use or research. Most often found in developing countries, these ancient crops and wild relatives are often tolerant to a variety of insect pests, bacterial and fungal infection, harsh environmental conditions such as drought, and other stresses due to thousands of years of genetic selection. Plant breeders are using ancient germplasm to introduce desirable traits such as insect resistance into modern crops by crossing susceptible plants with more resistant varieties. However, the abandonment of traditional farming practices, the clearing of old fields, and the displacement of old crops by modern varieties are increasing the rate at which valuable germplasm of wild relatives and ancient

landraces (different varieties of old crops) is being lost. Saving ancient crop varieties and their plant relatives for future use is very important to modern agriculture.

Gene banks, storage facilities for valuable germplasm, are located in many countries throughout the world and contain thousands of plant accessions (Figure 6.6; also see Table 1.1). The plant material housed within these facilities is so valuable that there is often a sophisticated security system to prevent theft, and the material is protected from destruction by fire, earthquake, or flood. *Ex situ* conservation involves collecting and storing plants in a central location, usually in the country that is a center of origin or diversity for that particular variety. Usually seeds are stored if possible; however, they must be maintained in an environment of controlled temperature and humidity and must be tested periodically to determine viability. Seeds sometimes must be germinated and plants grown so that seeds of the next generation can be harvested. Some seeds cannot be stored, and some plants are propagated by tubers or bulbs that cannot be stored for long periods of time. If seeds are not produced or the crop does not reproduce vegetatively, cultured tissue must be used. Cell and tissue culture in conjunction with cryopreservation may provide a way for some germplasm to be maintained.

Chemicals from Plants

In addition to the primary metabolites such as amino acids and nucleotides, plants produce a complex array of secondary metabolites or compounds not directly involved in the organism's growth and development—there are probably at least 50,000 such chemicals in the Plant Kingdom. Although these compounds are not required for photosynthesis and respiration or for

Figure 6.6 A cold-storage facility for seeds at the National Genebank of China, Bejing.

Table 6.2 Selected Plant Compounds and Estimated World Market Value

Compound		Approximate Retail Market Value (millions)
Medicinal		
Ajmalicine	U.S.	$ 5
Codeine		90–100
Corticosteroids		300
Ephedrine, pseudoephedrine		100
Quinine		50
Vinblastine, vincristine		50–75
Flavor and fragrance		
Cardamon		25
Cinnamon		4-5
Spearmint		85–90
Vanilla		15
Agriculture/insecticide		
Pyrethrins		20

biosynthetic processes, many protect plants against predation by mammals, insects, and pathogens. Other compounds are involved in attracting pollinators and fruit dispersers. Many plant-derived compounds, extracted primarily from field-grown plants, are used as flavorings, colorings or dyes, fragrances, pharmaceuticals, and insecticides (Table 6.2). Although many of these compounds can be synthesized in the laboratory, the synthesis is often sufficiently complex that yields are quite low. Many compounds cannot be or have not been reproduced in the laboratory.

Indigenous peoples have been using compounds extracted from plants as medicines for thousands of years. In fact, 75% of the world's population still relies on herbal medicines. Although in the United States herbal medicines are not widely accepted, up to 25% of our pharmaceuticals come directly from plants. Many of these plants were identified because of their use in traditional medicine. Some pharmaceuticals once came from plants but now are chemically synthesized. Pharmaceuticals that were originally isolated from plants include acetylsalicylic acid, or aspirin; digitoxin, a cardiac glycoside that increases heart muscle contractions; and anticancer drugs such as vinblastine, vincristine, and taxol. Typically the development of a drug begins with the identification of an herbal medicine that is widely used, often by indigenous people. Herbal extracts are collected and the active compound is iden-

tified. Sometimes the compound can be chemically synthesized and modified to increase its effectiveness. Clinical trials follow.

Usually plants of interest are grown on tropical and subtropical plantations and harvested for extraction. This method is very costly and energy intensive, poses a potential cost to the native biodiversity, and often is practiced in politically unstable regions; moreover, fertile land often is unavailable. The synthesis of desirable compounds by plant cell culture would alleviate these and many other associated problems. Suspension cultures of plant cells cultured in bioreactors may one day replace plantations in the same way that microorganisms have been used for years in fermentation processes to produce a wide variety of important compounds (described in Chapter 5).

Unfortunately, the production of secondary metabolites by plant cell culture poses many problems. Compounds may be produced in only one organ or, as is the case for many secondary plant compounds, even in only one cell type. Thus, the particular tissue cultured is extremely important. Another difficulty is that changes in the regulation of genes during culturing sometimes dramatically reduces the amount of compound synthesized. Usually many plant varieties or cultivars must be cultured to determine which one, if any, will produce the highest concentration of a particular chemical. In addition, individual putative clones must be screened since not all clones from a plant may produce the desired compound, or in high enough quantities to be profitable.

Many variables influence the success of chemical production during plant tissue culture. The environment must be carefully controlled and supplements such as sugar, vitamins, minerals, plant hormones, and even oxygen must be monitored. In fact, the amount of product may vary depending on the amount of specific supplements in the medium. Other variables include light, temperature, and culturing substrate or medium used. Sometimes different stages require different conditions. For example, rapid cell division may require one set of conditions and cellular production and accumulation of large quantities of metabolite another set. Testing identifies the conditions required. High-producing, stable clones are selected as a source of tissue culture material.

One benefit of using cell cultures is that they may offer much higher metabolite concentrations per cell and synthesis rates than whole plants. Usually in large-scale productions, a bioreactor or fermenter (described in Chapter 5) provides cells with a highly controlled environment. Cells are harvested after an appropriate time so that the metabolites can be extracted from vacuoles in the cells. When cells are induced to over-produce a particular product, the cost of isolation may be reduced. Also, cells in culture may provide an endless supply of product.

An example of an important industrial plant metabolite produced by cell culture is shikonin, a red pigment found in the purple roots of a perennial herb, *Lithospermum erythrorhizon*, which grows in Japan, China, and Korea. Shikonin was first commercially marketed in 1985; it is used in many creams, cosmetics, and dyes. The production process was developed by two Japanese scientists, M. Tabata and Y. Fujita, from Kyoto University. They demonstrated that single cells or protoplasts were required to start clones. Critical variables in the production of shikonin were the amounts and types of hormones, nutrients, and nitrogen sources as well as the temperature, light quality, and amount of oxygen supplied to the cultures. Even increasing the amount of exogenous calcium present was beneficial. Shikonin production was significantly increased after much experimentation. In the first culturing step, cells were multiplied in a 750 liter bioreactor, after which new medium was added. After two weeks, 1.2 kg of shikonin was produced, equivalent to 1.6 grams per liter of cell culture. Today the commercial output of shikonin is 1.5–4.0 grams per liter of cell culture.

Although a number of important secondary metabolites have been produced commercially from cell and root cultures, the costs of research and development keep culturing from being cost-effective in most cases. It is not unusual to invest 10 years into research and development for one product. Sales must recover costs, and usually can do so only if the price per kilogram of product and volume of market demand are high. Sometimes the costs of the compound may be high but the total market is low; for example, jasmine sells at $5,000/kg, but annual sales total only $500,000. On the other hand, there may be a very low unit price but a huge market; for example, spearmint oil sells at $30/kg, but annual sales total $100 million.

Plants also may someday produce compounds not normally found in plants: The gene or genes for a specific enzyme may be inserted into a plant to enable its cells to synthesize a particular metabolite if provided with a readily available precursor compound (see the section Plants as Bioreactors later in the chapter). Large-scale commercial use of both whole plants and plant cells soon may be an effective way to produce compounds and chemicals for which industrial screening programs have identified important uses, such as in pharmaceuticals and pesticides. *In vivo* processes may replace many laboratory synthetic reactions in the future.

The production of commercially valuable plant-derived compounds through new biotechnologies may have a negative impact on the economies of many developing countries. Tropical and subtropical farms produce the plants that make many of the world's commercial secondary metabolites. Since many metabolites must be directly extracted from plants, cultivation has been the only way to obtain them. However, the development of biotechnologies in industrialized nations to produce metabolites from cell cultures may eliminate the need to purchase them from developing countries. The routine use of plants as bioreactors to produce commercial products may adversely affect economies that currently rely on the production and export of plant extracts to industrialized nations.

Genetically Engineered Plants

For all cells in a plant to receive a newly inserted gene, the plant must be regenerated *in vitro* from a few transformed cells. A variety of plant transformation methods are described in Chapter 4; they include viral infection, microinjection, electroporation, and microprojectile bombardment using a particle gun. For many transformations (usually in dicots), explants are exposed to *Agrobacterium tumefaciens* harboring a foreign gene with a plant promoter and an antibiotic resistance gene for selection on the Ti plasmid (also discussed in Chapter 4). Since only some of the cells receive the new gene, the explants must be moved to a solidified medium containing either an antibiotic or chemical to select for the transformed explants. After selection, the explants are cultured on solid medium with nutrients and plant growth hormones to support the callus tissue. Shoot and root growth is promoted by different combinations of hormones as discussed earlier in this chapter. Plants are examined to establish whether the foreign gene has been inserted and expressed in the desired tissue. In this manner, many genetically engineered plants will express new, desirable traits.

Challenges of Foreign Gene Expression Two factors determine temporal and spatial regulation: the expression of genes at the right time and in the appropriate tissue. First, a gene must be delivered to all cells and stably maintained for transmission to progeny. In addition, the promoter region must be recognized by the host cell and either regulated or constitutively expressed, depending on the type of gene and the desired outcome. Termination and polyadenylation signals also must be provided.

For strong constitutive expression, the Cauliflower Mosaic Virus 35S (CaMV 35S) promoter often is used. A highly regulated promoter responds to a certain signal—heat, light, nutrients, etc.—by turning the gene on. The promoter from the gene, ribulose 1,5 bisphosphate carboxylase small subunit, is light-regulated (that is, it activates the gene when the plant is in the presence of light). If the gene product is to be produced primarily in a specific region of the plant, and tissue-specific regulatory regions (such as for leaves) are available, they must be included in the construct.

Also to be considered is codon usage—the codons and amino acids used by an organism; some codons are used more than others depending on the species of plant or other organism. Genes from nonplant sources especially may use amino acids that do not match the plant's tRNA and amino acid pools. Recently, researchers have been able to use **codon engineering**: They resynthesize the donor genes if the codons or amino acids do not match the host system. Codon engineering helps ensure that the host plant will have the appropriate tRNAs and amino acids to synthesize the protein. For example, to increase the production of *Bacillus thuringiensis* toxin by potato, tomato, rice, and cotton, the Bt toxin gene has been engineered to match the host plant's translation machinery and to remove incompatible sequences.

APPLICATIONS OF PLANT GENETIC ENGINEERING

Crop Improvement

A major goal of plant genetic engineering is to improve agronomically important crops by transforming them with foreign genes. The first plant biotechnology products are just beginning to emerge. Investments in crop biotechnology and transgenic plants have exceeded one billion dollars in the private sector of the United States.

Almost 50 crop varieties have been genetically modified, and many of them will be used commercially before the end of the 20th century.

Much research has focused on agronomic traits for controlling insects, viruses, bacterial and fungal plant diseases, and weeds. Other modifications will include developing new, more nutritious food products, manipulating petal color of flowers, and synthesizing biodegradable, organic plastics (Figure 6.7). A very promising area of research is the production of specialty oils for detergents, cosmetics, and lubricants, and the modification of the lipid composition of seed crops to reduce levels of saturated fats. Examples of modified oils include those with increased levels of laurate and myristate for shampoos and soaps and those that solidify at room temperature without hydrogenation. Partially hydrogenated oils (such as margarine), which are added to many foods, have been shown to be unhealthy. Other uses for specialty oils include cooking oil with reduced saturated fat, an oil substitute for cocoa butter in chocolate, oils for cosmetics, and liquid wax for lubricants. Long chain fatty acids (C_{20}-C_{24}) are used as lubricants and as solvents for some pesticides, while short chain fatty acids (C_6-C_{14}) are used in the manufacture of soap and detergents (a much used oil in detergents is lauric acid, obtained mostly from tropical oil palm). Since many oils are obtained from tropical plants, the production of oil subsitutes from genetically engineered oilseed crops grown in the United States, primarily rapeseed, soybean, and sunflower, would decrease the dependency on tropical oils.

The loss of prime farmland may one day require that crops be cultivated in areas that are less suitable or marginal for agriculture. Crops may have to be developed that can be grown in marginal soil or environmental conditions. Plants are a source of both monomeric and polymeric compounds such as sugars, fatty acids, celluloses,

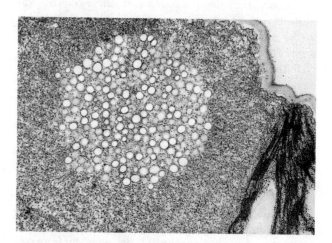

Figure 6.7 Transmission electron micrograph of an *Arabidopsis thaliana* leaf mesophyll cell with an accumulation of polyhydroxybutyrate (PHB) granules in the nucleus.

starches, waxes, and rubber. Plants may soon function as living bioreactors for the production of industrially important chemicals and pharmaceutically active compounds. The controversies that the genetic engineering of plants (and animals) has sometimes created are discussed briefly at the end the chapter.

Herbicide Resistance

Weeds are plants that effectively compete with crop plants for nutrients, water, light, and carbon dioxide and often significantly lower crop yields. Unfortunately, methods that promote crop growth also stimulate weed growth. Weeds make good competitors for several reasons: (1) They have very effective reproduction modes and highly evolved seed dispersal mechanisms. Each plant produces numerous seeds, which can remain dormant for long periods of time and do not always germinate at the same time. (2) Weeds produce seeds throughout the growing season and so re-seed themselves several times during the season. (3) Weeds are fast-growing and can tolerate drought and low nitrogen levels, while most crops cannot. (4) They are self-fertilizing, wind-pollinated, or pollinated by unspecialized insects. (5) They outgrow other plants and form large root systems that are difficult to uproot.

Herbicides are used annually in agricultural areas to decrease the impact of weeds on crops. Between 1966 and 1991, the use of herbicides in agriculture in the United States more than quadrupled, reaching an estimated 495 million pounds at a cost of $10 billion. Although farmers routinely apply more than 100 chemical herbicides, weeds still reduce crop productivity by approximately 12%. The sensitivity of crops to herbicides limits the types that can be used. Since many herbicides do not discriminate between crop and weed,

they must sometimes be applied early, before crop emergence. Often the chemical persists in fields when the crops are germinating and can kill them.

Plants such as corn, potato, soybean, tomato, and cotton have been genetically engineered to resist the toxic effects of chemical herbicides, allowing herbicides to be used where they otherwise would kill the crop plants. Most of the genetically engineered organisms being field tested in the United States are transgenic crops. These crops are also among the most controversial **transgenic organisms**.

Herbicide-resistant plants are produced by a variety of genetic manipulations. These affect the plant metabolism of the herbicide, the herbicide-binding to the target protein, and overproduction of the target protein herbicide. Most genetic engineering uses a single bacterial gene to confer resistance to a herbicide. For example, glyphosate-resistant plants have been engineered to readily degrade the herbicide (e.g., Monsanto's Roundup®) into nontoxic compounds (Figure 6.8). The herbicide normally inhibits an important enzyme (EPSPS or 5-enolpyruvylshikimate-3-phosphate synthase) in the aromatic amino acid synthesis pathway (the shikimate pathway) in both plants and bacteria. The gene encoding the EPSPS in a glyphosate-resistant *Escherichia coli* strain was isolated, placed under control of a plant promoter, and transferred into plant cells. Transgenic, glyphosate-resistant tomato, potato, petunia, cotton, and tobacco have been produced.

Other bacterial genes also are being used to confer herbicide resistance in plants. For example, bromoxynil inactivates photosynthesis. Bromoxynil-resistant plants are produced when a gene encoding the enzyme nitrilase is transferred into plants from the soil bacterium *Klebsiella ozaenae*. Nitrilase inactivates bromoxynil before this herbicide can kill the plant.

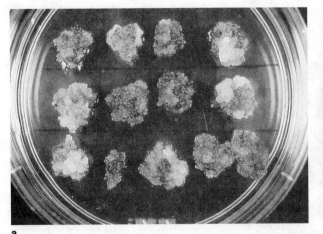

a b

Figure 6.8 (a) Glyphosate-resistant callus after transfer of a bacterial EPSPS gene to cells.
(b) Glyphostate-resistant soybeans before (left) and after (right) application of glyphosate (the product Roundup®). Note the absence of weeds in the row of soybeans treated with the herbicide.

Scientists speculate that transgenic plants that are tolerant have the potential to allow fewer toxic chemicals to be used in the environment. A larger selection of biodegradable or less-toxic herbicides could be made available to the farmer. The Monsanto Corporation already has produced Roundup Ready™ crops, such as corn, cotton, soybean, tomato, potato, and wheat, that are resistant to its readily biodegradable glyphosate herbicide Roundup®. Companies perhaps will adhere to a less toxic approach to herbicide resistance and not produce transgenic plants that tolerate toxic, non-biodegradable herbicides.

Companies that develop herbicides also are engineering herbicide-tolerant plants to accompany their chemicals. If a chemical approach to weed control continues in the future, chemical use will most likely increase. In this case, instead of creating a safer and cleaner environment, biotechnology will have perpetuated our dependence on toxic chemicals. This dependence may generate more herbicide-resistant weeds through increased abundance of chemical applications or perhaps by accidental crossing of herbicide-resistant crops with neighboring related wild plants.

Insect Resistance

In Chapter 5 the use of the Bt gene from *Bacillus thuringiensis* to generate insect-resistant plants was discussed. Other ways to confer insect resistance also are available. For example, plant **protease** inhibitors have been shown to be effective general pesticides. After they are ingested by insect larvae, they inhibit digestive enzymes, and starvation results (Figure 6.9). In laboratory experiments, the cowpea trypsin inhibitor gene has been successfully transferred into tobacco by *A. tumefaciens*. A more effective pesticide may be generated when plants produce a fusion protein that is encoded by portions of a specific protease inhibitor and Bt-protoxin gene (the protoxin must be activated by a trypsin-like enzyme).

Storage pests attack stored cereal grains, peas, and beans and result in huge losses. Recently, garden pea seeds have been engineered that resist attack by two species of weevils that damage crops in storage (Figure 6.10). A protein, α-amylase inhibitor, blocks the action of the enzyme α-amylase, which digests the starch in seeds. A seed-specific regulatory region was connected to the α-amylase inhibitor gene. In studies, most of the weevils that fed on peas either died or suffered inhibited development. Interestingly, the inhibitor protein is expressed only in the seeds of peas. Eventually, other legumes such as pinto, mung, and kidney beans, in addition to black-eyed peas (cowpeas) and chickpeas will be protected from insect attack.

Virus Resistance

Many crops are lost to viral diseases, resulting in losses of millions of dollars each year. Chemicals are widely used to help control the insect vectors that spread viral diseases, but genetic engineering may provide a more desirable alternative to chemicals. Research has focused on isolating genes involved in resistance to diseases caused by viruses, bacteria, and fungi. Scientists are generating many virus-resistant crops such as melons, tomatoes, squash, cucumbers, and potatoes. Plants are "immunized" by a "vaccine" that usually consists of the genes encoding viral coat proteins. A gene that encodes a viral coat protein is transferred to the plant, which then expresses a small amount of viral protein. Although the exact mechanism of protection is unknown at this time, the virus cannot replicate and spread in a virus-resistant plant. Numerous plant viruses have been identified, and many of them are host-specific. To date, the coat protein genes of a number of viruses are being used to genetically engineer resistance in crops. Among the viruses are cucumber mosaic virus, alfalfa mosaic virus, tobacco streak virus, tobacco etch virus, tobacco rattle virus, potato virus Y and potato virus X, and potato leafroll

Figure 6.9 Transgenic tobacco plant leaves with (right) and without (controls on left) a protease inhibitor gene were exposed to larvae. Extensive damage to controls occurred, while transgenic leaves were protected.

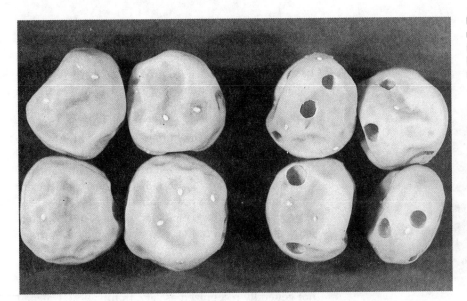

Figure 6.10 Pea seeds genetically modified with the α-amylase inhibitor gene resist attack by weevil larvae (left). Larvae have burrowed into the pea seeds that do not harbor the gene (right).

virus. They have been used on such crops as alfalfa, tomato, tobacco, and potato.

Resistance genes for a variety of diseases—not just viral infections—have been isolated from plants: A gene isolated from *Arabidopsis thaliana* (which serves as a model for the study of plant molecular genetics) and tomato confers resistance to the bacterial pathogen *Pseudomonas syringae*; a tomato gene is involved in resistance to the fungal pathogen *Cladosporium fulvum*; a flax gene has been isolated that confers resistance to fungal rust disease; and a tobacco gene protects against tobacco mosaic virus. One day these genes may be transferred to other agronomically important crops.

Plants as Bioreactors

One day whole plants may be used as living bioreactors (just as microbes already are being used) by inserting a foreign gene to produce proteins of therapeutic and industrial value. An inexpensive, readily available complex metabolite could be used as a nutrient source, and cells could then convert this compound into a valuable product, which accumulates until the cells are harvested. Whole plants would then be harvested and the product extracted, allowing producers to avoid complex *in vitro* biosynthetic reactions in the laboratory.

Recently, soybean plants have been used as bioreactors to produce a variety of monoclonal antibodies with therapeutic value (such as for treatment of colon cancer). The cost of producing foreign compounds in plants may be much lower than producing them in bacteria and mammalian cells for a number of reasons: (1) commercial scale-up involves simply planting seeds rather than using costlier fermenters; (2) proteins are produced in high quantity (e.g., soybeans have a protein content of 40–45%); (3) foreign proteins produced are active, unlike those produced by bacteria, which are

denatured and must be renatured after isolation; (4) foreign proteins in stored seeds are very stable (e.g., in experiments, they have remained active after five years); (5) pathogens, toxins, viruses, and other contaminants are not likely to be present. In addition, since plants are eukaryotes, post-translational processes (e.g., glycosylation, phosphorylation, proteolytic cleavage) are present.

The First Genetically Engineered Food

In the next decade we are likely to see many new agricultural products created through genetic engineering. The ones most likely to arouse public interest and concern will be genetically engineered foods. The first genetically modified food product to come to our stores is the Flavr Savr tomato developed by Calgene, Inc., a biotechnology company in California (Figure 6.11). The bioengineered tomato, which has generated much discussion and controversy, offers many benefits, including a garden-fresh taste all year round.

This feature is, of course, in contrast to the rather tasteless varieties that must be picked at the onset of ripening to avoid over-ripening and softening by the time they reach the market. Although the green, unripe tomatoes are gassed with ethylene to turn them red, time does not allow them to develop any flavor. The Flavr Savr has been genetically altered to soften more slowly so that the fruit can remain on the plant until ripe. Thus, the tomatoes stay fresh without rotting and deliver natural flavor all year. The shelf life is several days longer, and Calgene claims the Flavr Savr tomato retains the characteristics of a fresh-picked product. In addition, the tomato still remains firm for picking, packing, and shipping.

The Food and Drug Administration (FDA) approved the Flavr Savr tomato in May 1994, making it the first genetically engineered whole food to gain government approval in any country. The FDA ruled that the

Figure 6.11 Calgene's Flavr Savr tomatoes have been genetically modified to soften more slowly than conventional tomatoes so that they can remain on the vine for full flavor development before being harvested.

Calgene tomato is as safe as a conventional tomato and has not required labeling in markets to indicate which gene has been added or that it has been genetically modified. Calgene identifies their tomatoes with a small label that says "MacGregor's Grown from Flavr Savr seeds." The sale of fresh tomatoes declines by as much as 30% from October to June when they are not as tasty. The Flavr Savr can help increase sales during those months. Calgene contracts farmers to grow the tomatoes and will control the distribution of the Flavr Savr tomato.

The Flavr Savr tomatoes were produced by blocking the expression of the polygalacturonase (PG) gene. Fruit softens because the PG enzyme degrades pectin, a major component of plant cell walls. When the amount of PG is decreased, the fruit remains on the plant longer without becoming soft, and is redder and riper by the time it is picked for distribution.

Tomato plants were transformed with the antisense PG gene, which encodes mRNA that is complementary to PG mRNA. The two messenger RNAs (anti-PG mRNA and PG mRNA) may bind to one another so that they are degraded and translation of PG protein is prevented. Antisense technology is being developed to block the production of many other proteins that will have application in agriculture and medicine.

The controversy about Calgene's genetically engineered tomato arose over a marker gene that was used to identify plant cells that had been transformed with the antisense PG gene. The marker gene confers antibiotic resistance to kanamycin and is linked to the antisense gene. Opponents of genetically engineered foods feared that foods harboring antibiotic resistance genes could

exacerbate the problem of increasing resistance to antibiotics in humans. Calgene scientists have used the tomato in animal studies and demonstrated that most of the kanamycin resistance gene product is destroyed by the digestive system. Experimental results have indicated that the expression level of the marker gene in the tomato is extremely low, and the likelihood of gene transfer to *E. coli* of the human gut is highly unlikely.

Under current FDA policy developed in 1992 (Statement of Policy: Foods Derived from New Plant Varieties, *Federal Register* May 29, 1992, 57 FR 22984), genetically engineered foods are not required to be reviewed and approved unless they contain additives. The characteristics of the food product and not the method of production (e.g., genetic manipulation) are to be reviewed for safety. The FDA 1992 policy states that expression products are to be regulated as food additives if they are not generally recognized as safe. In July 1993, Calgene filed a food additive petition requesting the use of the kanamycin resistance gene product, aminoglycoside 3'-phosphotransferase II (APH(3')II), in new plant varieties of tomato, oilseed rape, and cotton. Calgene submitted the findings from studies that provided evidence that APH(3')II poses no threat to either humans or the environment. The summaries of the data and deliberations are presented in the FDA document (Docket No. 93F-0232, May 23, 1994), "Secondary Direct Food Additives Permitted in Food for Human Consumption; Food Additives Permitted in Feed and Drinking Water of Animals; Aminoglycoside 3'-Phosphotransferase II." The FDA amended the food additive regulations to allow for the safe use of APH(3')II as a processing aid in the development of new plant varieties. The FDA also determined that the Calgene tomato does not differ significantly from a nonengineered variety and maintains the essential characteristics of a normal tomato, thus making it as safe as a normal tomato.

Although producers are not required to inform the FDA that they plan to sell their engineered products to the public, they are encouraged to do so. For example, Calgene voluntarily submitted Flavr Savr data for FDA review and assessment. Since the tomato was the first bioengineered food to be marketed, Calgene hoped that an FDA safety endorsement would help diminish the controversy. (Calgene had to obtain approval from the United States Department of Agriculture (USDA) before it could grow the tomatoes in commercial fields.) Opponents of genetically engineered foods have expressed concerns that FDA approval of the tomato will open the door to many other genetically engineered foods in the near future.

Many more transgenic products are indeed likely to follow—a wide variety of genetically engineered livestock, fish, grains, fruits, vegetables, and agronomically important crops. Already pending approval from the USDA and FDA are virus-resistant squash, zucchini, soybeans, and watermelons. Over 50 bio-engineered foods are expected to be ready for the commercial market by the year 2000. One product that may already have a promising future is a potato variety in which the starch content has been increased by 40%, making a potato chip with better texture. Vegetables and fruits in the future may have a higher sugar content; controlled conversion of sugar to starch in peas and corn, for example, will help these foods remain sweeter after harvest. Future products also include a naturally decaffeinated coffee bean, high-nutrient corn, and new crops that can be grown with fewer toxic herbicides and pesticides and with less irrigation.

GENETIC ENGINEERING AND PUBLIC CONCERNS

The development of genetically engineered plants and animals has sometimes been controversial. Fears that these novel organisms might pose risks to the environment have led to federal policies requiring that thorough analyses of ecological risks be conducted. Before bioengineered plants and plant products can be commercialized, scientists must ensure that scientifically sound, experimental methods are developed to assess the ecological risks.

More than 2,000 U.S. field trials have been completed or are in progress on genetically engineered versions of cotton, maize, alfalfa, oilseed rape, cucumber, asparagus, apple, melon, cauliflower, and sugarbeet. Potatoes with their own insecticide and a virus-resistant yellow crookneck squash are two of seven genetically engineered crops recently approved by the FDA. The yellow squash has been approved by the USDA, which means that it can be treated like any other fruit. Other engineered plants include a cotton resistant to the herbicide bromoxynil, a soybean resistant to the herbicide glyphosate, two tomatoes that are engineered to delay ripening (they differ from the Clagene Flavr Savr).

After many years of studies, the regulatory procedures for field testing transgenic plants have been relaxed, although some individuals continue to voice concern. In 1987, each genetically modified plant had to be field tested (required by the USDA's Animal and Plant Health Inspection Service, or APHIS) to demonstrate that the plant would not disperse into the environment—that is, would not become a weed—and disrupt the natural plant diversity. Approval for field testing required extensive review. Today, the product and not the process is the determining factor. In 1993, the USDA modified the restrictions for approximately 85% of the genetically engineered plants—primarily tomatoes, corn, soybeans, tobacco, potatoes, and cotton. One only has to inform the USDA that field tests will be conducted. An in-depth review is conducted only if a product has been altered in such a way as to be completely different from a naturally occurring plant and poses a potential health or environmental risk. The FDA has adopted a similar policy.

As we saw, the release of the first genetically engineered food, the Flavr Savr tomato, generated much discussion about the potential risks of genetically engineered food. The primary public fear was that genetically engineering a plant might produce unexpected results. Plants produce secondary metabolites that may be toxic to humans or livestock or may alter the quality of foods. Although these metabolites are present in very low concentrations in our foods, one fear has been that a genetically engineered plant may produce them in higher quantities that may be toxic to humans. Many people have food allergies that vary from a mild reaction such as a rash to a more severe reaction such as anaphylactic shock. There has been concern that the allergenic properties of food from a donor plant might be conferred on the host plant, and that people may not be aware that the genetically engineered food contains the new protein. Others have questioned whether the nutritional quality of engineered foods could change unexpectedly (a disruption in the balance of nutrients). Perhaps a particular nutrient may change to a form that cannot be metabolized or absorbed properly. Of course, there also is the question of what constitutes true freshness. If a vegetable or fruit looks, feels, and tastes fresher and riper, is it really just as nutritious as its nonengineered counterpart? Calgene conducted extensive tests to confirm the nutrient level of its transgenic tomato as well as to address many of the concerns stated above.

A very different type of concern has focused on food choice. Some religious and ethnic groups avoid eating certain foods (for example, Muslims and Orthodox Jews do not eat pork). Vegetarians may also avoid some genetically engineered foods. How will engineered produce affect food selection when, for example, an apple contains an animal gene? What if a cucumber contains a human gene? Does the cucumber possess a human trait? Of course, these questions address hypothetical situations, but they represent the types of questions posed by concerned individuals and critics of genetically engineered foods.

As with the Flavr Savr tomato, another source of debate has encompassed the use of antibiotic resistance genes as selectable markers to indicate when an organism has been transformed. These enzymes have the potential to inactivate clinically invaluable antibiotics. Although evidence has not supported this concern, opponents of genetically engineered foods still consider antibiotic resistance an issue. Calgene conducted studies to demonstrate that the product of the kanamycin resistance gene, when ingested, did not inactivate the antibiotics kanamycin or neomycin, and that the APH(3')II was rapidly broken down during digestion.

Some fear that deleting genes also may introduce harmful side effects when the product is ingested. For example, plants produce secondary compounds that may protect them from fungal and bacterial infection. If secondary compounds are removed, people may be exposed to cancer-causing compounds produced by the fungus. An example often cited is decaffeinated coffee. Research has demonstrated that caffeine may inhibit fungal synthesis of aflatoxins; if a gene involved in caffeine biosynthesis is removed, fungal aflatoxins may then contaminate the coffee bean.

Other fears include the potential that uncharacterized DNA included with the gene and selectable markers will code for additional, unknown proteins or produce unexpected, harmful side effects or undesirable traits in the engineered plant. Fear also has been expressed that the introduced trait in genetically engineered crops may be spread through the pollen to wild plants by wind or insect pollinators. This dispersal may irreversibly alter the ecosystem and even affect wildlife that feed on these wild plants.

PLANT GENETIC ENGINEERING AND THE FUTURE

Because of their newness, genetically engineered foods and other recombinant plant products, such as herbicide resistant crops, have raised concerns about safety. However, as future research demonstrates their safety and still more genetically engineered plant products are introduced to the public, many concerns and fears will most likely diminish, as they did in the 1970s when cloning methods were first developed.

Exciting new agricultural biotechnologies will provide solutions to important problems in agriculture and ecology. The wise use of modern biotechnology should embrace safer, less-toxic agricultural practices as well as the conservation and use of germplasm. Genetically engineered plants can contribute in positive ways toward sustainable agriculture and should provide an alternative to poor agronomic practices that ultimately contribute to environmental degradation. Questions to be asked include how we can become less dependent on chemicals, hybrid seed, and other costly amendments, rather than encouraging "high-input" solutions that promote soil loss, increased use of chemicals, and the loss of germplasm diversity. Plant genetic engineering may offer sound alternatives that will help decrease our dependence on agronomic practices that promote environmental degradation. A bright, exciting future for agricultural biotechnology is on the horizon.

General Readings

S.S. Bhojwani, ed. 1990. *Plant Tissue Culture: Applications and Limitations.* Elsevier, Amsterdam.

I. Chet, ed. 1993. *Biotechnology in Plant Disease Control.* Wiley-Liss, New York, NY.

M.J. Chrispeels and D.E. Sadava. 1994. *Plants, Genes, and Agriculture.* Jones and Bartlett Publishers, Boston, Massachusetts.

J.R.S. Fincham and J.R. Ravetz. 1991. *Genetically Engineered Organisms: Benefits and Risks.* University of Toronto Press, Toronto, Canada.

L. Kim, ed. 1993. *Advanced Engineered Pesticides.* Marcel Dekker, Inc., New York.

D.F. Wetherell. 1982. *Introduction to* In Vitro *Propagation.* Avery Publishing Group Inc., Wayne, New Jersey.

Additional Readings

See series of articles in *Science,* 5 May 1995, volume 268, Emerging Plant Science: Frontiers in Biotechnology.

R.N. Beachy, S. Loesch-Fries, and N.E. Tumer. 1990. Coat protein-mediated resistance against virus infection. *Annu. Rev. Phytopathol.* 28:451-454.

K.J. Brunke and R.L. Meeusen. 1991. Insect control with genetically engineered crops. *Trends Biotechnol.* 9:197-200.

T.-T. Chang. 1994. Plant genetic resource conservation and utilization. In *Encyclopedia of Agricultural Science,* Vol. 3, Academic Press, San Diego, pp 295-304.

E. Chasseray and J. Duesing. 1992. Field trials of transgenic plants: An overview. *Agro. Ind. Hi-Tech* 2:5-10.

P. Christou. 1992. Genetic transformation of crop plants using microprojectile bombardment. *Plant J.* 2:275-281.

P.J. Dale. 1995. R & D regulation and field trialling of transgenic crops. *Trends Biotechnol.* 13:398-403.

E.L. Flamm. 1994. Plant biotechnology: Food safety and environmental issues. In *Encyclopedia of Agricultural Science,* Vol. 1, Academic Press, San Diego, pp 213-223.

R. Fraley. 1992. Sustaining the food supply. *Bio/Technology* 10:40-43.

C.S. Gasser and R.T. Fraley. 1992. Transgenic crops. *Sci. Am.* 266: 62-69.

O.J.M. Goddijn and J. Pen. 1995. Plants as bioreactors. *Trends Biotechnol.* 13:379-387.

R.J. Goldburg. 1992. Environmental concerns with the development of herbicide-tolerant plants. *Weed Technol.* 6:647-652.

U. Halfter, P.C. Morris, and L. Willmitzer. 1992. Gene targeting in *Arabidopsis thaliana. Mol. Gen. Genet.* 231:186-193.

K.K. Hatzios. 1994. Herbicides and herbicide resistance. In *Encyclopedia of Agricultural Science,* Vol. 2. Academic Press, pp 501-512.

H. Klee, R. Horsch, and S. Rogers. 1987. *Agrobacterium*-mediated plant transformation and its further applications to plant biology. *Annu. Rev. Plant Physiol.* 38:467-486.

B. Lambert and M. Peferoen. 1992. Insecticidal promise of *Bacillus thuringiensis*. *Bioscience* 42:112-122.

H.S. Mason and C.J. Arntzen. 1995. Transgenic plants as vaccine production systems. *Trends Biotechnol.* 13:388-392.

B.J. Mazur. 1995. Commercializing the products of plant biotechnology. *Plant Biotechnol.* 13:319-323.

L. Miele. 1997. Plants as bioreactors for biopharmaceuticals: regulatory considerations. *Trends Biotechnol.* 15:45-50.

J.N.M. Mol, T.A. Holton, and R.E. Koes. 1995. Floriculture: genetic engineering of commercial traits. *Trends Biotechnol.* 13:350-355.

R. Olembo. 1995. Biodiversity and its importance to the biotechnology industry. *Biotechnol. Appl. Biochem.* 21:1-6.

S.R. Padgette et al. 1996. New weed control opportunities: Development of soybeans with a Roundup Ready™ gene. In S. O. Duke, ed. *Herbicide-resistant crops: Agricultural, environmental, economic, regulatory, and technical aspects.* CRC Lewis Publisher, Boca Raton, Florida. pp 53-84.

Y. Poirier, D.E. Dennis, K. Klomparens, and C. Somerville. 1992. Polyhydroxybutyrate, a biodegradable thermoplastic, produced in transgenic plants. *Science* 256:520-523.

I. Potrykus. 1991. Gene transfer to plants: Assessment of published approaches and results. *Annu. Rev. Plant Physiol.* 42:205-225.

F. Powledge. 1995. The food supply's safety net. *Bioscience* 45:235-243.

W. Schuch. 1994. Improving tomato quality through biotechnology. *Food Technol.* 48:78-83.

D.M. Shah, C.M.T. Rommens, and R.N. Beachy. 1995. Resistance to diseases and insects in transgenic plants: progress and applications to agriculture. *Trends Biotechnol.* 13:362-368.

D.M. Stark, G.F. Barry, and G.M. Kishore. 1993. Impact of plant biotechnology on food and food ingredient production. In M. Yalpani, ed., *Science for the Food Industry of the 21st Century.* ATL Press, Mount Prospect, Illinois. pp 115-132.

J. Van-Brunt. 1985. Non-recombinant approaches to plant breeding. *Bio/Technology* 3:975-980.

J. Van Rie. 1991. Insect control with transgenic plants: Resistance proof? *Trends Biotechnol.* 9:177-179.

R. Walden and J. Schell. 1990. Techniques in plant molecular biology—progress and problems. *Eur. J. Biochem.* 192:563-576.

R. Walden and R. Wingender. 1995. Gene-transfer and plant-regeneration techniques. *Trends Biotechnol.* 13:324-331.

S. Williams, L. Friedrich, S. Dincher, N. Carozzi, H. Kessmann, E. Ward, and J. Ryals. 1992. Chemical regulation of *Bacillus thuringiensis* δ-endotoxin expression in transgenic plants. *Bio/Technology* 10:540-543.

Selective animal breeding has been practiced for nearly 10,000 years to produce desirable traits in livestock, as we saw in Chapter 1. Today, breeders are under increasing market pressure to produce livestock that grow faster and convert animal feed to lean tissue, with less total fat. Breeders have focused not only on manipulating the composition of animal products but also on increasing production, such as by enhancing reproduction rates. Increased milk production, decreased fat content, better wool quality, faster maturation rate, disease resistance, and the increase in frequency of egg-laying are but a few of the changes brought about by selective breeding. However, such methods of producing desirable traits take time; many generations of animals must be born and bred before an appropriate trait is finally expressed. Hundreds of gene pairs are involved in the modification of measurable traits such as egg and milk production, fat composition, and disease resistance. Consider the numbers the breeder is up against: If 20 heterozygous gene pairs are involved, the number of possible gene combinations is approximately 3.5 billion! Simply maintaining the rate of genetic improvements made in the past, let alone increasing the rate, requires intensive screening and selection programs.

As we saw in Chapter 6, on plant biotechnology, the availability of tools for genetically manipulating plants is revolutionizing agriculture. Recombinant DNA technology now allows us to introduce foreign genes into organisms for the expression of specific new traits. Animals also can be engineered for a variety of purposes, ranging from use as human disease models to the introduction of desirable traits into a variety of agronomically important animals—including fish, as we see in the next chapter.

An approach that combines breeding with molecular genetics and recombinant DNA technology will lead to greater progress in the future. Powerful genetic methods such as marker-assisted selection (MAS) technology will help identify chromosomal loci associated with physical traits. MAS identifies relationships between gene markers mapped to specific chromosomal regions or loci and measurable traits such as disease resistance, milk production, and fat content. MAS has identified some 75 genes that influence growth. Twenty-nine of these encode either growth factors or growth factor receptors. Now that more efficient transformation methods and vectors are available, specific genes are beginning to be cloned and transferred into many species to increase the production of animals with desirable traits.

Biotechnologies have affected animals in many significant ways. Genetic engineering has enabled scientists to transfer genes across species, families, and even kingdoms. These evolutionary boundaries, which were obstacles in the past, are now readily crossed with modern gene transfer methods that allow the transfer of DNA into living cells and fertilized eggs of animals. Transgenic animals not only provide invaluable research tools for studying gene regulation and disease, but they may be genetically modified for the production of pharmaceuticals, **vaccines**, and rare chemicals as well as for food production. This is an exciting yet sometimes controversial new frontier in animal biotechnology.

Major goals of biotechnology include the generation of livestock that are more economically produced and products that are more nutritious to the consumer. Livestock research has had two primary focuses: increasing growth rate and muscle development with growth-hormone genes from various sources, and producing antibodies and recombinant vaccines to increase disease resistance. A more distant goal is to develop animal "bioreactors" for producing rare pharmaceuticals and other medical compounds. Genetically engineered livestock will yield important products in milk or blood for treating a variety of human diseases and health needs. Among such products might be:

- Human hemoglobin, which could be used during trauma when much blood is lost. Hemoglobin is more desirable than whole blood or red blood cells for transfusions, since it requires no refrigeration and is compatible with all blood types, eliminating the need for blood typing.

- Human protein C, which helps prevent blood clotting.

- Human tissue plasminogen activator, which is used to treat patients after a heart attack.

- Human alpha-1-antitrypsin, which may be useful in treating the more than 20,000 people in the United States who have ha1AT-deficiency, which predisposes them to a life-threatening type of emphysema.

GENE TRANSFER METHODS IN ANIMALS

One challenge in creating transgenic animals is to ensure that the **transgene** turns on at the right time and in the right tissue. To be functional, the integrated gene must be expressed and regulated appropriately (for example, in a tissue-specific manner). Thus, the gene to be transferred must be accompanied by the appropriate promoters and regulatory sequences, as described in Chapter 3. Some genes require an enhancer that may be located far from the promoter. Gene regulation can be extremely complex during embryonic development, when many genes are regulated temporally and tissue-specifically. The engineering of organisms requires

fusion of the correct promoter/enhancers and gene coding sequence. The construct must then be incorporated into the chromosome where gene expression is regulated. Figure 7.1 shows the results of a gene-transfer experiment.

Since foreign genes were first introduced into mice by **pronuclear microinjection** of DNA in 1980, a tremendous effort has been expended to develop efficient methods of producing transgenic animals. In addition to microinjection, methods have included retrovirus-mediated and **embryonic stem cell**-mediated gene transfer into the germ line, the reproductive cell line. Other methods, described in Chapter 4, have included electroporation, virus infection, and DNA precipitation onto the cell surface using calcium phosphate.

Microinjection

Microinjection, first developed through use in the mouse, has been the method of choice for most transgenic research. Once a gene has been characterized and appropriately expressed in eukaryotic cells, a transgenic animal can be made by microinjection of the cloned gene into the fertilized eggs (ova) of a donor animal. The foreign gene must be injected before the first cell division, or cleavage, occurs so that all cells of the organism harbor the gene.

Donor females are superovulated (induced to produce exceptional numbers of ova) and mated, and the

Figure 7.1 Two 10-week-old sibling male mice, one of them the product of genetic engineering experiments. The mouse on the left harbors a new gene comprising the mouse metallothionein promoter fused to the rat growth hormone structural gene. This mouse weighs 44 grams, while its control sibling weighs 29 grams. The gene is passed on to offspring that also grow larger than controls. In general, mice that express the gene grow two to three times faster than controls and up to twice normal size.

fertilized eggs are removed and placed in a sterile dish with buffer. Before fusion occurs to make a diploid **zygote** with one nucleus, male and female pronuclei are separate. A very thin pipette or needle injects DNA into the large male pronucleus (refer to Figure 4.2). Microinjected eggs are implanted into the oviduct of a surrogate female made pseudopregnant with hormones. After birth, to identify which of the young are transgenic, tissue samples are subjected to DNA analysis by Southern blot hybridizations with the new gene as a probe or by PCR amplification of the new gene. Founder animals—those animals with the new gene integrated into their germ cell line and somatic cells—are bred to establish new genetic lines with the desired characteristics.

Some microinjected fertilized eggs (e.g., the morula or blastocyst in cows) must be cultured *in vitro* until they reach a certain stage, and others (such as those of pigs) are transferred to hosts almost immediately after microinjection. Other species have presented the problem that the opacity of the cytoplasm makes it difficult to observe the pronuclei. A special type of microscopy—**differential interference contrast microscopy**—has been used for rabbit, sheep, and goat eggs. However, cow and pig eggs must be centrifuged to stratify the cytoplasm so that the pronuclei are visible (Figure 7.2). The efficiency of experimental gene integration into the chromosome has been lower in livestock (ranging from 1 to 13% in sheep, pigs, and cattle) than in the mouse (approximately 27%). Although the difference in transformation efficiency is unknown, the microinjection parameters were developed for the mouse (e.g., age of ovum, buffer composition) and may not be optimal for other animals.

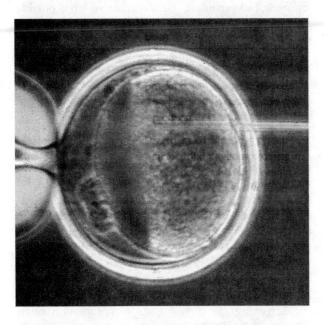

Figure 7.2 Microinjection of DNA into the pronucleus of a bovine egg after centrifugation has moved the cytoplasm to make the pronuclei more visible.

Microinjection is not without problems. Few injected eggs survive, and not all of those retain the new DNA. The new DNA is integrated at random if it is not targeted to a specific chromosomal locus. Unfortunately, **homologous recombination** and gene targeting efficiency are extremely low in animals. Microinjected genes often integrate into a single chromosomal locus as head-to-tail concatamers. In addition, many transgenic animals are mosaics because the gene has integrated into only some of the cells. This mixture of transformed and untransformed cells is especially problematic if the gene has not incorporated into the germ cells, or is in only a portion of the germ cells. All germ cells must be transformed to ensure that the gene is transmitted to progeny. There are also instances in which a gene is incorporated but expression is low or nonexistent. Further research will yield more efficient transfer and expression of foreign genes.

Embryonic Stem Cell Gene Transfer

Another method of producing transgenic animals uses cultured embryonic stem (ES) cells (ES cell lines are available in culture). ES cells are isolated from the inner cell mass of donor blastocysts of early embryos and can be cultured *in vitro* prior to **transfection** with a specific gene. A gene is targeted to specific areas by homologous recombination and the use of selectable markers; treated cells are then screened by either PCR or a selection method to determine whether the gene has been integrated. Transformed ES cells are microinjected into animal blastocysts so that they can become established in the somatic and germ line tissues. They are then passed on to successive generations by breeding founder animals (Figure 7.3).

At present, the ES method is most successful with mice because mouse ES cells are **pluripotent** (capable of differentiating into other cell types) and, when integrated into blastocysts, can divide and differentiate in the mouse embryo (compare this with plant regeneration, described in Chapter 6). However, with current technology, cow, pig, and chicken ES cells do not seem to be pluripotent when placed in blastocyst embryos.

The mouse is used as a model in the study of gene function, and is useful in many areas of medicine (see the section on lung diseases in Chapter 10, Medical Biotechnology). More than 5,000 human diseases are caused by genetic defects; thus, by using gene targeting to introduce mutations in experimental animals in order to produce human disease models, researchers may identify effective treatments and gene therapies. Specific genes are inactivated in mouse models to simulate human diseases such as Alzheimer's disease, cancer, atherosclerosis, amyotrophic lateral sclerosis (Lou Gehrig's disease) and cystic fibrosis. Methods are used to inactivate gene function and look for changes in the phenotype. Gene targeting also can be used to add genes encoding desirable characteristics. ES cell transformation technology has been used to inactivate specific

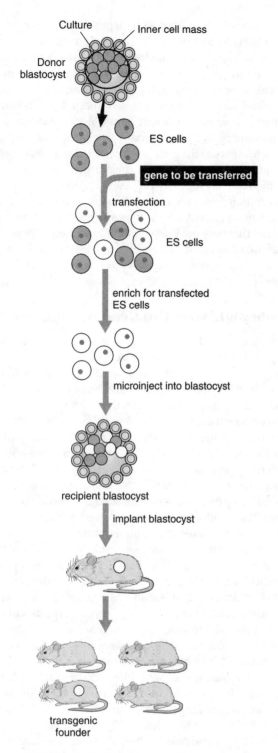

Culture
Inner cell mass
Donor
blastocyst

ES cells

gene to be transferred

transfection

ES cells

enrich for transfected
ES cells

microinject into blastocyst

recipient blastocyst

implant blastocyst

transgenic
founder

Figure 7.3 Production of transgenic mice by the transfer of embryonic stem cells containing a gene of interest. Stable transgenic lines are obtained by crossing founder animals that have the gene in their germ cells.

genes in order to determine their effect on growth and development or to develop effective treatments.

Gene Targeting in Mice Gene targeting is the insertion of DNA into a specific chromosomal location; it is often used to inactivate a specific gene in the genome. In mice, for example, gene inactivation is a common means of elucidating how a specific gene functions. In gene targeting, a cloning vector (called a targeting vector), which harbors the gene to be inserted, recombines with a region of the target chromosome that is homologous with a DNA region on the targeting vector. A selectable marker gene (for example, bacterial neomycin phosphotransferase, which confers resistance to the antibiotics neomycin and kanamycin) ensures that the desired recombination event is selected. Antibiotics included in the growth medium of cells transformed with the targeting vector and inserted gene allow only those cells that have the correct insertion into the chromosome to survive.

A gene is first isolated, altered *in vitro* according to the requirements of the experiment, and then targeted to its counterpart on a specific chromosome in cells. A portion of the gene is disrupted by inserting an antibiotic resistance gene, such as neomycin, into it, with regions of the gene flanking the marker remaining on either side. This marker enables cells that contain targeted DNA to be selected (positive selection). A second marker (a negative marker), located at the end of the targeting vector, is used to select for cells in which the transferred gene is targeted to the specific site (rather than to a random site, or for cells in which no insertion occurred).

The marker often is the herpes simplex virus type I thymidine kinase gene (HSV-*tk* gene). After the gene construct is assembled, the gene and its vector are transferred to embryo host cells, and homologous recombination allows the gene to insert into a specific chromosome region. The transferred DNA lines up next to its chromosomal counterpart in cells so that the identical regions are in alignment. When recombination occurs, there is an exchange of DNA in the identical regions of the gene, but excluding the *tk* marker, which is outside those regions (Figure 7.4).

To select for cells with the foreign gene and its mutation, cells are placed onto a medium containing antibiotic (for example, neomycin), which kills cells in which the vector has not inserted into the genome. A drug called ganciclovir selects for targeted insertion by killing any cells that contain the *tk* gene, specifically killing cells harboring the herpes virus *tk* gene. Thymidine kinase produces from ganciclovir a compound that is toxic to cells. If random insertion occurs, the cells will receive a copy of the *tk* gene and will be killed by ganciclovir. Cells that do not receive DNA will not receive the antibiotic resistance gene and will be sensitive to the antibiotic in the medium. The surviving cells contain the antibiotic resistance gene and the targeted gene region, but lack the *tk* gene. Therefore, only those cells that have undergone homologous recombination survive treatment with neomycin and ganciclovir.

Screening for homologous recombination often involves using PCR with two primers; one primer is complementary to the transferred DNA that is not present in the chromosomal DNA of host cells, and the other

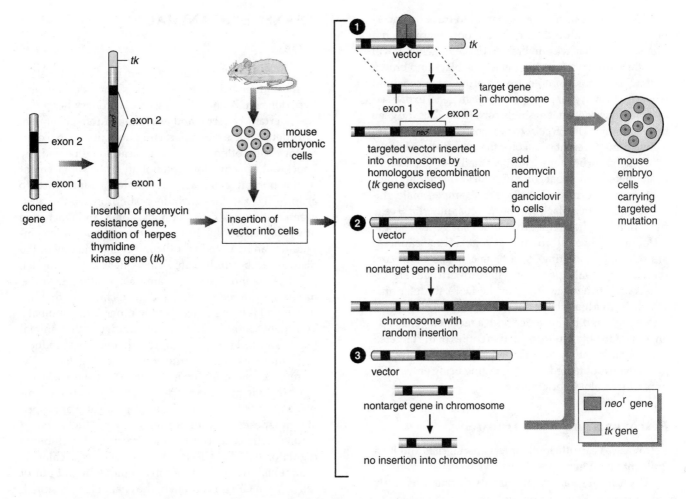

Figure 7.4 Targeted gene replacement in cultured mouse embryonic cells.

primer is complementary to a region of chromosomal DNA outside the targeted area. The predicted size of the amplified DNA fragment can be confirmed using gel electrophoresis; if a different size is obtained, a random integration occurred. This method is especially important if a selection method, such as the one described for the *tk* gene and ganciclovir, is not used.

ES cells from early mouse embryos are used in gene targeting experiments and can be altered as previously described for gene targeting experiments. ES cells are placed into mouse embryos, where they become incorporated (Figure 7.5). These genetically manipulated embryos are placed in surrogate mothers, where they develop.

Figure 7.5 Transfer of mouse ES cells into embryos. Coat color is a marker to determine incorporation of ES cells into embryos.

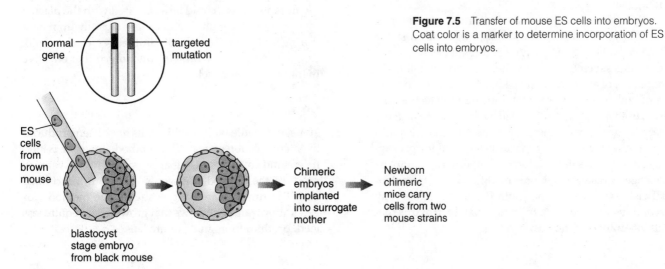

Coat color can serve as a marker to track the incorporation of introduced ES cells in mouse embryos. If a mouse has both black and brown coat color genes, it will be a **chimera** with a readily detectable black and brown coat. ES cells, for example, from a brown mouse would contain a brown coat gene (called *agouti* gene, expressed as a brown coat even if in single copy) as well as the gene to be targeted. The ES cells to be transferred will contain a gene for a different coat color than the host embryo's.

When newborn mice are screened for the presence of a targeted gene, coat color indicates whether the mice contain transformed ES cells. For example, male chimeric mice with a black and brown coat inherit ES cells from both the host embryo (black coat gene) and the introduced ES cells (brown coat gene). These chimeras are selected for matings to a female with a black coat (non-*agouti*). Only progeny with brown coats are selected for further screening of the DNA to determine which individuals have inherited the targeted gene. Those male and female mice with a targeted gene copy are mated to obtain mice with two copies of the targeted gene. Confirmation of mice **homozygous** for the targeted gene is obtained by DNA testing using either PCR or Southern blot hybridization.

Retrovirus and Gene Transfer

Retroviruses can efficiently infect animal cells and integrate into their genomes. They have been used to transfer DNA into a variety of animals during very early development. Retroviruses use RNA as their genome instead of DNA, and can infect and replicate in host cells without killing them. During infection, two copies of RNA and the enzyme reverse transcriptase are "injected" into the host cell. After infection, reverse transcriptase catalyzes the synthesis of a double-stranded DNA using the RNA as a template. The DNA then inserts into the host genome (provirus). The formation of new virus particles requires viral entry and integration of the DNA copy of the RNA genome and expression of viral genes.

Retroviruses can infect a wide variety of cell types, and a large number of those cells, each with a single DNA copy. However, the type of foreign DNA that can be used is severely restricted because only small DNA fragments of approximately 8 kb can be transferred into an organism. To ensure safety during transfection, the virus used as a vector is disabled by the deletion of genes that encode reverse transcription and viral particle packaging. A retrovirus **helper virus** is required for packaging. As an extra safety factor, the helper virus cannot be packaged because of genome modifications. One potential risk, however, is that both disabled vectors may recombine to generate a fully functional retrovirus that can integrate into the host genome.

TRANSGENIC ANIMALS

Mice

As discussed earlier, mice have been used as a model for many human diseases; examples include arthritis, hypertension, Alzheimer's disease, coronary heart disease, certain cancers, and various neurodegenerative disorders. If mice lack a particular human disease counterpart, they often can be genetically engineered, by knock-out (inactivation) technology or by replacement of the normal gene with a mutated counterpart, to acquire it; examples include cystic fibrosis, β-thalassaemia, atherosclerosis, retinoblastoma, and Duchenne muscular dystrophy. In this manner, therapeutic compounds can be tested and the molecular biology of the disease can be studied in a convenient animal model.

The potential for using animals as living factories has been explored in mice, and the technology that has developed is being transferred to other, larger animals. Transgenes introduced into the mouse or other animal (e.g., cows and goats) are expressed and their products (e.g., interleukin-2, α_1-antitrypsin, clotting factor IX) secreted in the milk for isolation. The gene construct must include mammary-specific promoters (for example, a promoter from the β-casein gene) and appropriate enhancers. One important protein that is being studied in this way is the cystic fibrosis transmembrane regulator (CFTR). CFTR normally acts as a chloride channel allowing chloride ions to move in and out of cells. When CFTR is mutated, the channel is disrupted. The resulting disruption of ion flow causes mucus to accumulate in organs, especially in the lungs and pancreas. The mucus attracts bacteria that later die and release their DNA, further inhibiting normal organ function. Because the CFTR protein is associated with membranes of transformed cells, it has not been successfully expressed *in vitro*. Secretion into milk facilitates isolation since the mammary glands produce fat globules in milk that are encapsulated by plasma membrane. In transgenic animals, a heterologous transmembrane protein could be associated with the plasma membrane of the globule and isolated readily from milk during lactation. The CFTR protein secreted in milk can be used to study protein function so that effective therapies can be developed.

Cows

Transgenic cattle with specific traits have been produced by microinjection of eggs. This method, however, is inefficient and costly, requiring approximately two years to produce the desired transgenic calf from a fertilized egg. The future may bring cost-effective methods for generating transgenic cows that grow faster, require less feed, produce more milk, or are leaner.

Genetic engineering will one day change the composition of milk products. One goal is to increase the protein κ-casein (a phosphoprotein in milk) in cows by over-expressing the κ-casein gene in animals. The end result would be the production of more cheese from milk. Another desirable genetic modification would be to remove lactose from milk. A lactose-free product would be beneficial to the millions of people who are lactose-intolerant and therefore cannot digest milk containing lactose.

Other potential modifications include increased resistance of animals to bacterial, viral, and other pathogenic diseases. For example, mastitis, a bacterial infection of the mammary glands, is a common affliction of cows. Resistance in cows would save farmers money, reduce the use of antibiotics, and prevent much suffering. *In vivo* immunization, transferring the antibody genes for specific antigens to animals, would protect livestock from many diseases and eliminate the need for vaccinations. Genetic engineering could eventually reduce the number of vaccinations, drugs, and veterinarian visits, and ultimately the cost of raising cattle.

The production of recombinant bovine somatotropin (rBST) for use in cows to increase milk production has created a recent controversy. Somatotropins occur naturally in the milk and meat of certain animals, and BST, a protein hormone produced in the cow's pituitary gland, is essential for milk production. When injected into cows, BST stimulates increased milk production and facilitates the efficient conversion of feed into milk rather than body fat. In the past, somatotropin has been obtained from cow pituitaries by extraction, an expensive and time-consuming method. Today, rBST can be produced in *E. coli* after transfer of the *bst* gene. After injection into cows, rBST increases milk production by approximately 25%.

Although extensive testing has shown that rBST is not toxic to humans and that cows injected with rBST do not have elevated levels of the protein in their bodies, rBST has remained controversial. The FDA has approved rBST as an animal drug and has formally stated that the milk and meat of treated cows are safe for human consumption. Nevertheless, opponents of rBST express many concerns that have focused primarily on economics and human health. They argue that increased milk production may be detrimental to small dairy farmers, since fewer cows would be required to supply the nation's dairy products; the entire dairy production industry could soon be dominated by large corporations. From a human health perspective, opponents give expression to a general fear of genetically engineered hormones or drugs (especially antibiotics) as well as a concern that an increase in rBST may cause cancer in human consumers. Others have indicated that cows may be more susceptible to mastitis (which may be a result of increased milk production), thereby requiring more frequent medical treatment and antibiotics that could select for antibiotic-resistant, pathogenic bacteria. The FDA, on the other hand, has concluded that rBST-treated cows will not suffer increased frequency of mastitis and that BST is not absorbed and metabolized by humans. Producers of somatotropins claim emphatically that the hormones do not affect humans and are readily digested and converted to amino acids when the milk or meat is consumed. Thus, the FDA has deemed products from cows treated with rBST safe for human consumption.

Pigs, Sheep, and Goats

As with cows, experiments with sheep and goats as bioreactors to produce important human proteins or pharmaceuticals seem quite promising. In the future, important compounds such as factor VIII and IX, interferon, interleukin, growth hormone, plasminogin activator, and a variety of antibodies may be mass produced in this way. The secretion of proteins from transgenes in the milk of sheep and goats seems to have no ill effects on most host animals or their progeny. Pigs that harbor a foreign gene, on the other hand, seem to have many problems, which include lameness, lethargy, thickened skin, kidney dysfunction, inflamed joints, peptic ulcers, pericarditis, severe osteoarthritis, and a propensity toward pneumonia. When pigs were transformed with human growth hormone (effects of bovine growth hormone are somewhat less severe), all founder animals expressing growth hormone had reduced fertility.

In experiments similar to those conducted with cows, the porcine somatotropin gene has been cloned into *E. coli* and the product used to treat pigs. Recombinant porcine somatropin increases an animal's growth rate, as well as feed efficiency and decreases fat deposition. This recombinant hormone may have potential for enhancing production of pigs in the future.

Birds

Genetic engineering will allow the production of healthier birds used for food in the near future. Chickens, ducks, geese, and small game birds could be made resistant to viral and bacterial diseases. Poultry products could be improved by decreasing the fat content of meat and the cholesterol in chicken eggs. Chicken or duck eggs also could serve as factories for producing valuable proteins. Typically, birds secrete a large amount of ovalbumin in their reproductive tracts that becomes part of the egg. If transgenic birds are engineered to also secrete other types of proteins into the egg, these products could readily be isolated.

Retrovirus vectors have been used to generate transgenic birds. After being disabled to prevent adverse side effects, the virus is used to infect cells at the blastoderm stage of development. As discussed earlier, problems associated with retrovirus vectors include the instability of the introduced gene and restrictions on the size of DNA inserts. (There is also, among some, the fear of using retrovirus DNA when the transgenic organism may be a source of food.)

In avian species, the mode of egg fertilization and the structure of the egg membranes prevent the microinjection of foreign genes. After fertilization, the egg becomes encased in a tough inner membrane as well as shell membranes that are difficult to penetrate. The male pronucleus that will fuse with the female pronucleus cannot be identified since more than one sperm enter the ovum. DNA microinjected into the cytoplasm will not be incorporated into the genome.

Alternative methods of transformation are being investigated. A promising technique involves introducing blastoderm cells, transformed with liposomes, into the subgerminal space of host embryos (Figure 7.6). A chimeric organism would be produced having some cells from the host and others derived from the transformed donor cells. If the introduced transformed cells become a part of the germ cell line, stable transgenic lines can be produced through matings of founder animals.

Insects

Research on the genetic engineering of insects is still in the early stages. However, studies have been aimed at the introduction of insecticide resistance into insects. Most insecticides are broad-spectrum chemicals and do not select specifically for harmful pests. If made resistant, beneficial insects, such as ladybugs, honeybees, preying mantis, and many species of ants, would not be killed by chemical applications in agricultural fields and small gardens. Arachnids (spiders), important insect predators, also are affected by chemical insecticides. Loss of agronomically important insects and other arthropods often leads to outbreaks of insect pests. Engineered insecticide-resistant insect predators could be used in combination with chemicals against insect pests. (Indeed, the disadvantage of developing such insects is that they potentially allow increased use of chemical pesticides.)

The first step in engineering these insects is to identify and isolate insecticide-resistant genes. Recently, a fruit fly gene for resistance to cyclodiene insecticides (such as Dieldrin) has been isolated. The most difficult steps will be transforming insects with a gene that is stably maintained and expressed at a high level.

ANIMAL DISEASES

Many virulent animal diseases, especially in livestock, are very contagious and can be economically devastating, especially in developing tropical countries. Livestock are susceptible to many types of dysentery and diseases such as African horse sickness, bovine leukosis, bovine infectious rhinotracheitis, brucellosis, and Rift Valley fever, to name just a few. Developing countries often have no way of coping with the widespread outbreak of disease. Even in the United States, diseases account for significant livestock losses.

Recombinant DNA technology may be the only way of preventing some of the more widespread and devastating animal diseases found in many developing countries. Monoclonal antibodies and recombinant vaccines produced by this technology are effective against deadly bacterial and viral diseases, many of which will be controlled or eradicated once low-cost vaccines are developed and in widespread use. Genetic engineering enables immunogenic proteins produced in bacterial and yeast cells to be used as vaccines that offer several advantages over traditionally produced vaccines (that is, pure preparations of viable pathogenic organisms that are attenuated—weakened—so they cannot cause disease but can stimulate the immune response against a virulent pathogen). The mass-production of antigens from bacteria or yeast cells can dramatically reduce production costs. A pure protein is often produced, eliminating the need for extensive and costly testing, since the disease-causing pathogen is not present. Recombinant DNA technology also eliminates the need for inactivated vaccine (attenuated live virus) by enabling the antigen to be synthesized separately from the pathogen.

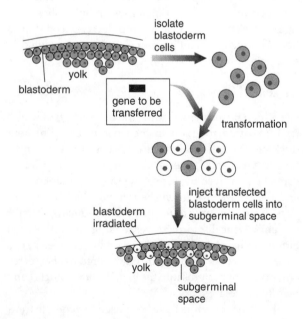

Figure 7.6 Production of transgenic birds by the transfer of transformed blastoderm cells into the subgerminal space of the embryo.

The genes of different proteins can be fused so that a variety of antigens are present in a single vaccine.

Traditional (attenuated virus) vaccines have serious drawbacks. First, many virus types and subtypes are specific to a particular region of the world. Thus, the most effective vaccines are polyvalent—specific for more than one virus. Some virus strains, however, cannot be grown in large quantities to provide antigens for vaccination. Second, after a period of time, vaccines may have to be modified to counter new virulent strains. Virus particles of traditional vaccines are unstable at less than pH 7 and must be refrigerated to remain potent. In the tropics, refrigeration may not be available where the vaccine is needed.

New diagnostic tests will increase the sensitivity with which parasites can be detected in animals. Serological tests of livestock populations have proven inadequate for detecting exposure to infectious disease-causing organisms. The presence of antibodies does not always indicate current infection, since antibodies can remain in an animal for long periods of time. Instead of detecting the antibodies produced in response to an infection, new diagnostic methods are needed for detecting parasite proteins or antigens and DNA.

Parasite antigen detection kits are available for detecting trypanosome proteins and tick-borne protozoan parasites. Mouse monoclonal antibodies raised against specific parasite proteins are used to identify parasite antigens in animals. These assays are highly sensitive and specific. Often, specific strains or species of an infectious organism produce different responses in an infected animal. The infecting strain determines which treatments are used. Not only can these assays detect current infections, they can also distinguish between different strains or species of parasite (for example, between the major tsetse-transmitted species of trypanosomes).

Parasite-specific DNA assays are desirable because of their sensitivity and because the DNA is stable. Samples can be collected in the field and delivered to a laboratory for DNA analysis. **Polymorphic** DNA sequences and their probes allow different parasites, such as tick-borne parasites and trypanosomes, to be detected and distinguished. These sequences are detected in both animals and vectors by hybridization or the polymerase chain reaction. These methods allow animal and vector populations (such as ticks) to be monitored to determine the extent of parasite infection.

Examples of current methods of treatment for a few important animal diseases are presented below. In many cases, biotechnology is only beginning to offer effective treatments. Developing countries in particular have special needs because of climate (often high temperatures), lack of modern facilities (e.g. refrigeration), and a need for low-cost treatments.

Foot-and-Mouth Disease. This highly contagious disease is one of the most devastating to livestock and very costly to the industry. Worldwide, approximately 30% of cloven-hoofed animals are affected. Although only about 5% of adult animals die from the disease, young cattle and pigs are most affected and approximately 50% of those afflicted die from myocarditis. Infected females abort their calves and rarely produce young. Usually infected animals are killed to avoid spreading the disease and their milk and meat cannot be used.

The most common way of protecting animals against foot-and-mouth disease is by vaccination with inactivated virus vaccine (when available). However, thermal instability, short-term protection (four to six months), and virus strain specificity present problems for treatment and prevention with vaccines. Experimental vaccines include a single inactivated antigen and a mixture of two strains of inactivated antigen for cross-protective activity to different viral strains.

Coccidiosis. Nine parasitic protozoan species in the genus *Eimeria* invade epithelial cells of the digestive tract and associated glands in livestock such as cattle, sheep, and poultry. Coccidiosis can be extremely costly to the poultry industry, where overcrowded, stressful conditions promote the spread of disease and make treatment very difficult. Developmental problems are found in the young, and 20% of those infected die. In industrialized countries, many feeds contain a coccidiostat to prevent the disease. Several drugs are available both for treatment and prevention of coccidiosis. A gene for a particular protein produced by the oocysts of the protozoan has been cloned so that purified antigen can be prepared for an effective vaccine. *In vivo* trials have been ongoing to determine the appropriate mixture of purified antigens for a vaccine that will boost immunity to coccidiosis.

Trypanosomiasis. More than 60 million cattle, as well as humans, pigs, domestic buffaloes, camels, horses, and small ruminants, in sub-Saharan Africa are susceptible to trypanosomiasis, a disease transmitted by the tsetse fly. Trypanosomiasis is difficult to treat since the single-celled trypanosomes continually change their surface antigens during infection and can escape detection by the host's immune system. The most common method of preventing this disease is the routine use of chemical sprays or dips. Research is focusing on ways to prevent infection by preparing vaccines from trypanosome components involved in the pathological process, such as the proteolytic enzymes used to degrade host molecules or proteins responsible for the suppresion of the cattle immune system. Researchers are trying to isolate trypanosome-resistance genes in West African cattle so that transgenic trypanotolerant cattle can be produced. Host genes that inhibit parasite cell division may one day be identified and, through their increased expression, used to prevent the spread of the parasite within the host.

Theileriosis. Theileriosis, or East Coast fever, a deadly disease of cattle prevalent in 12 countries of east, central and southern Africa, is caused by several strains of tick-transmitted protozoans. Losses to African farmers exceed $170 million U.S. dollars. Once ingested by the tick, the parasite develops and finally forms sporozoites in the tick's salivary gland that are released into cattle bitten by infested ticks. Infected lymphocytes in the cattle become leukemic and eventually lympholysis occurs, usually causing death within one month of infection. Ticks feeding on the blood of cattle infected with the protozoan transmit the disease to other cattle. At least 24 million cattle are at risk of infection, and high-yielding exotic breeds are especially susceptible. Spraying or dipping cattle in a chemical insecticide has traditionally been used to control the disease. However, insect resistance can develop, the milk and meat may become contaminated with the pesticide, and chemicals can harm the environment. A crude preparation of live, infected ticks with an antibiotic has been used to stimulate cattle immune systems, but immunity is not always obtained and the antibiotic may not inhibit development of the parasite. Recombinant DNA technology offers a means of producing new drugs for treatment, synthesizing monoclonal antibodies for typing strains in geographical areas, and producing sporozoite antigen for vaccination in large quantities. If specific proteins expressed in the sporozoite stage of the parasite can be cloned and expressed in bacteria and insect cultures, these recombinant proteins in vaccine form may offer immunity in cattle.

ANIMAL PROPAGATION

Artificial Insemination

Artificial insemination allows genetically desirable animals to be bred more efficiently. Before this method was developed, very few females produced offspring from a single, selected male. Now, artificial insemination allows a diluted sperm sample from one bull to inseminate 500 to 1,000 females. This process has been used in the beef and dairy industries for more than 40 years to increase the frequency of desirable characteristics.

Animal Clones

Embryo twinning—also called splitting or cloning—to produce identical twins has been used in cattle breeding. This technique produces two superior animals instead of one, and can increase the frequency of specific traits in an animal population. Medical research has used animal (specifically mice) embryo splitting to genetically manipulate one embryo while the other twin is left untouched as the control. This cloning method has recently been extended to human embryos. In preliminary experiments, cells of very early human embryos were separated and cultured *in*

vitro. Although these were developmentally defective embryos and were not transferred to a human host, the potential for increasing the number of eggs in human *in vitro* fertilization is very real. Embryo splitting may eventually increase the number of eggs available to women who produce few eggs during superovulation.

Nuclear transfer methods increase the number of offspring from a female animal to perhaps thousands of clones (Figure 7.7). Pluripotent nuclei (capable of forming any cell type in an individual) are taken from embryonic cells and transferred into donor oocytes and eggs (lacking a functional nucleus or pronucleus), which are subsequently transferred to a surrogate female to produce animals that are genetic clones of one another and the nuclei donor animal. Nuclear transfer is used primarily in cattle, sheep, pigs, and rabbits; however, commercialization of this procedure and extension to many other species will allow desirable traits to be readily maintained and propagated.

In a landmark experiment in 1997, Scottish scientists produced a clone of an adult sheep using a nucleus from a differentiated cell rather than from an embryonic cell. They named the clone Dolly. In this experiment, a mammary cell with an intact nucleus was taken from the udder of one sheep and fused with an enucleated egg cell from a different sheep. The fusion product was implanted into a surrogate sheep that was prepared hormonally for implantation. The result was a sheep clone of the host that provided the genetic material from its mammary cell. Although Dolly was the only result of 277 cloning attempts, the potential exists for the cloning of animals using the DNA from differentiated cells.

CONSERVATION BIOLOGY

With the world's human population nearing six billion and expected to reach more than nine billion in some 25 years, population pressures contribute to increasing pollution, loss and fragmentation of wildlife habitat, species extinction, and extensive environmental degradation. The most effective solutions to these problems include preservation of important habitat, increased control of the exotic animal trade, re-establishment of wild populations using captive animals, a shift in the distribution of wealth in developing countries, and human population control. However, these are not often practical or viable solutions, and other methods to conserve biological diversity must be investigated.

Preservation of genetic diversity presents a formidable challenge. Biotechnology is playing a more important role in conservation as biologists look toward new developments to maintain or even increase captive animal populations. In the past few years, sig-

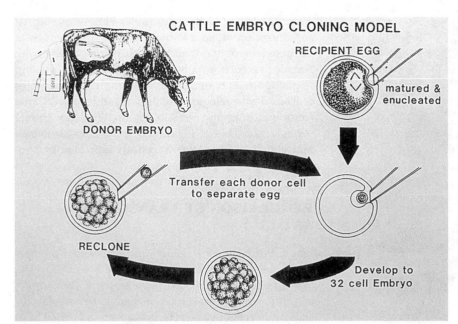

CATTLE EMBRYO CLONING MODEL

RECIPIENT EGG

matured & enucleated

DONOR EMBRYO

Transfer each donor cell to separate egg

RECLONE

Develop to 32 cell Embryo

Figure 7.7 Nuclear transfer enables many offspring with desirable traits to be generated from a single animal. Individual cells called blastomeres are obtained from embryos collected from a female and transferred into nucleated unfertilized eggs. Cells divide, and at the 32-cell stage the process can be repeated to obtain many additional embryos. Embryos also are transferred to host females for the completion of development and the production of calves.

nificant advances have been achieved in *in vitro* fertilization, in techniques for embryo transfer, oocyte maturation, control of the ovarian cycle, and in the production of transgenic laboratory and livestock animals. Controlled and assisted breeding, used in domestic livestock and in human reproduction, is being extended to conservation biology to help preserve endangered animals. Many zoos, for example, have instituted captive breeding programs for endangered species to ensure their *ex situ* survival.

Currently, frozen semen is in widespread use only for a small number of species, but in the future, cryopreservation of gametes and embryos will allow the creation of comprehensive frozen germplasm banks. Material could be collected and preserved to provide optimal genetic diversity in animal populations or to restore lost or severely reduced diversity in such populations. A coordinated international effort would be required to salvage some of the remaining genetic diversity of highly endangered species.

To sustain genetic diversity, careful genetic analysis of populations must be conducted. DNA "fingerprinting" (described in Chapter 11, Forensics and DNA Profiling) enables individuals to be readily distinguished, and is being used in the management of captive breeding programs. Fingerprinting is becoming an accepted way to resolve genetic differences between closely related individuals in breeding populations, and can identify relationships of highly inbred animals.

Embryo Transfer

Embryo transfer has been used in the cattle industry for many years. Females that have been superovulated are artificially inseminated with select bull sperm. Fertilized

eggs or embryos are retrieved from the female and frozen in liquid nitrogen at –196°C. These embryos can be implanted into surrogate mothers that are made pseudopregnant with hormones.

Embryo transfer has been practiced in some zoos in recent years. The embryos can easily be transported to various regions of the world for implantation. Genetic stocks can thus be transferred to regions lacking genetic diversity or requiring new genetic lines of livestock. Because it eliminates the use of natural reproductive processes, embryo transfer can increase the individual's annual production of progeny, and it allows surrogate mothers of a different, although closely related, species to be used. The ability to implant embryos of rare and endangered species in common surrogate hosts is important for the captive reproduction of these species.

The Center for Reproduction of Endangered Wildlife (CREW) at the Cincinnati Zoo in Ohio has been using embryo transfer technology since 1984, when Holstein cows first were used as surrogate mothers for the rare Malaysian ox, or Gaur (Figure 7.8). Holstein cows and elands also have served as hosts for embryos of the endangered bongo antelope (Figure 7.9).

Embryo transfer has been extended to natural populations: In Kenya, CREW used surrogate eland antelope to reintroduce bongo antelope in the hopes of increasing dwindling wild populations. The success rates of implantation in native animals versus those in captivity could be compared, as could the success rates of native- versus captive-collected embryos. The CREW Kenya Project is the first binational conservation effort involving inter- and intra-species embryo transfers. In the initial phases of the project, cryopreserved bongo embryos (in liquid nitrogen at –196°C) were transferred

Figure 7.8 Male Gaur calf born to a surrogate Holstein.

Figure 7.9 Bongo calf born to an eland surrogate.

from the Cincinnati Zoo to Kenya for implantation into eland surrogates. Cryopreserved embryos collected from wild bongo at the Mt. Kenya Game Ranch in Nanyuki were implanted into both eland and bongo surrogates at the Cincinnati Zoo. Successful transfers using embryo transfer technology under closely monitored conditions provided a knowledge base from which to increase bongos and their genetic diversity in the wild. Females taken from a herd of Mt. Kenya bongos and implanted with bongo embryos from the United States will one day be released into the wild and radio-collared for monitoring by Kenyan researchers and animal behaviorists. This is a landmark experiment in which biotechnology complements conservation biology.

REGULATION OF TRANSGENIC ANIMALS

The question whether genetically engineered animals should be regulated has stirred a hotbed of controversy. Several issues have elicited often heated debate among consumer groups, concerned citizens, governmental officials, and even scientists. The major fears expressed have been the potential detrimental effect on the environment, the health risk of consuming genetically engineered foods, and the health risk to the engineered animal.

At present, there is little federal control of environmental impacts of transgenic animals released into the environment. Since few transgenic animals existed in 1988, the federal framework published in that year said little about environmental risks of genetically engineered animals, focusing instead primarily on the risks of using engineered animals as food. Guidelines for USDA-funded research using genetically engineered organisms are voluntary; investigators can choose not to comply.

PATENTING GENETICALLY ENGINEERED ANIMALS

Winning approval for the patenting of genetically engineered animals is the most recent victory for biotechnology companies in what has been a long struggle to obtain patents on living organisms. In 1930, the U.S. Congress passed the Plant Patent Act, which allowed patents to be granted for innovations in horticulture involving asexually propagated plant varieties; 40 years later sexually propagated plants became patentable. In 1980, after numerous appeals, the U.S. Supreme Court delivered a landmark decision stating that an "invention" was not unpatentable simply because it was a living organism (Diamond *v.* Chakrabarty). Finally, on April 7, 1987, Donald J. Quigg, the Commissioner of Patents and Trademarks, announced that the Patent and Trademark Office considers non-naturally occurring nonhuman multicellular organisms (which of course includes animals) to be patentable.

On April 13, 1988, Harvard University obtained patent number 4,736,866 for a genetically altered mouse used as a model for studying how genes contributed to breast cancer. The "Harvard mouse" set off an emotional debate. Although patenting living organisms is not new—plant varieties, lower animals, and microorganisms have been patented—the idea of patenting mammals has far-reaching implications, and has caught the attention of animal-rights groups, farmers, and consumer groups.

The issue of patenting genetically engineered animals has also been addressed by lawmakers seeking ways to protect **intellectual property**, which includes patents, copyrights, trade secrets, and trademarks. The patent debate has been given added heat by dramatic developments in biotechnology in the past 15 years: recombinant DNA technology, cell fusion methods, monoclonal antibody technology, commercial bioprocessing, and the like. Although biotechnology, broadly defined, is very old (as we saw in Chapter 1), such recent innovations have ignited the controversy over patenting living organisms. Much of the debate has focused on patenting animals.

The debate has extended beyond the appropriateness of animal patents to encompass questions about the consequences of commercial uses of patented organisms and about the merits of the technology itself. Reluctance to embrace animal patents arises from a myriad of concerns. These include the environmental implications, the welfare of the engineered animal, and the ethics of creating a novel living organism and changing the course of evolution. Just a few of the many objections raised against animal patents are excessive interference in the "natural world," the increased suffering of animals in both the research and agricultural sectors, a decline in genetic diversity in commercialized animal species, and the demise of family farms. Some scientists fear that animal patents will encourage academic researchers to move away from basic research and conduct only applied research.

Farmers have been especially concerned about the long-term implications of animal patents. New technologies and patented livestock will most likely accelerate the trend toward larger and fewer farms: Small farmers are not equipped to adopt new, patented biotechnologies; nor could most afford them. The high cost of using patented animals would erode the profit margins of many farmers. Unable to compete effectively with larger operations that can adapt to biotechnological developments, small farms will decline still further in number, and fewer, larger farms will remain.

In the future, we will see many patented products, both living and nonliving, resulting from modern biotechnology. The open discussion of ethical, social, and legal implications will be required to ensure that the end-products of modern technology will contribute in a positive way toward a better world.

General Readings

L.A. Babiuk, J.P. Phillips, and M. Moo-Young, eds. 1989. *Animal Biotechnology*, Pergamon Press, Oxford, England.

R.B. Church, ed. 1990. *Transgenic Models in Medicine and Agriculture*, Wiley-Liss, New York, NY.

N. First and F.P. Haseltine, eds. 1988. *Transgenic Animals*, Butterworth-Heinemann, Boston, Massachusetts.

J.A.H. Murray, ed. 1992. *Transgenesis*, John Wiley & Sons, Inc., New York, NY.

M.A. Vega, ed. 1995. *Gene Targeting*, CRC Press, Inc., Boca Raton, Florida.

Additional Readings

M.R. Capecchi. 1994. Targeted Gene Replacement. *Sci. Am.* 270: 52-59.

J.G. Cloud. 1990. Strategies for introducing foreign DNA into the germ line of fish. *J. Reprod. Fertil. Suppl.* 41:107-118.

L.V. Cundiff, M.D. Bishop, and R.K. Johnson. 1993. Challenges and opportunities for integrating genetically modified animals into traditional animal breeding plans. *J. Anim. Sci.* 71 (Suppl. 3):20-25.

P.L. Davies, C.L. Hew, M.A. Shears, and G.L. Fletcher. 1990. Antifreeze protein expression in transgenic salmon. In R.B. Church, ed., *Transgenic Models in Medicine and Agriculture*, Wiley-Liss, New York, NY, pp 141-161.

R.H. Devlin et al. 1994. Extraordinary salmon growth. *Nature* 371:209-210.

P. DiTullio, S.H. Cheng, J. Marshall, R.J. Gregory, K. Ebert, H.M. Meade, and A.E. Smith. 1992. Production of cystic fibrosis transmembrane conductance regulator in the milk of transgenic mice. *Bio/Technology* 10:74-77.

K.M. Ebert. 1989. Gene transfer through embryo microinjection. In L.A. Babiuk, J. P. Phillips, and M. Moo-Young, eds., *Animal Biotechnology*, Pergamon Press, Oxford, England, pp 233-250.

J. Guenet. 1995. Animal models of human genetic disease. In M.A. Vega, ed., *Gene Targeting*, CRC Press, Inc., Boca Raton, Florida. pp 149-44.

L.M. Isola and J.W. Gordon. 1988. Transgenic animals: A new era in developmental biology and medicine. In N. First and F.P. Haseltine, eds., *Transgenic Animals*, Butterworth-Heinemann, Boston, Massachusetts, pp 3-20.

H.D.M. Moore. 1992. *In vitro* fertilization and the development of gene banks for wild animals. In H.D.M. Moore, W.V. Holt, and G.M. Mace, eds. *Biotechnology and the Conservation of Genetic Diversity*, Oxford University Press, Oxford, pp 89-99.

M. Müller and G. Brem. 1994. Transgenic strategies to increased disease resistance in livestock. *Reprod. Fertil. Dev.* 6:605-613.

E. Notarianni and M.J. Evans. 1992. Transgenesis and genetic engineering in domestic animals. In J.A.H. Murray, ed., *Transgenesis*, John Wiley & Sons, Inc., New York, NY, pp 251-281.

K.W. O'Connor. 1993. Patents for genetically modified animals. *J. Anim. Sci.* 71(Suppl. 3):24-40.

R.D. Palmiter and R.L. Brinster. 1986. Germ-line transformation of mice. *Annu. Rev. Genet.* 20:465-499.

D.J. Porteous and J.R. Dorin. 1993. How relevant are mouse models for human diseases to somatic gene therapy? *Trends Biotechnol.* 11:173-181.

D. Powell. 1995. Saftey in the contained use and release of transgenic animals and recombinant proteins. In G.T. Tzotzos, ed., *Genetically Modified Organisms: A Guide to Biosafety*, The University of Arizona Press, Tucson, Arizona.

D.A. Powers, T.T. Chen, and R.A. Dunham. 1992. Transgenic fish. In J.A.H. Murray, ed., *Transgenesis*, John Wiley & Sons Ltd., Chichester. pp 233-249.

R.S. Prather, F.L. Barnes, M.M. Sims, J.M. Robl, W.H. Eyestone, and N.L. First. 1987. Nuclear transplantation in the bovine embryo: Assessment of donor nuclei and recipient occyte. *Biol. Reprod.* 37:859-866.

V.G. Pursel and C.E. Rexroad, Jr. 1993. Status of research with transgenic farm animals. *J. Anim. Sci.* 71 (Suppl. 3):10-19.

R.M. Shuman. 1991. Production of transgenic birds. *Experientia* 47:897-905.

C. Stewart. 1997. Nuclear transplantation. An udder way of making lambs. *Nature* 385:769-771.

A. Teale. 1993. Improving control of livestock diseases. *Bioscience* 43:475-483.

P.B. Thompson. 1993. Genetically modified animals: ethical issues. *J. Anim. Sci.* 71 (Suppl. 3):51-56.

A.S. Waldman. 1995. Molecular mechanisms of homologous recombination. In M.A. Vega, ed., *Gene Targeting*, CRC Press, Inc., Boca Raton, Florida. pp 45-64.

I. Wilmut, A.L. Archibald, M. McClenghan, J.P. Simons, C.B.A. Whitelaw, and A.J. Clark. 1991. Production of pharmaceutical proteins in milk. *Experientia* 47:905-912.

I. Wilmut, A.E. Schnieke, J. McWhir, A.J. Kind, and K.H. Campbell. 1997. Viable offspring derived from fetal and adult mammalian cells. *Nature* 385:810-813.

I. Wilmut and C.B. Whitelaw. 1994. Strategies for production of pharmaceutical proteins in milk. *Reprod. Fertil. and Dev.* 6:625-630.

MARINE BIOTECHNOLOGY 8

The Office of Technology Assessment defined marine biotechnology in 1991 as "any technique that uses living marine organisms (or parts of these organisms) to make or modify products, to improve plants or animals, or to develop microorganisms for specific uses." The chief traditional biotechnologies have been **aquaculture** and the isolation of natural products from marine organisms for commercial applications. Few such products have been developed, perhaps in part because marine biological diversity is vast, and most of it remains unexplored. Modern marine biotechnology, although still in its infancy, encompasses such new technologies as protein engineering, recombinant DNA technology, and hybridoma and monoclonal production using marine organisms or their components. In recent years increasing funds have been devoted to marine biotechnology research; the United States, for example, spent some $55 million in 1995. In this chapter we examine the main areas of recent research, which range from aquaculture to transgenics.

The marine environment covers almost 71% of the earth's surface and contains 97% of the available water. The oceans play an important role in global ecology; they regulate global climate, moderate fluctuation of temperatures, produce a third of the global supply of oxygen, regulate levels of carbon dioxide, and recycle 90% of the world's carbon in the marine food web. A vast number of people depend on the biological diversity in the oceans. Approximately 100 million tons of seafood is harvested each year to feed a growing world population. The global marine harvest accounts for 16% of animal-protein consumption, and is especially important in developing countries. In Asia alone, fish is a main protein staple for one billion people. The world population is expected to almost double by the year 2025. Eventually, there may not be enough arable land or energy available to grow enough grain and raise enough livestock and poultry to sustain so many people. Fish may play a larger role in feeding this population (although we are placing this role at risk by depleting the oceans).

The oceans are an ancient ecosystem where life, in the form of bacteria, originated about four billion years ago. Marine organisms provide us with a valuable gene pool that is beginning to be tapped. To ensure a healthy and stable marine food web, pollution must be controlled, threatened and endangered species must be protected (and perhaps organisms cultured that are in high commercial demand), and the link between the ocean's primary producers (that is, plankton) and other organisms must be understood.

AQUACULTURE

In many regions of the world overfishing has dangerously reduced fish and **shellfish** stocks, and some areas now are closed to commercial fishing. People are therefore beginning to look to aquaculture: the propagation of aquatic animals and plants at high densities in fresh, brackish, or salt water. Salt water, or marine, culture (called **mariculture**) and freshwater aquaculture use similar methods.

Aquaculture has been practiced throughout the world for thousands of years. Ancient aquaculture, which was primarily freshwater, began in the Far East; Chinese aquaculture dates back probably at least 3,000 years. The earliest-known record, in 473 B.C., reports monoculture of the common carp. As methods were refined, polyculture was practiced, combining species with different food preferences; ponds contained varieties of finfish, shellfish, or **crustaceans**.

Today, high-density aquaculture is practiced in many countries. In the Philippines, for example, the giant tiger prawn *Penaeus monodon* is cultured in ponds at densities of 100,000 to 300,000 prawns per hectare. Significant advances have been made in the culturing of oysters, clams, abalone, scallops, brine shrimp, blue crab, and a variety of finfish.

Traditional aquaculture encompasses low-input technologies to propagate, harvest, and market fish, algae, crustaceans, and **mollusks**. An aquaculturist must know the chemical composition of the soil where the pond will be situated, the amount and type of water to be used, and the type and quantity of animal feed necessary for maximum production. Unfortunately, aquaculture ponds have the potential of destroying delicate habitats through pollution by wastes or by decimating, for example, mangroves to make room for shrimp ponds. Biotechnology may be able to help solve some of the problems created by intensive aquaculture practices.

Marine and freshwater biotechnologies are used today to increase the yield and quality of finfish, crustaceans, algae, and various **bivalves** such as clams and oysters. Recently, transgenic organisms with desired characteristics have been generated for future commercial production. Modern Japanese mariculture is more productive than freshwater aquaculture, and has recently accounted for over 92% of Japan's total aquaculture yield. The major food products have been the Japanese oyster, *Crassostrea gigas*, and a red alga called nori (*Porphyra*). Nori provides the largest seaweed harvest. Other products include the fish yellowtail jack, the red sea bream, as well as a scallop and two types of brown algae (wakame and kombu).

Although the primary products of aquaculture have been food for human and animal consumption, it has also yielded food supplements, natural products, pharmaceuticals and medicines, ornamental jewelry, and ornamental fish for home aquariums. Production of cultured organisms has increased dramatically in recent years (from almost 10 million tons in 1985 to 14 million tons in 1993, with projections of 22 million tons by the year 2000). The United Nations Food and Agricultural Organization (FAO) predicts that by the end of the century, products from aquaculture will account for 20 to 25% of the world's fisheries production by weight. The world fisheries have virtually reached the maximum sustainable yield of 100 million tons per year, and in some regions the fisheries have collapsed—which means that aquaculture will become increasingly important. According to a 1991 FAO report, the wild marine and freshwater fish catch accounts for 82% and 6% of the world total, respectively, while freshwater and marine aquaculture make up 7% and 5%, respectively, of the total fish catch.

Improved hatchery methods would increase the production of threatened or endangered fish and shellfish (examples include striped bass, oysters, and scallops), thereby enhancing marine fisheries for both recreation and commercial use. Improving marine fisheries could benefit other species that are used extensively for human consumption or sport, such as snook, tarpon, haddock, halibut, cod, red snapper, and flounder (Table 8.1).

The by-products of fish culture and agriculture are often used complementarily: Pond humus can be used to fertilize fields, and livestock waste can serve as pond fertilizer to stimulate the growth of plankton, which is food for the fish. Crop by-products also can be used as fish feed. Anaerobic digesters sometimes are used to produce methane gas as an energy source from aquacultural and agricultural wastes.

Gastropod, Bivalve, and Crustacean Production

The high worldwide demand for clams, oysters, mussels, abalone, crabs, shrimp, and lobsters favors the development of efficient culturing methods to increase their production. A variety of methods exist, from rearing in tanks to using floating platforms and substrates for the growth of organisms (Figure 8.1). Genetic manipulation in culture promotes faster growth and maturation, increased disease resistance, and triploidy. Normal **diploid** oysters spawn in the summer and lose their flavor because they form a massive amount of reproductive tissue. Instead of the two sets of chromosomes that diploids have, triploid Pacific oysters have three (two from the female and one from the male). **Triploid** oysters are obtained in culture by treating eggs with cytochalasin B. This inhibitor of normal cell division doubles the number of chromosomes so that, when the eggs are fertilized with normal sperm, zygotes have three sets of chromosomes. Since triploid oysters are sterile and do not form reproductive organs, they are more flavorful and meatier year round. They also grow

Table 8.1 Examples of Aquacrops (all have potential for culture in the U.S.)

Common Name (scientific name)	Culture and Importance
Alligator (*Alligator mississippiensis*)	Grown in southern U.S., primarily Louisiana and Texas; flesh is used for food and the skin is used for leather.
Bullfrog (*Rana catesbiana*)	Cultured for food and laboratory use; mostly imported to U.S.
Carp (common) (*Cyprinus carpio*)	Many nations produce and consume; not particularly popular in the U.S.
Carp (grass or white amur) (*Ctenopharyngodon idellus*)	Consumes huge amounts of vegetation; produced in China.
Catfish (channel) (*Ictalurus punctatus*)	Widely grown in southern U.S.
Chinese waterchestnut (*Eleocharis dulcis*)	Plant grown for corms; production is experimental in the U.S.; popular and successful in China; imported food to U.S.
Crawfish (red swamp) (*Procambarus clarkii*)	Primarily grown in Louisiana and Texas for food.
Eel (Japanese) (*Anguilla japonica*)	Popular food in Italy, Japan, and other countries; cultured in Taiwan, Japan, and other countries.
Goldfish (*Carassius auratus*)	Popular ornamental fish in U.S.; grown in Arkansas and Missouri.
Freshwater prawn (*Macrobrachium rosenbergii*)	Experimental in U.S.; successful production in Asian and other countries.
Minnow (fathead) (*Pimephales promelas*)	Popular baitfish in U.S.; grown in Arkansas and other areas.
Oyster (American) (*Crassostrea virginica*)	Limited culture success in Chesapeake Bay area and Eastern Shore of U.S.; successful culture in Japan, Taiwan, and other areas.
Salmon (chinook) (*Oncorhynchus tshawytscha*)	Popular food fish; harvested in several locations.
Shrimp (Chinese white) (*Penaeus chinensis*)	China is world leader in Shrimp production; other species produced elsewhere.
Striped bass (*Morone saxatilis*)	Thought to hold much potential for culture, particularly its hybrids, still experimental.
Tilapia (blue) (*Tilapia aurea*)	Grown in warm water or intensive systems; sometimes considered a pest.
Trout (rainbow) (*Salmo gairdneri*)	Popular; produced in northern U.S.
Watercress (*Nasturtium officinale*)	Grown in temperate and subtropical climates, such as in Hawaii; the only aquatic foliage consumed on a regular basis in U.S. in raw salads or cooked.

Compiled from J. E. Bardach, J. H. Ryther, and W. O. McLarney, *Aquaculture: The Farming and Husbandry of Freshwater and Marine Organisms*. New York: John Wiley & Sons, Inc., 1972; H. K. Dupree and J. V. Huner, *Third Report to the Fish Farmers*, Washington D.C.: U.S. Fish and Wildlife Service, 1984; B. Rosenberry, *World Shrimp Farming*, San Diego, *Aquaculture Digest*, 1990; and S. W. Waite, B. C. Kinnett, and A. J. Roberts, *The Illinois Aquaculture Industry: Its Status and Potential*, Springfield, State of Illinois, Department of Agriculture, 1988.

J. S. Lee and M. E. Newman, *Aquaculture: An Introduction*, Table 1-1. Interstate Publishers, Inc., 1992. Used by permission. See this book for original references.

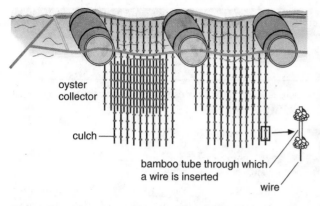

Figure 8.1 (above) A common method of culturing oysters: A raft floated by buoys holds strings of culches with oyster larvae attached. (right) Aquaculture facility where shrimp are cultured in fiberglass tanks. An enclosure similar to a greenhouse helps regulate the temperature of the aquatic environment.

larger and faster than diploid oysters and can be harvested sooner. Triploid oysters represent a significant portion of the total oyster production in the United States (making up almost 50% of the Pacific Northwest oyster hatchery). Concerns about the safety of using cytochalasin B may soon lead to the production of triploid oysters by mating tetraploids (having four sets of chromosomes) and normal diploid oysters.

Bivalves such as oysters and **gastropods** such as abalone (which has a commercial value of $20 to $30 per pound) are cultured by the manipulation of the reproductive cycle. Hydrogen peroxide added to the seawater induces synthesis of prostaglandin, a type of hormone that triggers spawning. Larvae are then induced to settle by the addition of the amino acid aminobutyric acid (GABA), an important neurotransmitter in animals. Larvae exposed to GABA settle onto the substrate and begin developmental metamorphosis and cellular differentiation. Growth-accelerating genes are being identified and cloned for the eventual production of compounds from these genes, which are important in the efficient, controlled culturing of abalone and other shellfish. Productivity is also being increased by crossing specific genetic lines to increase, for example, the growth rate of oysters (by up to 40%). The use of recombinant fish growth hormone may also produce faster-growing shellfish.

MARINE ANIMAL HEALTH

Like terrestrial animals, marine animals are susceptible to **protozoan**, bacterial, viral, and fungal diseases. Marine pollution, which is increasing dramatically worldwide, often promotes growth of **pathogens** (such as the fungus *Saprolegnia* and the common marine bacterial pathogen *Vibrio*) that infect marine organisms.

Ninety percent of the marine catch comes from coastal waters, which are the areas most susceptible to pollution.

High densities of animals in aquaculture also increase the chance of disease. In fact, disease contributes to a significant loss of production each year, with devastating economic consequences. Shrimp aquaculture provides an important source of revenue to many Asian and Latin American countries, primarily through export. For example, because of a virus that decimated black tiger shrimp, Taiwan's production of these shrimp dropped from 114,000 metric tons in 1987 to 50,000 in 1988 and 30,000 in 1991 (of a world total, in 1991, of 690,100 metric tons). At this time there are no vaccines to protect shrimp and other shellfish. If disease-resistant organisms could be cultivated, they would lower production costs and dramatically increase the amount of biomass available for human consumption.

Methods of bacterial control include the use of antibacterial agents such as disinfectants and antibiotics (tetracycline, chloramphenicol, penicillin, etc.) in ponds or culturing pens. However, the heavy use of antibiotics —Norway, for example, used 8,000 kilograms in salmon aquaculture facilities in 1987—poses an ultimate risk to human health: Antibiotic resistance could be transferred from bacteria in culturing facilities to human bacterial pathogens that cause serious diseases such as typhoid, dysentery, and cholera, normally treated with antibiotics. Antibiotic residues also could remain in the fish, crustacean, or bivalve consumed by humans.

Salmon in aquaculture facilities sometimes become infected with viruses. Two of the most deadly viral diseases are infectious hematopoietic necrosis and infectious pancreatic necrosis. Infectious hematopoietic necrosis was first found in 1953, when a tremendous number of salmon died in Washington state; it has since spread worldwide. Protozoans such as flagellates and ciliates may not kill the fish, but can cause extensive damage by feeding on parts of the fish so that it can no longer be used for food. Viral diseases pose a particularly difficult problem for fish aquaculture facilities. Infected animals often become carriers, and must be destroyed. Sometimes all the residents of a pond must be destroyed and the pond decontaminated. The lost revenue is recovered through higher prices to the consumer. Efficacious fish vaccines would greatly benefit the aquaculture industry, but very few are available.

Since few marine disease-causing organisms have been identified, additional research is needed before vaccines can be generated. Moreover, fish are not easily vaccinated. The usual method is injection or immersion in water containing the vaccine. However, the aquatic environment dilutes concentrations. A new method involves immersing fish and using ultrasound for approximately 10 minutes, which somehow facilitates entry of the vaccine into the fish.

Researchers are seeking a vaccine that is effective against infectious hematopoietic necrosis. Such a vaccine might also protect wild populations of salmon that are susceptible to this disease. Several types of vaccines have been tested that confer resistance. A recombinant vaccine, produced by isolating and expressing genes encoding viral proteins, induces the formation of antibodies in fish. The development of new vaccines and efficient delivery systems will enhance productivity from the egg through larval stages (a risky part of the life cycle). Marine metabolites and other constituents also may prove to be invaluable in the fight against disease. For example, a shellfish extract has been shown to increase immunity against infection in blue crabs and prawns as well as to protect eels from *Aeromonas* infection.

Biotechnology can also be used against the spread of disease in aquaculture through the development of diagnostics. If diseases can be detected early, before they spread, timely preventive measures can be taken and infected organisms can be removed.

ALGAL PRODUCTS

Algae are a diverse group of photosynthetic eukaryotes that are used throughout the world for various products, including food. Of the two types of algae, microalgae and macroalgae, the latter are the most economically important. The three major groups of eukaryotic macroalgae are green (*Chlorophyta*), red (*Rhodophyta*), and brown (*Phaeophyta*).

Algae are harvested in the wild as well as produced in culture. The world algae harvest is about four million tons a year, which is worth approximately one billion dollars. Most of the production is from Japan, China, and Korea, although the United States (especially California, where brown algae, or kelp, is harvested), Canada, France, the United Kingdom, Indonesia, Chile, and the Philippines also contribute.

Wild seaweeds were collected for food and medicines as early as 900 B.C., and are still being collected in some parts of the world. The California kelp forests (*Macrocystis*, sometimes called giant kelp because they can grow to lengths of 5 m) have been harvested since the early 1900s (Figure 8.2). Kelp was used as fertilizer and

a   b

Figure 8.2 (a) Sunlight filters through a giant kelp forest. One of the fastest-growing plants, giant kelp can grow as much as two feet a day (Coronado Islands, Baja California). (b) Large ships like this kelp harvester prune the kelp canopy, leaving enough submerged fronds to grow new canopy (Cedros Island, Baja California).

as a source of potash and acetone for the production of explosives. The Kelco Company of San Diego, founded in 1929, was the world's first producer of algin products; the first product was kelp meal for livestock feed. Later, algin was extracted. Today approximately 70 algin products are manufactured for many applications (Table 8.2).

Kelp has been used for many years as a food supplement, and is an important source of potassium, iodine, and other essential minerals; carbohydrates; and vitamins. Alginates are the main structural components of the cell wall and intercellular matrix of brown seaweeds. They are used as food thickeners and stabilizers (they retain moisture and assure smooth texture and uniform thawing of frozen foods). Algin is added to desserts, dairy products (just check your favorite ice cream), canned foods, salad dressings, cake mixes, and even beer for foam stabilization. Their industrial applications include paper coatings and textile printing, and they are used in pharmaceuticals (e.g., antacids, pill coatings, and capsules) and cosmetics.

The red alga *Porphyra,* or nori, has been cultured as a food source in Japan since 1570 (Figure 8.3). The culturing process relied on the natural dispersal of spores (actually propagules called conchospores) from wild

Table 8.2 Some Applications of Alginates in Food and Industrial Products

Property	Product	Performance
Food Applications		
Water-holding	Frozen foods	Maintains texture during freeze-thaw cycle.
	Pastry fillings	Produces smooth, soft texture and body.
	Syrups	Suspends solids, controls pouring consistency.
	Bakery icings	Counteracts stickiness and cracking.
	Dry mixes	Quickly absorbs water or milk in reconstitution.
	Meringues	Stabilizes meringue bodies.
	Frozen desserts	Provides heat-shock protection, improved flavor release, and superior meltdown.
	Relish	Stabilizes brine, allowing uniform filling.
Gelling	Instant puddings	Produces firm pudding with excellent body and texture; better flavor release.
	Cooked puddings	Stabilizes pudding system, firms body, and reduces weeping.
	Chiffons	Provides tender gel body that stabilizes instant (cold make-up) chiffons.
	Pie and pastry fillings	Cold-water gel base for instant bakery jellies and instant lemon pie fillings. Develops soft gel body with broad temperature tolerance; improved flavor release.
	Dessert gels	Produces clear, firm, quick-setting gels with hot or cold water.
	Fabricated foods	Provides a unique binding system that gels rapidly under a wide range of conditions.
Emulsifying	Salad dressings	Emulsifies and stabilizes various types.
	Meat and flavor sauces	Emulsifies oil and suspends solids.
Stabilizing	Beer	Maintains beer foam under adverse conditions.
	Fruit juice	Stabilizes pulp in concentrates and finished drinks.
	Fountain syrups and toppings	Suspends solids, produces uniform body.
	Whipped toppings	Aids in developing overrun, stabilizes fat dispersion, and prevents freeze-thaw breakdown.
	Sauces and gravies	Thickens and stabilizes for a broad range of applications.
	Milkshakes	Controls overrun and provides smooth, creamy body.

Table 8.2 *continued*

Property	Product	Performance
Industrial Applications		
Water-holding	Paper coating	Controls rheology of coatings; prevents dilatancy at high shear.
	Paper sizings	Improves surface
	Adhesives	Controls penetration to improve adhesion and application.
	Textile printing	Produces very fine line prints with good definition and excellent washout.
	Textile dyeing	Prevents migration of dyestuffs in pad dyeing operations. (Algin is also compatible with most fiber-reactive dyes.)
Gelling	Air freshener gel	Firm, stable gels are produced from cold-water systems.
	Explosives	Rubbery, elastic gels are formed by reaction with borates.
	Toys	Safe, nontoxic materials are made for impressions or putty-like compounds.
	Hydro-mulching	Holds mulch to inclined surfaces, promotes seed germination.
	Boiler compounds	Produces soft, voluminous flocs easily separated from boiler water.
Emulsifying	Polishes	Emulsifies oils and suspends solids.
	Antifoams	Emulsifies and stabilizes various types.
	Lattices	Stabilizes latex emulsions, provides viscosity.
Stabilizing	Ceramics	Imparts plasticity and suspends solids.
	Welding rods	Improves extrusion characteristics and green strength.
	Cleaners	Suspends and stabilizes insoluble solids.

Courtesy of The NutraSweet Kelco Company, a unit of Monsanto Company.

populations in the ocean. In the beginning, the substrate for algal spores was plant debris placed in the ocean; much later, horizontal nets were used (Figure 8.4). The spores germinate and grow into fronds that are harvested, dried, and processed. A significant amount of research has been devoted to this economically important alga, and much is known about spore formation, storage of inoculum, strain improvement, disease resistance, and nutrient composition. With the elucidation of the *Porphyra* life cycle, large sheets of nori are produced from controlled seeding of culture nets.

Other algae have been similarly cultured. The brown algae *Undaria* (wakame, shown in Figure 8.5) and *Laminaria* (kombu), are grown off the coasts of Japan and China. Ropes are seeded with spores, the germinated organisms are propagated in environmentally controlled culturing tanks, and the mature plants are grown in coastal waters on large floating rafts and then harvested and dried. Wakame and kombu have a variety of commercial applications; they are used in noodles, soups, and salads, or with meats.) Annual production

Figure 8.3 Nori has been cultivated as a food source for hundreds of years.

Figure 8.4 Nori is cultured on horizontal nets. The blades are dried and processed for human consumption.

of wakame exceeds 20,000 tons, and the annual market value of both algae is $600 million.

Alginic acids (alginates) from brown algae and phycocolloid polysaccharides (agars, carrageenans) from red algae were commercially produced early in the twentieth century. In seventeenth-century France, soda ash was obtained from wild kelp (brown algae). By the mid-nineteenth century, iodine was extracted. Agar production originated in China and Japan, probably in the seventeenth century, while carrageenans were produced in the 1830s, first in Ireland and then in the United States.

Today, alginic acids and phycocolloid polysaccharides are used in food, industrial products, fertilizer, and energy production. Alginates are in large demand by several industries: Of the more than 35,000 tons pro-

duced, the textile industry uses 50%; foods use 30%; pharmaceuticals use 5%; and paper uses 6%. Carrageenan is used extensively as an extender in foods such as evaporated milk and ice cream, in toothpaste, and in a variety of cosmetics. Agars are used primarily in foods but also in pharmaceuticals (e.g., as a component in capsules holding medication) and in scientific laboratories for making gels (e.g., for gel electrophoresis) and solidified culture media. The annual production of agar products is approximately 11,000 tons, with a value of approximately $160 million, according to 1990 figures.

The demand for products like carrageenans and agars far exceeds the red algae harvest. Although the process of isolating these compounds is straightforward, genetic manipulation of cultured algal strains might enhance growth and hydrocolloid production. Genetically improved strains might decrease the cost of products—agar and agarose can be expensive, ranging from $2 to $200 and $250 to $40,000 per kilogram, respectively, depending on purity.

Experiments using algal cell culture are in progress to increase the amounts synthesized of such products as agar. Macroalgae, or seaweeds, can be cultured *in vitro* by producing protoplasts and callus tissue from which these algae can be regenerated. Cell and tissue culture allows new traits to be selected or genes for specific characteristics to be transferred. Protoplast fusion allows desirable characteristics from two different organisms to be mixed. *In vitro* propagation and selection may one day lead to the development of disease resistance; faster growth; tolerance to variations in light, temperature, and nutrients; and increased production of metabolites and nutrients.

The microalgae comprise a diverse group of both eukaryotic algae (e.g., green algae) and prokaryotic photosynthetic bacteria (e.g. **cyanobacteria**, sometimes called blue-green algae). Although used primarily as food, microalgae are also a source of pigments such as

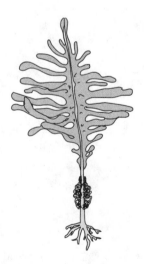

Figure 8.5 Wakame is cultured in Japan and China and has a large market value.

phycoerythrin, phycocyanin, β-carotene, and zeaxanthin (Table 8.3). Five types of microalgae have been most exploited: *Dunaliella*, *Scenedesmus*, *Spirulina* (a cyanobacterium), *Porphyridium*, and *Chlorella*. The last three in particular are sources of protein, vitamins, etc., especially in developing countries and in health food stores (Tables 8.4 and 8.5).

Mass culturing enables large quantities of microalgae to be grown and harvested in outdoor ponds. The potential exists for lowering costs so that microalgae products can compete commercially with other compounds. For example, arachidonic acid is an essential dietary fatty acid and precursor to prostaglandins and other important compounds. One of the few natural sources, the red unicellular algae *Porphyridium*, has one of the highest concentrations of arachidonic acid—36% of the total fatty acids. This organism also can be mass cultivated in open ponds.

Phycobiliproteins are pigments associated with the photosynthetic apparatus of red algae, cyanobacteria, and cryptomonads. *Poryphryidium* is an excellent source of phycobiliproteins, since compounds can easily be

Table 8.3 Microalgal Products and their Approximate Market Value

Algal Products	Uses and Examples	Approx. Value, $/kg	Approx. Market*	Algae Genus or Type	Product Content	Production System	Current Status
Isotopic compounds	Medicine Research	>1000	Small	Many	>5%	Tubular, indoors	Commercial
Phycobili-proteins	Diagnostics Food colorings	>10,000 >100	Small Medium	Red Cyanobacteria	1–5%	Tubular, indoors	Commercial Commercial
Pharmaceuticals	Anticancer Antibiotics	Unknown (very high)	Large Large	Cyanobacteria Other	0.1–1%	Indoor ponds, fermenters	Research
β-Carotene	Food supplement Food colorings	>500 300	Small Medium	*Dunaliella*	5%	Lined ponds	Commercial
Xanthophylls	Chicken feeds Fish feeds	200–500 1000	Medium Medium	Greens, diatoms, etc.	0.5%	Unlined ponds Lined ponds	Research Commercial (?)
Vitamins C & E	Natural vitamins	10–50	Medium	Greens, unknown	<1%	Fermenters	Research
Health foods	Supplements	10–20	Medium to large	*Chlorella, Spirulina*	100%	Lined ponds	Commercial
Polysaccharides	Viscocifiers, gums	5–10	Medium to large	*Porphyridium*	50%	Lined ponds	Research
	Ion exchangers		Large	Greens, others			Commercial
Bivalve feeds	Seed production Aquaculture	20–100 1–10	Small Large	Diatoms, Chrysophytes	100%	Lined ponds	Commercial Research
Soil inocula	Conditioner, Fertilizers	>100	Unknown Unknown	*Chlamydomonas* N-fixing species	100%	Indoor ponds Lined ponds	Commercial Research
Amino acids	Proline Arginine, aspartate	5–50 5–50	Small Small	*Chlorella* Cyanobacteria	10% 10%	Lined ponds Lined ponds	Research Conceptual
Single cell protein	Animal feeds	0.3–0.5	Large	Greens, others	100%	Unlined ponds	Research
Vegetable oils	Foods, feeds	0.3–0.6	Large	Greens, others	30–50%	Unlined ponds	Research
Marine oils Waste treatment	Supplements Municipal, industrial	1–30 1 per kg algae	Small Large	Diatoms, others Greens, others	15–30% n.a.	Lined ponds Unlined ponds	Research Commercial
Methane, H$_2$, liquid fuels	General uses	0.1–0.2	Large	Cyanobacteria, greens, diatoms	30–50%	Unlined ponds	Research

*Market sizes ($ million); small <$10; medium $10–100; large >$100.

J. R. Benemann in *Algal and Cyanobacterial Biotechnology*, Cresswell et al., eds., Table 11.1. Addison Wesley Longman, London. Used by permission of the publisher.

Table 8.4   Comparison of the Gross Chemical Composition of Human Food Sources and Different Algae (percentage of dry matter)

Commodity	Protein	Carbohydrates	Lipids	Nucleic Acid
Baker's yeast	39	38	1	—
Rice	8	77	2	—
Egg	47	4	41	—
Milk	26	38	28	—
Meat muscle	43	1	34	—
Soya	37	30	20	—
Scenedesmus obliquus	50–56	10–17	12–14	3–6
Scenedesmus quadricauda	47	—	1.9	—
Scenedesmus dimorphus	8–18	21–52	16–40	—
Chlamydomonas rheinhardii	48	17	21	—
Chlorella vulgaris	51–58	12–17	14–22	4–5
Chlorella pyrenoidosa	57	26	2	—
Spirogyra sp.	6–20	33–64	11–21	—
Dunaliella bioculata	49	4	8	—
Dunaliella salina	57	32	6	—
Euglena gracilis	39–61	14–18	14–20	—
Prymnesium parvum	28–45	25–33	22–38	1–2
Tetraselmis maculata	52	15	3	—
Porphyridium cruentum	28–39	40–57	9–14	—
Spirulina platensis	46–63	8–14	4–9	2–5
Spirulina maxima	60–71	13–16	6–7	3–4.5
Synechococcus sp.	63	15	11	5
Anabaena cylindrica	43–56	25–30	4–7	—

E. W. Becker, *Microalgae: Biotechnology and Microbiology*, Table 12.1. Cambridge University Press, 1994. Used by permission of the publisher and the authors.

extracted and cells are readily cultured. *Poryphyridium* phycobiliproteins (e.g., phycoerythrin in red algae) is a potential source of phycofluors, used to label or tag biologically active molecules such as immunoglobulin, protein A, and biotin.

Records indicate that indigenous people in Mexico were culturing the cyanobacterium *Spirulina* for food in 1524. People living around Lake Chad in West Africa also were consuming *Spirulina* centuries ago. This cyanobacterium is a desirable food source because it is readily harvested, its cell wall constituents are easily digested, and it is approximately 70% protein by dry weight. The largest culturing facility is near Mexico City, at Lake Texcoco, an ancient lake, now dry, that was an Aztec site. At this lake, the Indians harvested *Spirulina*

and another cyanobacterium, *Oscillatoria*. Today, *Spirulina* is propagated in ponds at this lake by pumping underground water into a spiral evaporation system created by dikes (Figure 8.6). At a certain site in the spiral, the water has enough salinity so that *Spirulina* can grow. There are culturing facilities in Thailand, Israel, Japan, Taiwan, and the United States; total worldwide production is approximately 850 tons a year. This alga is marketed as dried flakes, and is found primarily in health food stores, in fish food, and in Japanese cuisine. Current cost is about $10/kilogram, but it has been as high as $150/kilogram—most likely because of an earlier interest in its potential therapeutic value.

The cost of culturing algae as a protein source is much higher than the cost of soybean production ($2 to

Table 8.5 Comparison of the Recommended Daily Intake of Vitamins with the Content of Human Food Sources and Different Algae

Vitamin	RDI (mg d^{-1})	Beef liver (fresh)	Spinach (fresh)	mg kg^{-1} dry matter					
				1	2	3	4	5	6
Vitamin A	1.7	360	130	840	225	—	230	554	480
Thiamine (Vit B$_1$)	1.5	3	0.9	44	14	55	8	11.5	10
Riboflavin (Vit. B$_2$)	2.0	29	1.8	37	28.5	40	36.6	27	36
Pyridoxine (Vit. B$_6$)	2.5	7	1.8	3	1.3	3	2.5	—	23
Cobalamine (Vit. B$_{12}$)	0.005	0.65	—	7	0.3	2	0.4	1.1	0.02
Vitamin C	50	310	470	80	103	—	20	396	—
Vitamin E	30	10	—	120	—	190	—	—	—
Nicotinate	18	136	5.5	—	—	118	120	108	240
Biotin	—	1	0.07	0.3	—	0.4	0.2	—	0.15
Folic acid	0.6	2.9	0.7	0.4	—	0.5	0.7	—	—
Pantothenate	8	73	2.8	13	—	11	16.5	46	20

Notes:

RDI, Recommended daily intake (adults).

1, *Spirulina platensis* (Faggi, 1980); 2, *Spirulina maxima* (Jaya, Scarsino & Spadoni, 1980); 3, Spirulina sp. (Durand-Chastel, 1980); 4, *Scenedesmus obliquus* (Becker, 1984); 5, *Scenedesmus quadricauda* (Cook, 1962); 6, *Chlorella pyrenoidosa* (Fisher & Burlew, 1953).

E. W. Becker, *Microalgae: Biotechnology and Microbiology,* Table 12.5. Cambridge University Press, 1994. Used by permission of the publisher and the author. See this book for original references.

$10/kilogram vs. $0.20/kilogram). However, in developing countries protein is desperately needed, fertile farmland is in short supply, and high-technology amendments to the soil such as chemical fertilizers and herbicides are not affordable. Thus, the low-technology cultivation of high-protein algae like *Spirulina* may be attractive. (Single-cell protein is discussed in Chapter 5, Microbial Technology.)

Modern algal biotechnology involves genetic manipulation, chiefly through mutation and selection, to produce algae that grow faster in culture, are more disease resistant, synthesize more of a particular metabolite, and even produce new, unique products. Recombinant DNA technology has not been used extensively, since gene transfer methods must be established for each type of organism, and specific genes and promot-

Figure 8.6 An aerial view of the spiral algal pond at Lake Texcoco, in Mexico.

ers must be isolated that would allow the expression of a desired trait or product.

Over the past 15 years, algae have received much attention for their pharmacological potential, their use as agricultural fertilizers and as energy biomass, and their use in cell cultures for producing unusual and rare chemicals. The variety of potential products is extremely broad—encompassing polysaccharides, proteins, lipids, pigments, carotenoids, sterols, vitamins, antibiotics, enzymes, pharmaceuticals, fine chemicals, as well as biofuels such as hydrocarbons, methane, and alcohol. Moreover, commercially available marine polysaccharides have antiviral, antiulcer, antitumor, anticoagulant, and cholesterol-lowering activities. Algae are an efficient, renewable, environmentally friendly source of chemicals, pigments, and energy.

Many unusual compounds with unique structures have been isolated from algae. Unfortunately, some marine pharmacology research programs have declined in recent years because of the tremendous (and costly) screening effort required to identify bioactive compounds, difficulties of collecting and culturing, and the small amounts in which many algal compounds are isolated. Most algal products do not seem to show much economic potential, especially in competition with chemical syntheses. Two exceptions have been β-carotene and glycerol. The pigment β-carotene is found in green plants, but some algae such as the halotolerant, unicellular green alga *Dunaliella salina* produce very high amounts, and can be mass cultured in shallow brackish ponds. Glycerol, an osmoregulatory product of *Dunaliella,* has been commercially produced. However, since glycerol can also be produced during petroleum extraction, the future of algal glycerol is uncertain.

Currently algal biotechnology represents only a small fraction of modern biotechnology. Commercial interests and the bulk of investments have focused on the development of genetically engineered plants and animals, diagnostics, new pharmaceuticals from plants and synthetic compounds, and industrial fermentations using animal, plant, and bacterial cells. In the future, however, increased awareness of the importance of algae should stimulate research activities, and algal biotechnology will most likely represent a larger portion of modern biotechnology.

FUELS FROM ALGAE

Nonrenewable fossil fuels provide most of the world's energy (78%). In 1992, oil provided 26% of domestic energy production in the United States and 41% of energy consumption; coal has been the largest source of domestic energy (33%), and is the least expensive fossil fuel per energy unit ($1.42 per million BTU). Although fossil fuel stores are not inexhaustible, no one knows exactly how much remains. Readily accessible reserves are being depleted, and new sources must be identified. Continued extraction and combustion of fossil fuels also pose environmental problems. For example, coal has been a major source of air pollution, contributing 66% of total U.S. sulfur dioxide emissions and 36% of greenhouse gases (human-generated carbon emissions). Projections of the worldwide market for clean power technologies range from $270 to 750 billion during the next 20 years (1996 figures). In 1996, renewable (such as wind and solar) and nuclear power represented only 18% and 4%, respectively, of global energy production.

A potential alternative to fossil fuels is photosynthetically generated biomass, a resource that is renewable and will not damage the environment. Biomass can be converted to methane by any of several methods: catalytic hydrogenation, thermal gasification, and biological gasification. Biological gasification is the best method for converting algal biomass. A mixed microbial population digests the biomass, generating methane and carbon dioxide as end products. The process has two steps: an acid-forming step and a methane-producing step. In the first step, acid-forming bacteria (nonmethanogens) catalyze the production of simple molecules from complex biological macromolecules. These simple molecules are further modified to generate carbon dioxide, hydrogen gas, hydrogen sulfide, ammonia, and acids (fatty acids and acetic acids). The short-chain acids (having from one to six carbons) produced, of which acetic acid is the major constituent, decrease the pH of the reaction.

In the second step, methanogens (bacteria that produce methane) use the products from the first step to produce methane and carbon dioxide. Either short-chain acids are converted to methane and carbon dioxide, or carbon dioxide and hydrogen gas form methane and water.

The bacteria that the second step requires are sensitive to nutrient availability, pH, and oxygen levels. The type of bacteria (for example, thermophiles or mesophiles) present determines the temperature range of the reactions. The two steps need not be performed in immediate succession; the acids produced in the first step can be stored for later conversion to methane. Acid storage might be desirable if there is a continuous demand for methane because the supply of algal biomass is seasonal. From 47 to 65% of the available carbon is converted, and the gas produced varies from 60 to 72% methane, with carbon dioxide and trace amounts of nitrogen making up the rest.

Bacteria used in the bioconversion are obtained from agricultural wastes, high saline environments, and estuaries. Biomass conversion to methane can use a variety of algae, which provide all the nutrients that the bacterial culture requires. The amount of methane produced depends in part on the composition of the algal culture, since the amounts of polysaccharides, lignin, protein, and other components that influence the efficiency of the bioconversion vary greatly among species. At the present time, biological gasification is not competitive with other sources of natural gas. Algal harvesting and production account for 70% of the total cost of the methane.

Although *Dunaliella* produces up to 85% of its dry weight in glycerol, this alcohol is not a good fuel because its high viscosity and oxygen content give it a very low energy value. However, bacterial fermentation and catalytic conversion can convert glycerol into more useful products. Several bacteria, such as *Klebsiella* sp., *Clostridium pasteurianum*, and *Bacillus* sp. can ferment glycerol from *Dunaliella* biomass to higher-energy, nonviscous compounds. *Klebsiella* produces 86% 1,3-propanediol, 9% acetate, and 3% ethanol; *Clostridium*-produces 72% n-butanol, 21% ethanol, 1% acetate, and trace amounts of 1,3-propanediol; and *Bacillus* produces 92% ethanol and 8% acetate. Ethanol, butanol, and 1,3-propanediol can be used as liquid fuels and have higher energy contents than glycerol. Gasohol, a gasoline with an ethanol additive produced from the bacterial fermentation of corn, is available in the Midwest of the United States. A simple chemical conversion can further convert ethanol to alkenes, alkanes, and aromatics similar to those found in gasoline.

Algae can potentially produce fuel directly. Algae—primarily microalgae—can produce large amounts of hydrocarbons derived from either fatty acid or isoprenoid biosynthesis. Resulting hydrocarbons are high in energy, having two times the energy of any other biological compound per comparable mass. Small isoprenoids with fewer than 16 carbons can be used directly as liquid fuels. Larger molecules can be chemically converted to liquid fuels, although economical conversion methods still must be developed. Naturally occurring, high-yielding microalgae that redirect a large portion of fixed carbon into lipid production must be identified and mass cultured in saltwater ponds for use as fuel. The lack of technologies may be due to the fact that some hydrocarbons such as triglycerides are more valuable to the food industry than as a source of energy.

In the future algae may be genetically modified so as to synthesize gasoline-type fuels such as acyclic hydrocarbons (among them, alkanes and alkenes). Some brown algae (*Macrocystis*) and cyanobacteria (e.g. *Anacystis nidulans*) already synthesize small amounts of "fuels" from fatty acids. Genetic engineering may increase production.

For the time being, algal fuels will not replace fossil fuels. Research first must find ways to increase culturing efficiency and increase biofuel yields. However, even if biofuels account for only 2 to 3% of the total global energy use, this will represent an enormous amount of energy and would decrease the impact of fossil fuel use on the environment.

ALGAL CELL CULTURE

Algal culture is used chiefly to generate biomass from which cells and metabolites can be isolated. Culturing often takes place in large ponds or raceways (Figure 8.7 shows several facilities). Cultured algal cells also are used to maximize the production of high-cost or rare compounds. This type of culturing usually takes place in a fermenter or bioreactor rather than in ponds. Culture conditions are highly controlled, and products undergo significant downstream processing of product collection, purification, and packaging. Currently, microalgae are the main organisms used in culture, although macroalgae yield valuable agars and agarose (polysaccharide-containing polymers) that are used in research and diagnostic laboratories.

The world demand for agars far exceeds production because of seasonal variations in yield and quality, inadequate production methods, and the scarcity of agar-producing algae. Cell culture technology may be able to contribute to agar production. International collaborators are spending many millions of dollars to develop adequate cell culture production technologies. Red algal cells are being isolated and selected for efficient growth in culture, as well as being genetically engineered for high rates of agar and agarose production.

Specialty chemicals are produced from microalgal cultures. For example, a particular strain of *Chlorella* (green algae that produce a large quantity of amino acids) has been developed that produces 30% more proline than other strains produce. The proline can be readily extracted from the culture for commercial use (such as for food supplements and cell culture media). A variety of hydrocarbons (including important lipids), polysaccharides, and other important compounds also can be produced. Gene transfer from other organisms can yield new algal products; for example, after genes are transferred from bacteria, the green alga *Chlamydomonas* can produce the rare amino acid octopamine.

Although algal cell culture is a high-cost operation, the products that are isolated can be very profitable. For example, amino acids fetch $5 to $100/kilogram; food coloring phycobiliproteins, $100/kilogram; β-carotenes and other pigments, $300 to $500/kilogram; and medical phycobiliproteins, $10,000/kilogram. Other industrial

Figure 8.7 A variety of algal culture facilities: (a) Circular algal ponds with rotating agitator in Taiwan. (b) Concentric algal pond at Auroville, Indiana. (c) A sloping algal pond in Peru has floating injectors. (d) A primitive algal pond with a polyethylene lining in Madras, India.

compounds, such as dihydroxyacetone, gluconic acid, hydrogen, and acetic acid, are produced by immobilized microalgal cells. One algal culture in a large fermenter potentially can generate thousands of dollars of specialty chemicals each month.

MEDICAL APPLICATIONS

Marine organisms also make important contributions to medical research. Sharks do not form cancerous cells, and so are being used to study the immune system; compounds in shark-cell extracts might hold the key to cancer suppression. Other marine organisms serve as models for the study of human disease. For example,

because the sea cucumber's abdominal cavity is filled with bacteria, the creature is being used to study resistance to peritonitis, and because the icefish lacks red blood and hemoglobin, it is a model for anemia.

Marine Natural Products and Their Medical Potential

Marine organisms produce many metabolites, which are usually referred to as natural products. Chemists and pharmacologists are studying these natural products because of their unusual and sometimes unique chemical structures. Unlike primary and intermediate metabolites and cofactors, secondary metabolites are not essential for growth and reproduction, but most probably confer some selective advantage on the organism (or did so at some time during the organism's evolution). The many diverse compounds provide much diversity of chemical structure and function.

Researchers are evaluating these marine natural products for their usefulness. In the past, studies were limited to cataloging the metabolites and other chemicals isolated from marine organisms. Today, investigators focus on identifying applications in medicine and agriculture. Compounds that are pharmaceutically active or that can help control insect pests destructive to agriculture are being studied. Chemists also use these compounds to synthesize new derivatives or analogs to generate more compounds which have potential value. Secondary metabolites exhibit several types of activities: antitumor, antiviral, anti-inflammatory; enzymatic; insecticidal and herbicidal (a highly neglected area); antibacterial (antibiotic); and toxic. Some secondary metabolites may one day be developed commercially.

For the past 100 years, much energy has been devoted to screening the world's organisms for useful chemicals. Approximately 20,000 chemicals have been characterized, and many of these come from marine organisms such as bacteria, algae, **sponges**, corals, jellyfish, bryozoans, mollusks, ascidians, and **echinoderms**. Most of the new natural products identified each year have come from terrestrial organisms: Annual sales of pharmaceuticals from plants exceeds $10 billion in the United States. However, plants were virtually untapped until recently, so natural products from marine organisms may hold similar promise. Indeed, marine organisms have been shown to be an excellent source of new compounds which exhibit a variety of biological activities. From 1977 to 1987, approximately 2,500 new metabolites from a variety of marine organisms were reported. In 1992 alone, some 200 published articles characterized over 500 marine natural products. Most of these compounds came, in descending order of abundance, from sponges, ascidians (**tunicates**), algae, mollusks, and soft and gorgonian corals. Sponges have recently become a primary object of investigation because of their wide range of biosynthetic capabilities. Metabolites from echinoderms (such as starfish and sea stars) also have been isolated for natural products. Microbial organisms such as bacteria, cyanobacteria, and actinomycetes also are important sources of pharmaceutical compounds.

Anticancer and Antiviral Compounds

Many compounds from marine organisms are extremely toxic to humans; however, controlling toxicity by diluting concentrations exploits their therapeutic value. The National Cancer Institute tests marine extracts and compounds against many *in vitro* cell lines that represent various cancers (such as kidney, brain, colon, lung, skin, blood cells, and ovary, as well as a lymphoblastic cell line infected with the AIDS virus).

A variety of compounds isolated from marine organisms show promise as anticancer drugs. Didemnin B (Figure 8.8a), a cyclic peptide from a Caribbean tunicate (*Tridedemnum solidum*), was shown experimentally to be very effective against leukemia and melanoma cells in mice. Although in human clinical trials, the compound did not exhibit significant anticancer properties, didemnin derivatives may nevertheless provide cancer-fighting properties. Didemnin B is also an effective immunosuppressive agent—it is approximately 1,000 times more effctive than the standard cyclosporin A. Further study may prove it useful in repressing rejections of transplanted organs.

The Romans knew of the toxic effects of crude extracts from sea hares (a type of mollusk) containing dolastatins, and by the year 150 were using such extracts in the treatment of various diseases. However, not until the 1970s were dolastatins known to be effective against lymphocytic leukemia and melanoma. Different structural forms of dolastatins that show antitumor properties have since been isolated. Dolastatin 10 (Figure 8.8b) is a linear peptide from the sea hare *Dolatella auricularia*, found in the Indian Ocean. This compound is the most active of the isolates, and is a potent antimitotic (inhibitor of cell division) because it inhibits the polymerization of tubulins into microtubules, which are involved in cell division. This compound is undergoing clinical trials and is being compared to other antitubulin polymerization drugs, such as vinblastine, which is isolated from the Madagascar periwinkle.

Marine sponges seem to be an excellent source of natural products. Alkaloids, terpenoids, and sterols are but a few of the hundreds of compounds isolated from them. Some compounds may actually be produced by symbiotic organisms living within the sponge, such as cyanobacteria, **dinoflagellates**, microalgae, and an assortment of bacteria. The South Pacific marine sponge *Luffariella variablis* is the source of manoalide, a member of the terpene group of chemicals. Pharmacological studies have demonstrated that it has anti-inflammatory, analgesic, and antifungal properties, in addition to being effective against leukemia. Luffariellin A (Figure 8.8c) and B, also isolated from *Luffariella*, may be anti-inflammatory agents. A variety of potentially important chemicals from several marine sponges have exhibited antitumor activities: halichondrins (Figure 8.8d), mycalamides, onnamide A, calyculins, swinholides (Figure 8.8e), and misakinolides (Figure 8.8f). Halichondrins in particular have generated special interest by demonstrating potent activity against leukemia and melanoma cell cultures.

Other organisms such as nudibranchs (molluscs that are essentially snails lacking a shell) extract potent compounds from invertebrate prey and use them for

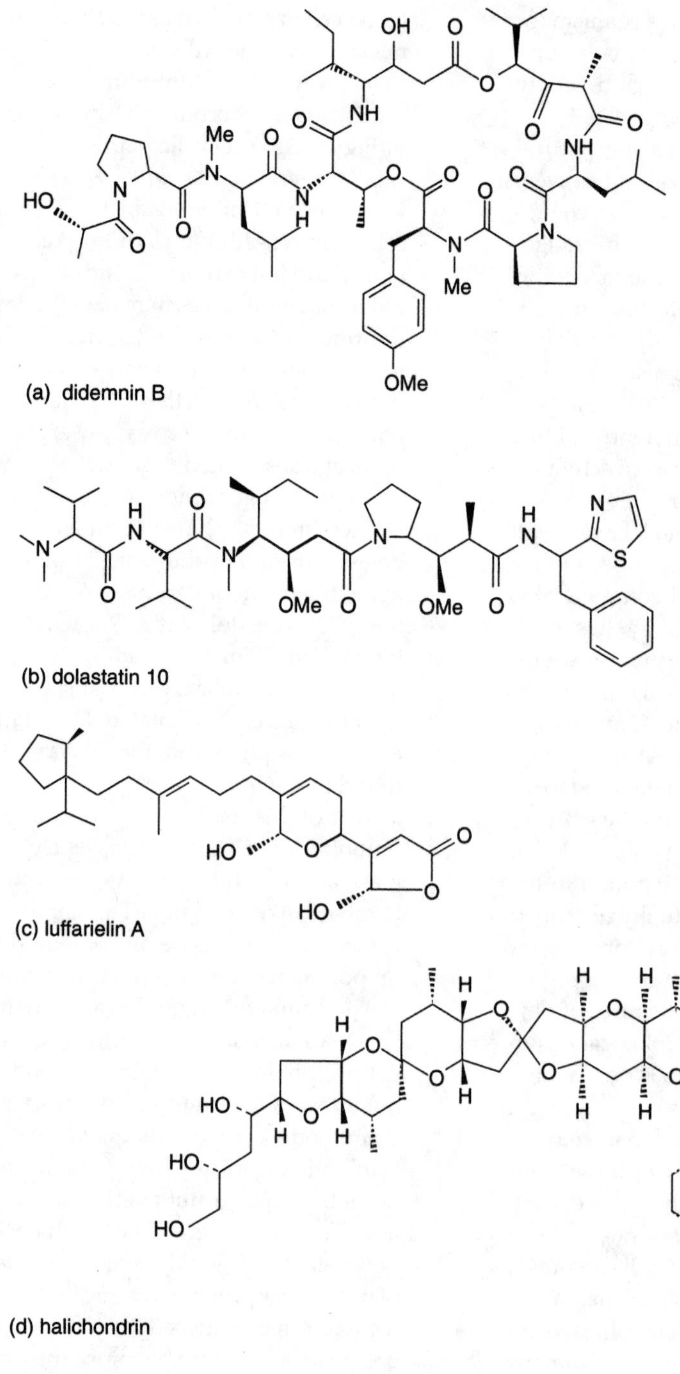

Figure 8.8 Chemical structures of (a) didemnin B, (b) dolastatin 10, (c) luffariellin A, and (d) halichondrin B.

protection. The ulapualides, compounds demonstrating activity against both leukemia and fungi, can be transmitted to the nudibranch's eggs to protect them from predators seeking a readily available meal.

Useful antiviral drugs (such as aciclovir and AZT) are extremely rare. However, marine organisms may be a source of antiviral agents. Researchers worldwide are screening compounds from marine organisms for antiviral activities. Potent antiviral agents, such as eudistomins derived from tunicates (e.g., *Eudistoma olivaceum*), show promise, and are being evaluated in clinical trials for their antiviral and antitumor activities.

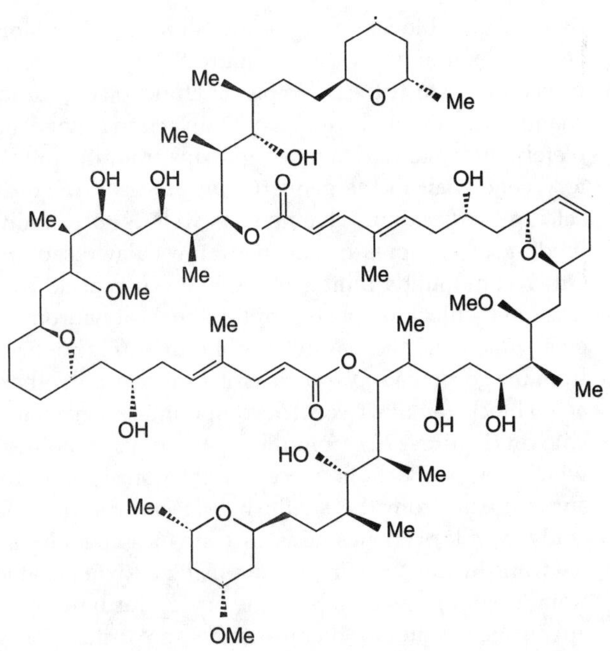

(e) swinholide A

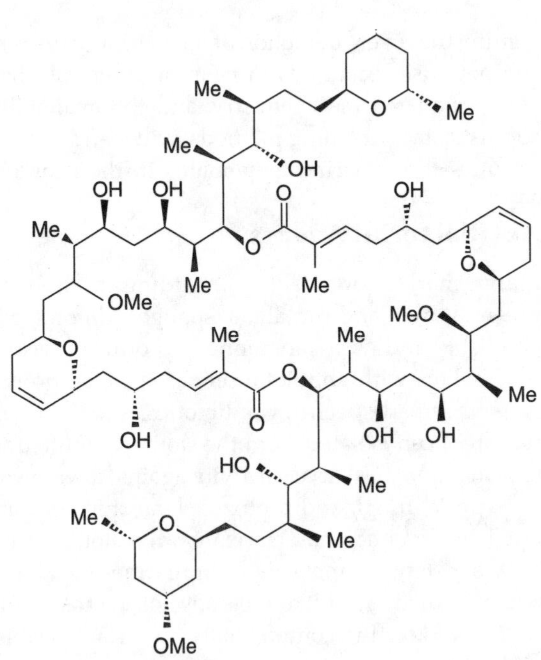

(f) misakinolide A

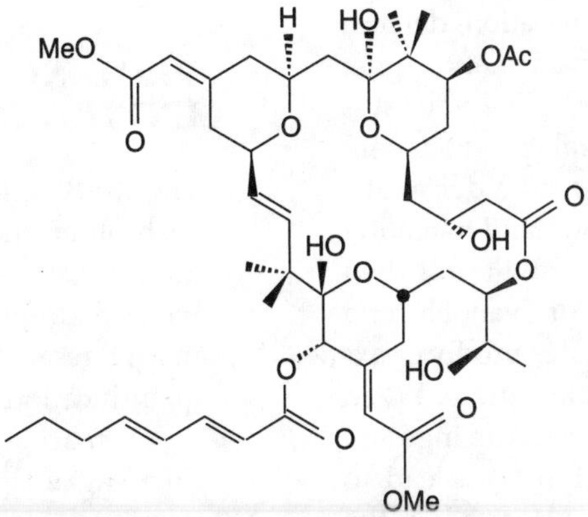

(g) bryostatin 1

Figure 8.8 Chemical structures of (e) swinholides, (f) misakinolide A, and (g) bryostatin 1. Me = methyl group, CH_3.

Cyanobacterial extracts (active agents composed of sulfolipids) also show activity against the AIDS virus.

Marine bryozoans, also known as false corals, are colonial filter-feeders that produce many potentially important compounds. One important class of compounds is the bryostatins (Figure 8.8g); one source is the symbiotic bryozoan *Bugula neritina*, found in association with a yellow sponge in the Gulf of California, Mexico. The bryostatins seem to be effective against a variety of cancers such as ovarian carcinoma and lymphocytic leukemia. Bryostatins inhibit RNA synthesis, bind to certain enzymes such as protein kinase C, stimulate protein phosphorylation, activate certain human leukocytes, and induce the synthesis of interleukin-2.

Unfortunately, the amount of any marine natural product that is available is limited by how many organisms are harvested that produce it. Studies that focus on the chemical synthesis of pharmaceutically active marine compounds will be indispensable not only for developing new medical treatments but also for preventing the severe depletion of the marine sources of the compounds. Recent advances in biochemical pharmacology and marine chemistry, and the availability of sophisticated screening methods, will lead to new therapeutics from marine metabolites in the near future.

Antibacterial Agents

Many marine organisms—including cyanobacteria, green, brown, and red algae, sponges, dinoflagellates, jellyfish, and sea anemones—produce secondary metabolites with antibiotic properties. One promising class of broad-spectrum antibiotics, squalamine, has recently been isolated from the stomach of the dogfish (*Squalus acanthias*); it has activity against a wide variety of bacteria, fungi, and protozoa. Unfortunately, antibiotic isolation and characterization has not been extensive, and few compounds from marine organisms are being studied, most likely because of the costs of identifying and testing commercially important antibiotics.

Marine Toxins

Toxins are thought to be produced by marine organisms for a variety of purposes, including predation, defense from predators and pathogenic organisms, and signal transduction in the autonomic and central nervous system. Many of these marine natural products are extremely toxic; their effects can range from dermatitis to paralysis, kidney failure, convulsions, and haemolysis. Livestock and humans have become ill and died when specific toxin-producing algae or cyanobacteria have contaminated water. Dinoflagellates produce saxitoxins that are 50 times more toxic than curare. When bivalves that have incorporated saxitoxins are ingested by humans, paralytic shellfish poisoning can lead to severe illness or death. Dinoflagellates also produce the potent ciguatoxin. When small grazing fish that have ingested dinoflagellates while feeding on algae are then eaten by larger fish, ciguatoxin is transferred, and can poison humans who consume the fish.

Marine toxins are being studied for their antitumor, anticancer, and antiviral compounds, tumor promoters, and anti-inflammatory properties. When concentrations are controlled, marine toxins show promise as antitumor agents, analgesics, and muscle relaxants. They are useful neurophysiology and neuropharmacology research tools for studying such processes as signal transduction in the nervous system. Even compounds that are too toxic to be used therapeutically are being used as models for new, chemically synthesized drugs.

Many neurotoxins act only on specific receptors and ion channels. Each toxin has its own mode of action and so can be used to probe a different step in signal transmission. Ion channels and membrane excitation are being studied by means of specific sodium channel blockers; an example is tetrodotoxin, from the puffer fish, which blocks peripheral nerve transmission. Tetrodotoxin and saxitoxin (Figure 8.9 a and b) have been used to study the assembly and function of sodium channels, although the mode of toxin action is not completely understood. Potent conotoxins from the predatory cone snail in the genus *Conus* are used to study calcium and sodium channels, as well as to examine binding to vasopressin and neuromuscular receptors. Other compounds inhibit nerve-induced muscle contraction by blocking postsynaptic signals at neuromuscular junctions; the paralytic lophotoxin (Figure 8.9c) from the gorgonia *Lophogorgia* is an example. Still others act to block ganglia; two such compounds are neosurugatoxin (Figure 8.9d), from the Japanese ivory mollusc, which is an acetylcholine receptor antagonist , and prosurugatoxin, from the shellfish *Babylonia japonica*. To study the role of adenosine as a second messenger in signal transduction, researchers are using a toxin isolated from a red alga as an adenosine analog for binding to adenosine receptors. Other researchers are using sponge dysidenin (Figure 8.9e), an inhibitor of iodide transfer in thyroid cells, to study iodide transport.

PROBING THE MARINE ENVIRONMENT

The study of the interaction of marine organisms with each other and with their environment is an exciting area of marine biology. Modern biotechnology methods have made important contributions in many areas of marine research, including the rapid identification and quantification of new species and populations of microscopic marine organisms, such as bacteria, protozoans, microalgae, and viruses; the tracking of commercially important organisms; the bioremediation of marine niches; the development of diagnostics to detect contaminants, pollutants, and pathogens in seafood and the environment; and the use of marine organisms in biomedical research.

Biotechnology methods are increasing our understanding of the abundance, distribution, and rates of growth of various marine organisms. Specific DNA probes and the polymerase chain reaction (PCR, described in Chapter 4) allow different organisms to be detected and identified in marine environments without their having to be isolated and cultured. Larval invertebrates and microbes, which are often difficult to isolate, can readily be tracked in the environment. The country of origin of important fish stock populations can be established as well. Molecular techniques can determine the level of genetic variation in organisms. Genetic diversity in a group of organisms is an indicator of a healthy population, and one that is more likely than an inbred one to tolerate environmental fluctuations and be resistant to disease.

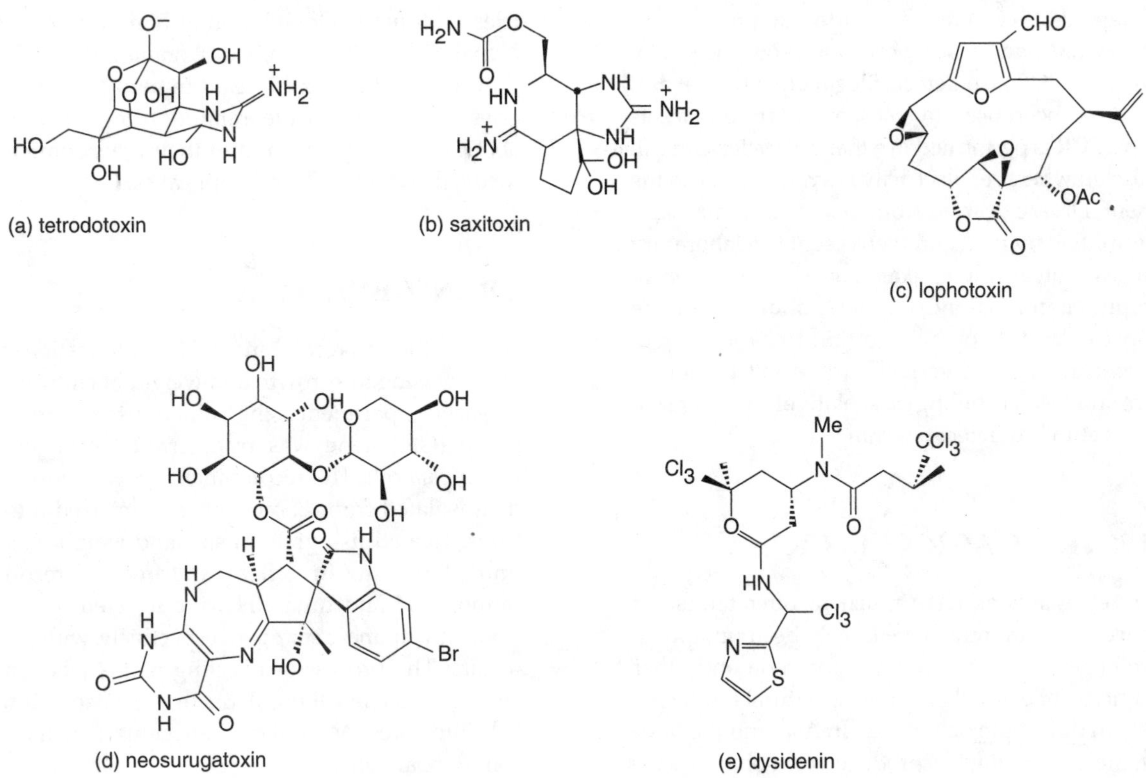

Figure 8.9 Chemical structures of (a) tetrodotoxin, (b) saxitoxin, (c) lophotoxin, (d) neosurugatoxin, and (e) dysidenin. Me = Methyl group, CH_3.

DNA probes or PCR can identify currently unknown intermediate life stages of important marine organisms such as shellfish. Once these intermediates are identified for study, researchers may be able to isolate and characterize the chemical signals that induce settling and metamorphosis. Often such research findings can be applied to the culture and propagation of economically important crustaceans, bivalves, and gastropods.

Microplankton have been shown to be the base of the food chain that supports most of the other marine animal species in the oceans. Microplankton can be identified by fluorescent DNA probes (probes labeled by fluorescent tags rather than radioisotopes; detection is by fluorescence). Microbial primary producers are being identified and quantified by PCR and flow cytometry (which distinguishes different types of cells by defined characteristics). Thus, ecological processes as critical as primary production in the oceans can be studied more easily with the help of biotechnology. In polluted areas, such as areas of industrial waste and raw sewage outfalls, the numbers and types of microbes that cause disease can be readily identified. For example, hepatitis A and E and cholera affect large numbers of people exposed to polluted water through swimming, washing, or harvesting food. Mussels, oysters, and clams, often consumed by humans, reside in offshore areas where ocean dumping occurs and can harbor pathogens.

Many of the pathogens released into the marine environment remain unidentified. Diagnostic kits using DNA probes and monoclonal antibodies will help researchers track pathogens along polluted and contaminated shores and enable health officials to determine what areas should be closed to the public. These detection kits will also be invaluable for detecting pathogens and toxins in commercial seafood.

CONSERVATION

Biotechnology has begun to play an important role in the conservation of threatened and endangered marine fish and mammals. By using the DNA from animals, scientists can determine whether protected animals are illegally being used for food, jewelry, or traditional medicines and potions. Makeshift DNA laboratories are being used in countries worldwide to determine whether markets and restaurants are selling endangered species in violation of international law. Most recently, DNA methods identified unlabeled meat from the humpback (protected since 1966) and fin whale (protected since

1989) in Japan; other samples included dolphin, northern minke whale, and a beaked whale about the size of a killer whale. PCR also detects illegal animal meat and parts, and has been used to test samples from markets and shops. PCR is performed in a thermal cycler, and can be conducted wherever electricity is available to run the instrument. DNA extracted from food and animal samples is amplified by PCR and then is sent to a laboratory for analysis. For example, examination of a region of PCR-amplified mitochondrial DNA allows both the whale species and its ocean of origin to be identified. This "fingerprinting" (described further in Chapter 11, Forensics and DNA Profiling) is a relatively inexpensive and easy method to detect poaching.

TERRESTRIAL AGRICULTURE

Marine organisms have much to offer terrestrial agriculture, but until now have largely been untapped. Soil-dwelling and freshwater cyanobacteria and other aquatic microalgae have been used as natural fertilizers, especially in developing countries. In Asia and the Middle East, mass culturing has produced cyanobacteria for fertilizer (usually distributed as dried flakes) and for use in rejuvenating nutrient-exhausted soils. Marine plants, macroalgae and microalgae, and animals have characteristics that, through the application of modern biotechnology, can benefit agriculture. Marine organisms are salt-tolerant (the most tolerant aquatic plant is *Dunaliella*, from the Dead Sea), and many, like flounder, survive at very low temperatures. Genes encoding adaptive traits are being experimentally transferred to agronomically important crop plants to determine whether they can express these traits. Salt tolerance might extend the range of important crops, or sustain ranges that would otherwise be lost. Intensive agricultural practices increase the salinity of soil, often to such a level that crops can no longer grow. Salt-tolerant crops could grow in these regions and would increase agricultural productivity substantially, which would be especially important to very poor, malnourished populations. Many highly populated regions of the world lack adequate fresh water but have plenty of water high in saline. Such water might be used for irrigation—reserving scarce fresh water for drinking—if salt-tolerant crops were introduced.

A terrestrial halophyte (an organism requiring salt water) was recently discovered in the Sonora desert in Mexico, where extremes of both temperature and dryness are found. The halophyte, *Salicornia bigelovii*, is a potential high-yielding oilseed crop from which locals can obtain an unsaturated vegetable oil for commercial production. Cold- or freeze-tolerant plants can grow at high altitudes, extending the range of important crops. These may be especially beneficial to developing coun-

tries with undeveloped land at high altitudes, such as Nepal, Peru, Chile, Bolivia, Afganistan, and Equador. Future research should result in many exciting discoveries in the next decade, ranging from new knowledge about marine organisms and their environment to new medical and agricultural applications.

TRANSGENIC FISH

Some of the first recombinant DNA applications focused on increasing the growth and weight of cultured finfish. In early experiments, the rainbow trout growth hormone (GH) gene was transferred to the bacterium *Escherichia coli*. The recombinant growth hormone was then isolated from *E. coli* cells and injected into young trout. Treated fish grew in size and weight faster than untreated fish. In other experiments, recombinant salmon GH and tuna GH were injected into rainbow trout and Japanese snapper, respectively, with promising results. The process of injecting fish is labor-intensive and poses technical problems that the use of transgenic fish alleviates. Moreover, transgenic fish have much commercial value.

Direct gene transfer allows fish to harbor the desired gene and express the appropriate product (such as GH). Fish are excellent organisms for gene transfer since their eggs are large and transparent, making the developmental process readily visible. Gene transfer by microinjection into the nucleus is straightforward, and has been the method of choice for fish transgenics. In addition to transgenic fish with genes for GH, fish with genes for "antifreeze" protein are being field tested, in the hope that economically important fish can extend their range of survival into colder waters.

Transgenic techniques are being developed to introduce desirable traits into many species of fish. For fish transgenics to be successful, appropriate promoters, enhancers, and tissue-specific DNA sequences must be identified. To regulate transgenes (for example, to turn them on and off and to ensure expression in specific tissues), light, temperature, and diet may be used as inducers of transcription. For example, a light-regulated promoter would be activated if fish were exposed to light of a certain color or intensity.

Methods of gene transfer include electroporation, direct pronuclear injection of the egg, and injection through the micropyle (the opening in the egg where sperm enter). Fish pronuclei are often too small to be seen clearly under a microscope and are obscured by an opaque chorion and large volume of cytoplasm. Thus, the gene is often transferred by microinjection into the cytoplasm of fertilized eggs (already released from the female). This method is easier than pronuclear injection of unfertilized eggs (which must be isolated, cultured, and fertilized *in vitro*). However, since the first cell division occurs only 60 minutes after the egg is fertilized, not much time is available to inject eggs. Staggered fish matings enable more freshly fertilized eggs to be injected.

Salmonid eggs (from, for example, Atlantic salmon, rainbow trout, Arctic char, or brown trout) have a hard chorion (Figure 8.10). For these, injection through the micropyle with very fine glass needles (2–3 μm in outside diameter) inserts the transgene very near the male pronucleus, which is just under the micropyle. Another procedure uses sturdy glass needles (15 μm or less in diameter) to penetrate the chorion and insert the DNA into the **blastodisc** just after fertilization but before the chorion completely hardens. Another successful method is to cut a minute hole in the egg after the chorion hardens for insertion of a glass needle (approximately 10 μm in diameter) into the blastodisc just before the first cell division.

Embryos can be left to develop in temperature-controlled tanks. The survival rates of microinjected fish embryos are much higher (from 35 to 80%) than for mammals. From 10 to 70% of the surviving embryos are transgenic. To determine which fish have received the foreign gene, DNA is isolated from fish scales and the nucleated red blood cells are analyzed by PCR. The founder fish are mated to produce stable transgenic lines.

Many important fish species cannot be cultured because of low tolerance to certain environmental conditions or diseases. A variety of environmental conditions may limit the home range. For example, rainbow trout and salmon cannot be raised in warm water; on the other hand, Atlantic salmon, a highly desirable commercial fish, is susceptible to freezing at temperatures below –0.7°C. (The body fluids of most fish freeze between –0.5 and –0.8°C.; seawater, with a much higher salt concentration than fish have, freezes between –1.7 and –2.0°C.) The waters of the Atlantic off the coast of Canada and the polar oceans reach subzero temperatures, and floating ice is abundant during the winter months. Fish such as winter flounder that survive these freezing temperatures do so by synthesizing antifreeze proteins in the liver that circulate in the blood. These fish are protected from freezing at temperatures as low as the freezing temperature of seawater. Research is underway to produce freeze-resistant salmon by introducing the winter flounder antifreeze gene. Stable transgenic lines of Atlantic salmon have been produced that may eventually be cultured in sea pens along the Atlantic coast.

Fish are being genetically modified to make them grow faster and larger and to allow them to flourish through aquaculture. Most studies have focused on the effect of growth hormone genes. With the success of recent research, fast-growing transgenic fish strains may become an important future aquaculture commodity. In one study, engineered catfish containing a

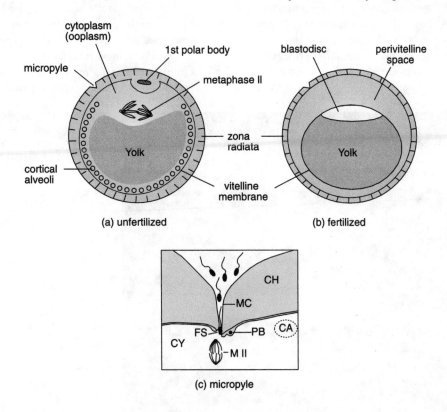

Figure 8.10 Schematic of unfertilized (a) and fertilized (b) salmonid eggs. Sperm enter the micropyle (c); fertilizing sperm (FS), chorion (CH), first polar body (PBI), cytoplasm (CY), cortical alveoli (CA), metaphase II (M II).

trout growth hormone are being studied in experimental ponds. Researchers have achieved mixed results, since only three of five engineered catfish families grew 40% larger than normal. Nevertheless, this area of research looks promising.

Recently, researchers produced transgenic Pacific salmon that harbor a metallothionein-B promoter (regulated by metals and glucocorticoid) and the sockeye salmon GH gene and regulatory elements. The construct was microinjected into the blastodisc of salmon eggs. Out of 3,000 injected eggs, approximately 6.2% transgenic fish survived to one year of age. As Figure 8.11 shows, these fish were more than 11 times larger than their nontransgenic counterparts (one fish was 37 times larger!).

Since males grow faster than females, methods for producing only male fish are being identified. Recent experiments using classical breeding methods have produced genetically homogeneous all-male tilapia (a very important fish in Asian aquaculture), channel catfish, and salmon. To produce these fish, homogeneous male and female breeding stocks must first be generated.

Another area of intense interest is the production of tropical fish and temperate-zone food fish that are resistant to the viral, fungal, and bacterial diseases that can decimate aquaculture stocks.

Concerns about environmental safety have been raised that focus on the containment of transgenic fish. Case studies demonstrate that engineered fish might not remain isolated in commercial facilities, but could accidentally end up in ponds, lakes, streams, and even oceans, where they could decimate already threatened wild fish populations. They could disrupt trophic food webs, increase competition for food and spawning sites, and disturb natural mating systems. Examples of exotic fish introductions that have decimated native fish populations and their ecosystems abound (to give just two, the sea lamprey entered Lake Erie through the Welland Canal and spread to all the Great Lakes, where it destroyed 97% of the trout population; the Nile perch was intentionally introduced into Africa's Lake Victoria and is eating indigenous fish species—by the 1980s 300 native species had become extinct). Fish that have been engineered to occupy an extended home range incorporating previously inhospitable habitats might displace their wild counterparts in those habitats; extending home range can be equivalent to introducing an exotic species into a particular habitat (for example, lake trout introduced into Yellowstone Lake, in Yellowstone National Park, are displacing the native cutthroat trout). Even small genetic modifications that dramatically alter an animal's home range could be introduced into wild populations through breeding.

Both physical and biological controls are needed to prevent release of transgenic fish into the environment. Because physical containment is always subject to compromise by such factors as weather and human error, a biological barrier must ensure that mating between cultured and wild fish does not occur. The elimination of reproductive capabilities can be engineered into transgenic stock; methods include triploid females, sterilization by high doses of **androgens**, and genetically engineered sterility.

Currently, there is no federal regulatory program for examining the risks of transgenic fish. The guidelines that apply to U.S. Department of Agriculture-funded

Figure 8.11 Comparison of 1-month-old coho salmon siblings; nonengineered fish are at left, transgenic fish are at right. The largest fish (top right) is 41.8 cm in length.

research using genetically engineered organisms are voluntary, and private institutions or industries not using federal funds can release genetically engineered fish without governmental approval. Federal agencies will eventually have to address any potential dangers from the use of genetically engineered fish (and other such animals as well). Environmental impact studies are necessary to help alleviate fears, assess risks, and ensure the safety of potential introductions.

General Readings

E.W. Becker. 1994. *Microalgae: Biotechnology and Microbiology*. Cambridge University Press, Cambridge, England.

Committee on Assessment of Technology and Opportunities for Marine Aquaculture in the United States. 1992. *Marine Aquaculture: Opportunities for Growth*. National Academy Press, Washington, D.C.

J.S. Lee and M.E. Newman. 1992. *Aquaculture—An Introduction*. Interstate Publishers, Inc. Danville, Illinois.

R.A. Zilinskas, R.R. Colwell, D.W. Lipton, and R.T. Hill. 1995. *The Global Challenge of Marine Biotechnology: A Status Report of the United States, Japan, Australia, and Norway*. Maryland Sea Grant Publication, College Park, Maryland.

Additional Readings

B. Baker. 1996. Building a better oyster. *Bioscience* 46:240-244.

A. Ben-Amotz and M. Avron. 1989. The biotechnology of mass culturing Dunaliella for products of commercial interests. In R..C. Cresswell, T.A.V. Rees, eds., *Algal and Cyanobacterial Biotechnology*, Longman Scientific and Technical, Essex, England, pp 91-114.

V.S. Bernan. 1997. Marine microorganisms as a source of new natural products. *Adv. Appl. Microbiol.* 43:57-90.

L.J. Borowitzka and M.A. Borowitzka. 1989. Industrial production: methods and economics. In R.C. Cresswell, T.A.V. Rees, eds., *Algal and Cyanobacterial Biotechnology*, Longman Scientific and Technical, Essex, England, pp 294-316.

B.K Carté. 1993. Marine natural products as a source of novel pharmacological agents. *Curr. Opin. Biotechnol.* 4:275-279.

B.K. Carte. 1996. Biomedical potential of marine natural products. *Bioscience* 46:271-286.

D.J. Chapman and K.W. Gellenbeck. 1989. An historical perspective of algal biotechnology. In R.C. Cresswell, T.A.V. Rees, eds., *Algal and Cyanobacterial Biotechnology*, Longman Scientific and Technical, Essex, England, pp 1-27.

G. Cimino, S. De Rosa, and S. De Stefano. 1984. Antiviral agents from a gorgonian, *Eunicella cavolini*. *Experientia* 40:339-340.

R.R. Colwell. 1984. The industrial potential of marine biotechnology. *Oceanus* 27:3-12.

D.J. Faulkner. 1992. Biomedical uses for natural marine chemicals. *Oceanus* 35:29-35.

D.J. Faulkner. 1995. Marine natural products. *Nat. Prod. Rep.* 12:223-269.

W. Fenical and P.R. Jensen. 1993. Marine microorganisms: A new biomedical resource. In D. H. Attaway and O.R. Zaborsky, eds., *Marine Biotechnology: Pharmaceutical and Bioactive Natural Products*. Vol. I, Plenum Press, New York, pp 419-457.

Food and Agriculture Organization of the United Nations. 1991. *Environment and Sustainability in Fisheries*. Committee on Fisheries Nineteenth Session, Rome, Italy. pp 1-23.

A. Fuji. 1987. Aquaculture and mariculture. *Oceanus* 30:19-23.

K.-W. Glombitza and M. Koch. 1989. Secondary metabolites of pharmaceutical potential. In R.C. Cresswell, T.A.V. Rees, eds., *Algal and Cyanobacterial Biotechnology*, Longman Scientific and Technical, Essex, England, pp 161-238.

M.E. Gonzalez, B. Alarcón, and L. Carasco. 1987. Polysaccharides as antiviral agents: Antiviral activity of carrageenan. *Antimicrob. Agents Chemother.* 31:1388-1393.

F. Gusovsky and J.W. Daly. 1990. Maitotoxin: A unique pharmacological tool for research on calcium-dependent mechanisms. *Biochem. Pharmacol.* 39:1633-1639.

Y. Hirata and D. Uemura. 1986. Halichondrins—Antitumor polyether macrolides from a marine sponge. *Pure Appl. Chem.* 58:701-710.

H. Ikenoue and T. Kafuku. 1992. *Modern Methods of Aquaculture in Japan*. Kodansha Ltd., Tokyo.

M. Indergaard and K. Østgaard. 1991. Polysaccharides for food and pharmaceutical uses. In M.D. Guiry and G. Blunden, eds, *Seaweed Resources in Europe: Uses and Potential*, John Wiley & Sons, Chichester, pp 169-184.

C.M. Ireland, B.R. Copp, M.P. Foster, L.A. McDonald, D.C. Radisky, and J.C. Swersey. 1993. Biomedical potential of marine natural products. In D. H. Attaway and O.R. Zaborsky, eds., *Marine Biotechnology: Pharmaceutical and Bioactive Natural Products*. Vol. I, Plenum Press, New York, pp 1-43.

I. Kitagawa and M. Kobayashi. 1992. Anti-cancer drugs from marine organisms. In S. Baba, O. Akerele, and Y. Kawaguchi, eds., *Natural Resources and Human Health—Plants of Medicinal and Nutritional Value*, Elsevier, Amsterdam, pp 123-132.

G. Liles. 1996. Gambling on marine biotechnology. 1996. *Bioscience* 46:250-253.

R.H. McPeak and D.A. Glantz. 1984. Harvesting California's kelp forests. *Oceanus* 27:19-26.

E. Pennisi. 1996. Sorcerers of the sea. *Bioscience* 46:236-239.

R.J. Radmer. 1996. Algal diversity and commercial alagal products. *Bioscience* 46:263-270.

R.J. Radmer and B.C. Parker. 1994. Commercial applications of algae: opportunities and constraints. *J. Appl. Phycol.* 6:93-98.

G. Skjåk-Bræk and A. Martinsen. 1991. Applications of some algal polysaccharides in biotechnology. In M.D. Guiry and G. Blunden, eds, *Seaweed Resources in Europe: Uses and Potential*, John Wiley & Sons, Chichester, pp 219-258.

L. Tangley. 1996. Ground rules emerge for marine bioprospectors. *Bioscience* 46:245-249.

H. Yamamoto. 1995. Marine adhesive proteins and some biotechnological applications. *Biotechnol. Genet. Eng. Rev.* 13:133-165.

T. Yasumoto and M. Murata. 1993. Marine toxins. *Chem. Rev.* 93:1897-1909.

The Human Genome Project is a long-term coordinated international effort to develop detailed genetic and **physical maps** of the human genome; researchers are engaged in locating and identifying all of its genes and establishing the sequence of the genes and all other components—introns, etc.—of the genome. This monumental task has the potential to dramatically increase our understanding of human evolution and variation, gene regulation, human development, and human disease. Knowledge of the human genome will revolutionize the medical field, leading to new areas of research, diagnostics, and treatments for genetic disorders. Areas such as genetic counseling and human genetics are expected to demand additional professionals. The Human Genome Project has stimulated the development of new technologies for organizing the vast quantities of data being generated. These technologies are available to the entire scientific research community, just as new DNA sequencing technologies are useful to investigators conducting molecular biology research in other areas.

The success of the Human Genome Project also poses potential societal problems. Some genetic disorders will be detectable long before there will be treatments for them. Controversy could also arise over reproductive issues, if, for example, individuals consider terminating pregnancies for reasons of genetic makeup, or if there is social pressure to limit reproductive rights on genetic grounds. Issues of genetic discrimination and confidentiality in the insurance industry and in employment also must be addressed.

In 1988, a committee organized by the National Institutes of Health (NIH) and the Department of Energy (DOE) developed an action plan for the Human Genome Project. In 1990, a five-year joint research proposal was submitted to Congress, and in October 1990, the Human Genome Project officially began. The project has been organized and supported primarily by the DOE and the NIH, which established working groups to address genome mapping, computational analysis to handle databases, and the social, legal, and ethical implications of human genome research. Congress provides funds for the project to the National Center for Human Genome Research at the NIH, which in turn awards grants and contracts to U.S. investigators. Additional funds from Congress go to the DOE, which conducts research on human genetics at three national laboratories and funds independent investigators. About 50% of the funding supports study of the human genome, 25% supports study of other model organisms, another 20% supports conferences, training, and program administration, and 5% supports consideration of ethical issues. The Human Genome Organization (HUGO), established in 1988, facilitates the international scientific effort; Canada, Japan, France, the United Kingdom, and Italy now have genome research programs, and the United Nations Educational, Scientific, and Cultural Organization (UNESCO) facilitates and promotes the inclusion of developing countries in international human genome initiatives.

The scope of the Human Genome Project is enormous. Mapping and sequencing the genome was projected to take 15 years and be completed by the year 2005, at a cost of three billion dollars—although reports indicate that progress has been more rapid than expected. After mapping and sequencing are complete, many years will be needed to completely identify all the genes and determine the mechanisms of gene expression and the genetic basis of diseases.

The Human Genome Project supports research on nonhuman organisms because smaller, less complex genomes are easier to study, and correlations can be made to the human genome. Sequencing information for nonhuman genomes generates invaluable information about gene organization and interspecies relationships. During the evolution of organisms, many DNA regions, including some genes, have remained almost unchanged. Comparisons of DNA sequences from different organisms will establish evolutionarily conserved areas and help identify functional regions. Genetic and physical maps are being constructed for organisms such as the mouse, the nematode *Caenorhabditis elegans*, the plant *Arabidopsis thaliana*, and yeast. A detailed mouse genome map will provide valuable information for identifying and isolating genes that cause diseases in humans, especially since animal models are used for studying human disease.

Researchers are making such rapid progress that genome maps and sequences are impossible to accurately summarize at any given time. However, a summary of some nonhuman genome research gives an idea of the progress made in recent years and the importance of these maps to the medical field.

Recently, work was completed on sequencing and determining the gene content of three bacterial genomes: *Mycoplasma genitalium* (589 kb, and one of the smallest genomes of any free-living organism), *Haemophilus influenzae* (1,830 kb), and *Methanococcus jannaschii* (1,660 kb). A novel approach was used— random, or "shotgun," sequencing—and the sequences were assembled into **contigs** with the aid of sophisticated computing and positioning on the physical map. Sequencing with oligonucleotide primers made to contig ends filled in sequence gaps. Interestingly, 56% of *M. jannaschii* genes (a total of 1,738 genes) are new to science. Also, most genes for energy production, metabolism, and cell division are similar to those in bacteria, while genes for replication, transcription, and translation are more like those in eukaryotes. The research on all three bacteria will provide information important to the study of the evolution of genes and genomes.

The yeast genome, comprising 16 chromosomes and 12,500 kb in size (12.5 Mb), has been completely sequenced (see *Science*, 1996, 272:491). The project began in 1989 and progressed rapidly because the genome is compact (it is mostly coding sequences), a large number of scientists participated, and sequencing was automated. Rearchers must next determine the functions of the many genes that seem not to have homologs in other organisms.

Construction of the physical map of the nematode *Caenorhabditis elegans* genome began in 1984. The genome comprises one sex chromosome and five **autosomal chromosomes**, 100 Mb or 100 million base pairs per haploid genome. By mid-1995, both yeast artificial chromosome (YAC) and cosmid libraries have yielded a complete X chromosome map (with the exception of telomeric regions) and a map of one autosomal chromosome with some small gaps. Approximately 17,500 cosmids and 3,000 YACs have been mapped, encompassing most of the estimated 13,100 genes and more than 95% of the total genome. Only small gaps remain. One-fifth, or 21.3 Mb, of the genome (primarily from chromosome II, III and X) has been sequenced. Updated information is available from the Genome Sequencing Center, St. Louis, Missouri (http://genome.wustl.edu/gsc/gschmpg.html), the *Caenorhabditis* Genetics Center, University of Minnesota, Minnaepolis (gopher://elegans.cbs.umn.edu:70), and the Sanger Centre in Cambridge, England (http://www.sanger.ac.uk).

Genomes of both plants and animals are being intensively mapped and sequenced. By mid-1994, scientists at the Whitehead Institute of the Massachusetts Institute of Technology had obtained more than 4,000 markers for the mouse genome (one goal is to obtain more than 6,000 markers). Both marker and sequence data for the mouse genome will provide invaluable information about the human genome and disease.

The small genome (100 Mb) of the plant *Arabidopsis*, with five chromosomes, is amenable to both mapping and sequencing. Researchers have recently focused on developing a detailed physical map of chromosomes 4 and 5 (http://nasc.nott.ac.uk/JIC-contigs/JIC-contigs.html).

THE HUMAN GENOME

The human genome harbors an estimated 100,000 genes: 23 pairs of chromosomes, one set from each parent, made up of one sex chromosome pair (XX or XY) and 22 autosomal chromosome pairs. The three billion base pairs in the haploid genome contain an amount of information equivalent to 200 telephone books of 1,000 pages each (see Figure 9.1). The smallest chromosome (Y) is about 50 million base pairs, the largest (#1) is 250 million. If all the chromosomes in one human cell were removed, unwound, and placed end to end, the DNA would stretch more than five feet (although it would be only 50 trillionths of an inch wide).

Chromosomes are distinguished from one another by banding patterns and size differences (see Figure 9.2). Chromosomes can be detected with a light microscope

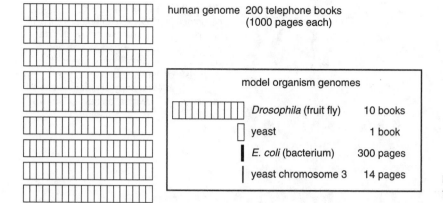

human genome 200 telephone books
(1000 pages each)

model organism genomes

	Drosophila (fruit fly)	10 books
	yeast	1 book
	E. coli (bacterium)	300 pages
	yeast chromosome 3	14 pages

Figure 9.1 Compiling the DNA sequence from the human genome into books would require 200 volumes, each the size of the 1,000 page Manhattan telephone book.

and special stains, and are condensed only during mitosis, a specific stage in the cell cycle. A pattern of alternating light and dark bands is observed in **metaphase chromosomes** that reflects the differences in the ratio of the bases adenine and thymine to guanine and cytosine. A picture of all the chromosomes grouped by size and type (1 through 22 with X and/or Y) is called a **karyotype** (an example is in Figure 9.3). Karyotyping detects some chromosomal abnormalities, such as broken and missing chromosomes or extra copies (for example, Down's Syndrome, with three #21 chromosomes, as the figure shows), but does not detect most genetic diseases such as cystic fibrosis, Huntington's disease, or certain cancers. Although karyotyping detects large changes in the complement of chromosomes, such as chromosome breakage, duplications, and translocations, it does not detect

small mutations, such as base changes and small insertions or deletions.

GOALS OF THE HUMAN GENOME PROJECT

A primary goal of the Human Genome Project is to generate detailed maps of the human genome. These maps will aid in determining the location of genes within the human genome—more specifically, will assign genes to their chromosomes. Two types of maps are being generated: **Genetic linkage maps** determine the relative arrangement and approximate distances between genes and markers on the chromosomes; physical maps specify the physical location (in base pairs) and distance between genes or DNA fragments with unknown functions that are mapped to specific regions of the chromosomes (Figure 9.4).

Maps have different levels of resolution, ranging from low to high, and the degree of resolution that is appropriate depends on whether, for example, a large fragment of DNA is to be studied or a more detailed picture of a small DNA region is needed. A human genomic library consists of random DNA fragments (such libraries are described in Chapter 4), and is used to establish sets of ordered, overlapping cloned DNA fragments or contigs for each chromosome of the genome—in other words, high-resolution maps.

After mapping is complete, the DNA must be sequenced to determine the order of all the nucleotide bases of the chromosomes, and the genes in the DNA sequence must be identified. In all phases of the project, a major focus has been on developing instrumentation to increase the speed of data collection and analysis. New, automated technologies are significantly increasing the speed and accuracy of DNA sequencing while decreasing the cost. Software and database systems manage the data generated from mapping and sequencing projects, and database management systems store and aid in distributing genomic information.

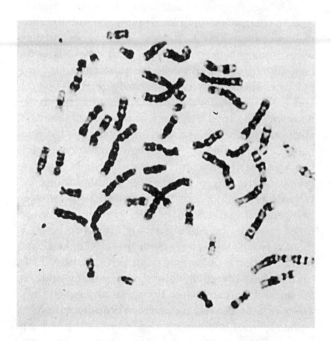

Figure 9.2 A microscopic preparation (metaphase squash before a karyotype is made) of human chromosomes shows the differences in size and banding patterns of the chromosomes.

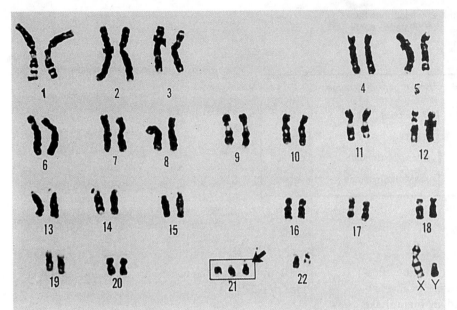

Figure 9.3 An individual's 46 chromosomes (22 pairs of autosomes and one pair of sex chromosomes, in this case, X and Y) can be arranged by size and banding pattern into a karyotype for the diagnosis of genetic disease. In this example, an extra copy of chromosome 21 identifies the individual as having Down's syndrome.

GENETIC LINKAGE MAPS

Genetic linkage mapping is analagous to assigning city locations and other landmarks along a highway on a road map. The more information—detail—that the map presents, the more useful it is. A map of the United States showing only Boston and San Francisco as end points on a coast-to-coast highway would be of little value. Adding other cities—Pittsburgh, Chicago,

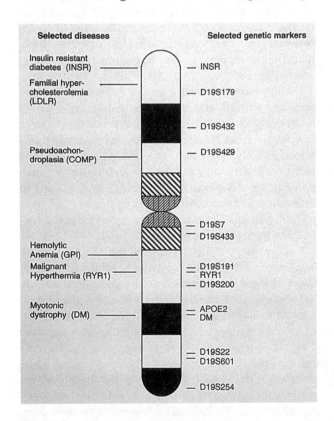

Figure 9.4 Diagram of human chromosome 19 with the locations of selected disease genes and genetic markers.

Denver, Salt Lake City—would increase its utility, as would adding national and state parks, landmarks, monuments of interest, and the intersections of other highways.

Genetic linkage maps show the order and genetic distance between pairs of **linked** genes—that is, genes on the same chromosome—that determine variable **phenotypic traits**. (The difference between genetic distance and physical distance is explained below.) Genetic linkage maps enable geneticists to follow the inheritance of specific traits (that is, genes) as they are passed from generation to generation within families.

Linkage maps also determine the arrangement of genes or markers with unknown functions on the chromosomes; they show the order of linked genes and pairwise distances between their loci. During **meiosis**, as the haploid egg and sperm cells form, homologous chromosomes (maternally and paternally derived) line up, and DNA segments can be exchanged between the homologs. Figure 9.5 shows that new combinations of alleles result from this process of homologous recombination. During meiosis, each human chromosome pair is involved, on average, in 1.5 crossover events. The likelihood of **crossing over** increases as the distance between the two loci increases. Crossing over between two genes or markers on the same chromosome sometimes can occur if there is enough distance between them. If two genes are very close, they are "linked," and recombination is unlikely to occur between them. Thus, the frequency of recombination is a quantitative index of the linear distance between two genes on a genetic linkage map. Distances are measured in **centimorgans** (cM), named for the famous geneticist Thomas Hunt Morgan. If genes (for example, A and B) are separated by recombination 1% of the time—that is, if one out of 100 products of meiosis is recombinant—they are 1 cM apart. A genetic distance of 1 cM represents a physical distance

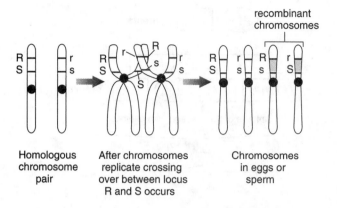

Figure 9.5 During meiosis (the formation of eggs and sperm), homologous chromosomes sometimes cross over and exchange DNA. Crossing over produces genetic diversity, since new combinations of alleles are established within the gametes.

of approximately one million base pairs (1 Mb). Genetic maps are very powerful: An inherited disease gene can be located on the map if a second gene or DNA reference marker is also inherited in individuals with the disease but is not found in individuals who do not have that disease. Exact chromosomal locations have already been found for many disease genes, including fragile X syndrome, cystic fibrosis, and Huntington's disease. Figure 9.6 shows the locations of disease numerous genes and markers on chromosome 16.

Many inherited diseases are caused by single genes, and thus can be studied by genetic linkage analysis. Almost 5,000 genetic disorders have been studied in this way. These maps, however, do not relate directly to the physical structure of DNA, and the gene of interest cannot be isolated on the basis of information from genetic linkage maps alone.

For human genome mapping, linkage analysis involves the study of family members carrying a particular trait for an inherited disorder. Often, several generations of one family are studied to obtain enough information with which to infer linkage. Some family members must express the trait (gene) or genetic disorder, and the trait must vary among individuals (that is, there must be different alleles or forms of the gene). Analysis also requires that there be individuals who are **heterozygous** for DNA reference markers or who have a second gene linked to the gene in question. Heterozygous family members (members carrying two different forms of the trait or gene—one dominant and one recessive allele) enable geneticists to determine which chromosome of the homologous pair carries the allele for the genetic disorder and whether it is passed on to the offspring.

POLYMORPHIC DNA MARKERS

Specific physical regions of the DNA can be identified by noting actual variations in DNA sequence within a specific region for different individuals. These sequence

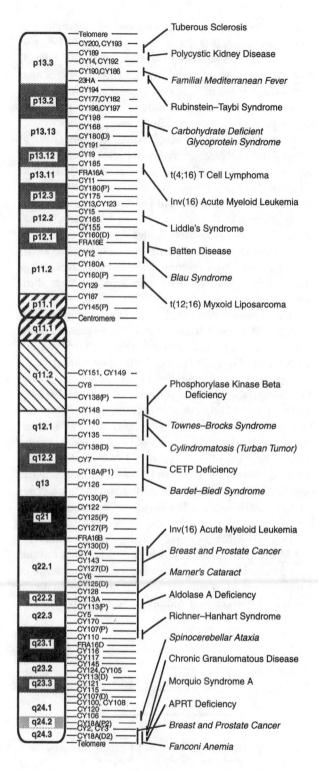

Figure 9.6 Location of disease genes on chromosome 16. Italicized genes have not yet been cloned.

variations are called DNA polymorphisms. DNA polymorphisms that occur within a coding region or gene lead to changes in phenotypic or observable characteristics (required for genetic linkage mapping), such as eye and hair color or in disease susceptibility. Variations that occur in an intron or other noncoding region lead to little or no observable change in physical traits. Because these polymorphisms are detectable at the DNA level, they can serve as markers. Polymorphic markers also can be short, tandem, repeated sequences varying in the number of repeat units (that is, length). This type of linkage mapping is revolutionizing human genetics.

Restriction fragment length polymorphisms (RFLPs) reflect actual DNA sequence variations, which can be detected with restriction endonucleases (described in Chapter 4). Restriction enzymes recognize specific nucleotide sequences and cut DNA at those sites. Fragments differ in length and number according to the location and number of restriction sites. A single base change within the recognition sequence can result in the gain or loss of a restriction endonuclease cut site. RFLPs are detected by gel electrophoresis and Southern blot hybridization.

Differences among individual family members can be detected when variations in DNA sequences exist. These sequence polymorphisms are passed from parent to offspring (Figure 9.7). These reference marker "alleles" are useful in studying the human genome. Because they are variable, they can be used just like variable phenotypic traits. Co-inheritance of DNA marker pairs can be used to determine the genetic distances between them, and thus a linkage map of DNA markers can be constructed. Co-inheritance of a DNA marker and a variable phenotypic trait can be used to determine the genetic distance between the marker and the gene. Polymorphic markers provide a way to relate loci on genetic linkage maps with the physical regions of human chromosomes (see Integration of Genetic Linkage and Physical Maps later in the chapter). In addition, a mutant gene can be located by examining the co-inheritance of a reference marker and a disease in families—that is, a linkage of marker and mutant gene is established, as

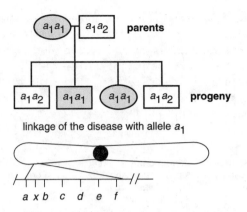

Figure 9.8 Demonstration of the co-inheritance or linkage of a marker (*a*) and a disease-causing allele (*x*) is a powerful way to determine the location of a disease gene. Shading identifies the individual with the disease *x* allele. When *a* and *x* are co-inherited, they are located so close together that a recombination event between them is rare. This linkage indicates that they are physically located close to one another. Southern blot hybridization with a probe for *a* is used to identify individuals (parents and progeny) with and without the disease and the *a* alleles they carry. Hybridization with a DNA probe for *b* also is necessary to establish linkage of *b* and the disease-causing allele *x*. A linkage map confirms the location of the disease-causing allele *x*.

Figure 9.8 shows. The physical location of the DNA marker on a chromosome can then be found by using the marker sequence as a DNA probe (as described in the next section, in the discussion of *in situ* hybridization). **Polymorphic DNA markers** serve as reference points or landmarks to help find a region of DNA that contains the gene of interest. If a gene is found between two DNA markers, that DNA region can be isolated for further study.

An early goal of Human Genome Project investigators was to generate linkage maps with polymorphic DNA markers spaced 2 to 5 cM along each chromosome; this goal was reached in 1995. Such a map helps scientists find genes of interest relative to about 1,500 markers within the genome. Once linkage maps have some 3,300 polymorphic DNA markers each separated by only 1 cM, gene hunting will be much easier. Thus, for polymorphic DNA markers to be valuable, their linkage with a gene must be established, and their physical locations must be identified through the use of probes. Several large scientific groups working on the human genome are identifying markers to generate comprehensive genetic linkage maps.

PHYSICAL MAPS

Physical maps provide information about the physical organization of the DNA; examples are the location of restriction enzyme sites and the order of chromosomal

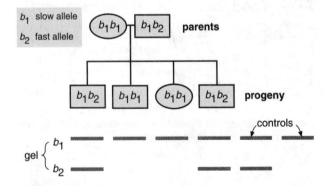

Figure 9.7 Schematic of Southern blot hybridization of DNA from parents and progeny cut with *Eco*RI and probed with DNA from locus *b*. b_1 is the "slow" allele, b_2 the "fast" allele. The inheritance of the restriction fragment length polymorphism (RFLP) at locus *b* in individuals can be demonstrated by the bandng pattern of the autoradiogram.

restriction fragments. Entire genomes can be studied with a library of genomic DNA. These clones are uncharacterized, random fragments, and are not placed in order as they would be on the chromosome.

Since the human genome is so large, very large DNA fragments must be cloned into vectors to maintain a manageable number of clones in the library. Yeast artificial chromosomes (YACs) are being used as cloning vectors for the human genome, since a DNA can be up to one million base pairs in length. Human DNA is attached to the yeast DNA and transferred into yeast host cells for replication. Only a small portion of the yeast's total DNA—origin of replication, telomere, and centromere—is required for replication, so most of the YAC is the foreign DNA insert.

The average insert used in YAC libraries is 200,000 to 400,000 base pairs in length. This range is 10 times larger than inserts used in other libraries, such as for bacteriophage and cosmids, where up to 20,000 to 40,000 base pairs, respectively, can be cloned. The human genome can be represented by 7,500 YAC clones and is maintained and amplified in yeast host cells. YACs and their inserts are cut into smaller fragments and recloned or subcloned (for example, into cosmids) so that a detailed map of a YAC clone is obtained (described below in the discussion of contig mapping).

YAC clones are screened by PCR to isolate specific genes of interest. DNA inserts also are analyzed by obtaining restriction maps, identifying polymorphic markers, and/or DNA sequencing. However, without an ordered physical map—one that refers to actual physical distances in base pairs between landmarks—the location of particular clones cannot be identified.

Fluorescence *in situ* hybridization (FISH) of probes to metaphase chromosomes provides information for constructing low-resolution chromosomal maps. Chromosomal maps are actual physical maps because distances are measured in base pairs. Metaphase chromosomes are spread out on a microscope slide, and a solution containing a fluorescent-tagged DNA probe is added. Under the appropriate conditions, the probe hybridizes to its DNA complement on the chromosome and is detected with a fluorescence microscope (Figure 9.9). The relative orientation of genes and DNA fragments can be assigned to specific chromosomes, and the gaps between mapped cosmids can be bridged. Chromosomal mapping is used to locate genetic markers that are associated with observable traits.

Another type of physical map is the cDNA map, which localizes coding regions (exons, for example) to specific chromosome regions or bands. The cDNA molecules are synthesized from an mRNA template. The cDNA map is probably one of the most important types of map, since it can identify the chromosomal location of specific genes, whether their functions are known or not. Researchers searching for a specific disease-causing gene can use cDNA maps to help locate it after

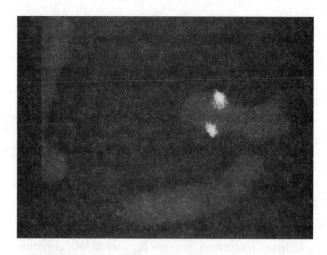

Figure 9.9 Fluoresence *in situ* hybridization (FISH) of a cosmid probe to chromosome 16. The light spots show where the probe hybridized to the chromosome. FISH helps assign DNA to specific chromosomes and fill in gaps between mapped cosmids. Metaphase chromosomes comprise two sister chromatids; therefore, a probe binds in two locations on each chromosome.

having established a general location by genetic linkage methods.

High-resolution physical maps can be generated by a method that is sometimes called bottom-up mapping. The chromosome is cut into small overlapping fragments, each is cloned, and the order is determined. These fragments form continuous DNA blocks called contigs. The bottom-up method generates a detailed map called a contig map. A library of clones ranging from 10,000 base pairs to 1 Mb is used for mapping. Each clone can be localized to specific regions within chromosomal bands. This "linked" library of overlapping clones comprises a chromosomal segment.

Human contig maps require several steps to produce. First, a library must be made that represents the human genome—either the entire genome or a segment—in cloned DNA fragments. The DNA fragments within each clone must overlap other fragments. Overlap is accomplished by cutting the DNA with a specific restriction enzyme. If every restriction site on the DNA were cut, no fragments would overlap. Therefore, enzyme digestion is conducted in such a way that not every DNA restriction site is cut. This partial digestion randomly leaves many sites uncut so that overlapping DNA fragments are produced and the order along the chromosomes can be determined.

The order of the clones or contigs can be determined by identifying the overlaps in the DNA fragments. Overlap can be detected when some of the DNA bands are the same—that is, two clones have bands in common. This method of assembling pairs of clones into contigs is difficult and time-consuming. Automation and sophisticated computer algorithms may increase efficiency.

Different approaches may be used to fill in the gaps that are likely to be present even after researchers generate detailed physical maps. For example, as Figure 9.10 shows, microdissection is used to physically cut a piece of DNA from a specific region of a chromosome. This chromosomal piece can be cut into smaller fragments by restriction enzymes, then cloned, mapped, and sequenced by standard methods.

Alternatively, chromosome walking is sometimes used. A small region at the end of the DNA fragment is used as a probe to screen the library for the adjacent clone. A DNA piece at the end of this second cloned fragment is used as the next probe. This process continues until a complete physical map has been obtained.

Since the human genome is divided into chromosomes, chromosome-specific libraries can be constructed so that each chromosome has a contig map. Mapping is simplified if each chromosome is separated from the others before being cut by restriction enzymes and cloned to make libraries. Twenty four libraries are required: 22 autosomal libraries and one each for the X and Y chromosomes.

Chromosomes can be separated by size using a process called **flow cytometry**. Cell sorting machines called flow cytometers (shown schematically in Figure 9.11) can analyze up to 2,000 chromosomes a second and sort 50 chromosomes a second. The separation method uses the fluorescence of chromosomes that have been stained with two dyes, each one preferentially binding to DNA that is rich in either adenine and thymine (AT) or guanine and cytosine (GC). Two laser beams excite the fluorescence of each dye bound to the chromosomes. Since the AT and GC content of chromosome regions differ, each chromosome will emit different light intensities for each of the two dyes. A computer analyzes the fluorescence characteristics for each chromosome and uses this information to sort the chromosomes. Some chromosomes, such as numbers 9 through 12, cannot be separated because they give very similar fluorescence patterns. A specific chromosome can be separated from the rest by placing an electric charge on it and using charged deflection plates to channel it into a collection tube.

To summarize, the several types of maps range from coarse to fine resolution. The map with the lowest resolution is the genetic map, which measures the frequency of recombination between linked markers (which can be genes or noncoding DNA). The next level of resolution is the restriction map, on which DNA restriction fragments ranging from 1 to 2 Mb are separated and mapped. The next higher level of resolution is achieved by placing in order 400,000 to 1,000,000 base pair fragments of overlapping clones from libraries of YAC clones. These clones then are further subcloned (with insert sizes of 20,000 to 40,000 base pairs) into other vectors to produce contig maps. Finally, the DNA base sequence map, having the finest resolution, is determined.

Sequence-Tagged Sites

A coordinated effort like the Human Genome Project requires that collected genome information be shared. A major problem is that investigators from different laboratories use a variety of methods for generating and mapping DNA fragments, thus making correlations difficult when data from different laboratories must be compared. A universal reference system has been developed. Unique regions of 200 to 500 base pairs of partially sequenced DNA are used to identify clones, contigs, and long stretches of DNA. These **sequence-tagged sites** (STSs) are standard markers that are used for physical mapping. An STS also can be a region of cDNA—that is, an expressed

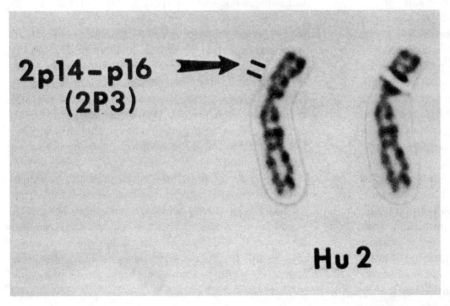

Figure 9.10 Microdissection of a DNA fragment (2p14-p16) from the short arm of human chromosome 2. A library of the DNA is constructed by cloning into a suitable vector. Clones from the library can be used to isolate larger regions of DNA from cosmid and YAC clones for the establishment of contigs and high-resolution physical maps. Left is before dissection, right is after dissection.

FLOW SORTING

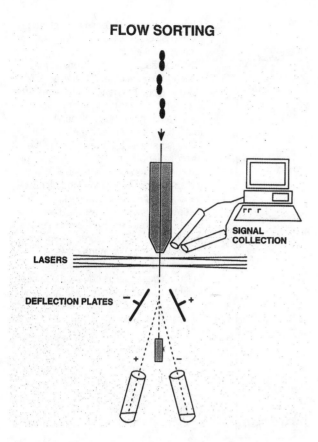

Figure 9.11 A simple diagram of a dual laser flow cytometer and sorter that rapidly sorts chromosomes. Chromosomes fluoresce after staining with AT- and GC-specific dyes (each emitting light at a different wavelength). Laser beams excite the fluorescence of the dyes bound to the chromosomes so that each chromosome type emits different light intensities at each wavelength (which depends on the GC:AT content). This enables discrete chromosomes to be identified (with the exception of chromosomes 9 through 12 because of their similar fluorescence patterns). A particular chromosome can be separated by giving it an electric charge so that charged deflection plates deflect it into a collection tube.

sequence, called an expressed-sequence tag (EST). ESTs are used to represent landmarks along the map, thus helping to identify where pairs of clones overlap. These special sequences constitute a "universal mapping language," enabling everyone to refer to a specific region of the genome by the same name, and enabling investigators to share information and construct compatible maps. A goal is to generate STSs every 100,000 base pairs for each human chromosome. Each chromosome would comprise contigs that are one to two million base pairs in length and cover over 95% of the chromosome. Thus, approximately 30,000 STSs must be identified.

To identify an STS, a short DNA fragment is isolated from a chromosome-specific library of clones or from a small fragment within a clone from a contig. A small region 200 to 400 base pairs in length is sequenced and compared to all known repeated sequences. Sophisticated computer analysis programs help identify unique

sequences. PCR can be used to screen a library for clones containing STSs. PCR amplifies the unique DNA region from the total human genome. Gel electrophoresis will yield one DNA band if the STS is unique and more bands on the gel if additional regions are amplified, indicating that the STS is not unique.

An STS also can include a repeated sequence (examples: unique sequenceGAGAGAGA.......GAunique sequence or unique sequence (GA_n) unique sequence) if unique sequences flank the repeat on either side. STSs then would be polymorphic in the region of the repeat (that is, repeated n times) and have many alleles in a population. Each individual carries two copies of the STS marker—one on each homologous chromosome. Thus, the inheritance of the STS can be determined. Once located on the physical map, polymorphic STSs aid in aligning the genetic linkage map with the physical map.

The use of STSs as probes for contig mapping allows clone overlap to be established; this method is called STS-content mapping. Clones that share an STS overlap belong in the same contig. To be most useful, STSs should be spaced every few thousand base pairs. STS information is stored in computer databases analogous to, for example, the large DNA sequence database called Genbank. As progress is made, information will include PCR primer sequences, PCR reaction conditions, and the size of the fragment amplified. The amplified PCR fragment then can be used as a probe in hybridizations to isolate clones containing the STS marker.

INTEGRATING GENETIC LINKAGE AND PHYSICAL MAPS

The use of polymorphic DNA markers for genetic linkage analysis is a powerful method for finding the approximate location of genes causing inherited diseases. If the polymorphic DNA markers are co-inherited or linked closely to the gene of interest, the markers and gene must be physically close to one another. To be amenable to further analysis, genetic linkage markers must be within two million base pairs (preferably no more than 1 cM) from the disease gene. To isolate the disease gene, a method called chromosome walking is used (Figure 9.12). DNA markers are used to link the genetic linkage map to the physical map. Flanking DNA markers are used as a starting point in hybridizations to clones comprising contigs in the disease gene region. Once the two types of maps are integrated, genes for inherited disorders can be identified, isolated, and sequenced.

Human Genome Progress

A high-resolution map has one reference point per 100 kilobases (100,000 bases). The next stage is to have overlapping fragments 10 kilobases (10,000 bases) in length

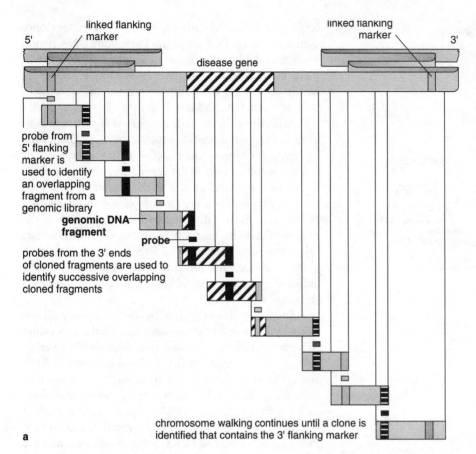

Figure 9.12 (a) Disease genes can be cloned by chromosome walking. After a marker is linked to within 1 cM of a disease-causing gene, chromosome walking from the marker to the gene is used to clone the gene. A probe from the 5′ flanking marker region identifies the next overlapping library clone. A restriction fragment from the end of that clone is used as a probe to isolate the next overlapping clone. This process of "walking" is repeated until the 3′ flanking marker on the other side of the disease-gene locus has been reached. The disease gene will be contained within one or more clones. (b) A section of the contig map of chromosome 19 spanning the region defective in the disease myotonic dystrophy. A variable-length polymorphism (VLP) sequence has been identified by sequencing DNA from one of the cosmid clones within the myotonin kinase gene region.

linked flanking marker

5′

linked flanking marker

3′

disease gene

probe from 5′ flanking marker is used to identify an overlapping fragment from a genomic library

genomic DNA fragment

probe

probes from the 3′ ends of cloned fragments are used to identify successive overlapping cloned fragments

chromosome walking continues until a clone is identified that contains the 3′ flanking marker

a

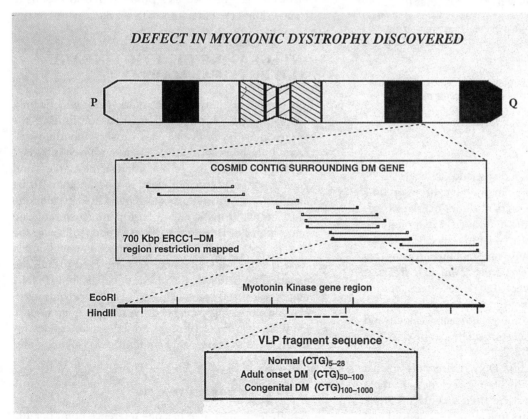

DEFECT IN MYOTONIC DYSTROPHY DISCOVERED

P

Q

COSMID CONTIG SURROUNDING DM GENE

700 Kbp ERCC1–DM region restriction mapped

Myotonin Kinase gene region

EcoRI

HindIII

VLP fragment sequence

Normal (CTG)$_{5-28}$
Adult onset DM (CTG)$_{50-100}$
Congenital DM (CTG)$_{100-1000}$

b

between reference points. This ratio has been achieved for chromosomes 16 and 19; these two most likely will be the first chromosomes to be sequenced in their entirety. Figure 9.13 shows the different levels of map resolution for chromosome 16 to illustrate the levels of detail obtained in mapping studies. Detailed integrated maps of chromosomes 3, 11, 12, 13, 16, 19, 21, and 22 also have been obtained.

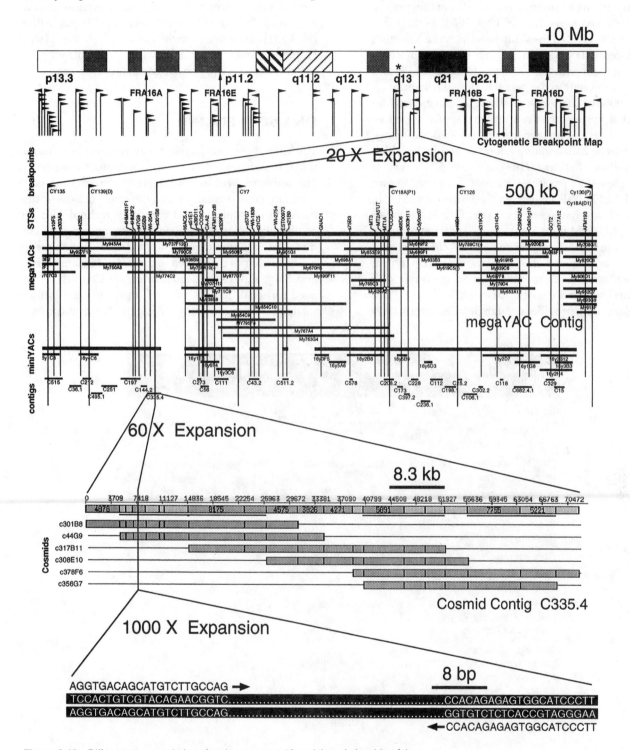

Figure 9.13 Different map resolutions for chromosome 16 and the relationship of the cytogenetic map to YAC contigs, cosmid clones, and sequence-tagged sites (STSs). STS markers can aid in integrating the genetic map (top map) and physical maps. The location of an STS (s301B8) is shown from the bottom up in an ordered set of cosmid clones (contig C335.4), both mini and mega YAC clones, and on a cytogenetic map within q13 (top).

Several major laboratories have constructed detailed genetic and physical maps of the human genome. Investigators from the Centre d'Étude du Polymorphisme Humain (CEPH) have obtained both physical and linkage maps covering 75% of the genome with 33,000 YACs (see *Nature*, 1993, 366:698–701; *Gene*, 1993, 135:275–278). At the Whitehead Institute, scientists have placed 10,000 STSs in 50 groups per chromosome (matching them with the YACs produced by CEPH and ordering them within the genome) to make up approximately 90% of the genome. These scientists used an automated system that can screen rapidly for STSs. Généthon's genetic linkage map includes 5,300 markers.

In a large collaborative effort the results of which were published in *Science* (1994, 265:2049–2054), several laboratories—Cooperative Human Linkage Center (CHLC), CEPH, Yale University, the University of Utah, and Généthon—have generated a detailed linkage map covering 4,000 cM with 5,840 loci, 970 of them ordered. Markers are located approximately 0.7 cM apart and consist of both genes (427) and short tandem repeat polymorphisms (3,617, providing a marker every 1×10^6 base pairs), which are useful for amplification by PCR. This linkage map provides enough detailed information for integration with physical maps. Current genome mapping data for humans and mouse genomes can be obtained through the Internet on World Wide Web, FTP, and e-mail (Table 9.1 lists several web sites). The number of genes assigned to human chromosomes as approved by the Human Genome Organization (HUGO) is available from the Genome Data Base, Johns Hopkins University School of Medicine (Table 9.2 gives the number of genes per assigned chromosome).

DNA SEQUENCING

The human genome physical map with the best resolution is the complete DNA sequence of all chromosomes. The sequence will be invaluable in determining gene function, gene regulation, and the cause of genetic diseases. More sophisticated computer technologies are continually being developed for organizing and handling the vast amount of generated data. Two sequencing technologies developed in the 1970s, Maxam-Gilbert and Sanger dideoxy sequencing (see Chapter 4), are used for small regions of the human genome but are inadequate for large-scale projects. New, automated

Table 9.1 Useful World Wide Web Sites for Genome Maps and Related Genetic Information	
Center	WWW Site
NIH Human Genome Program (general information)	http://www.nhgr.nih.gov/HGP/
DOE Human Genome Program (general information)	http://www.er.doe.gov/production/ober/hug_top.html
Oak Ridge National Laboratory Human Genome Management Information System (HGMIS) (general information	http://www.ornl.gov/TechResources/Human_Genome/home.html
Genome Data Base (GDB) (human map, loci data, general information)	http://www.gdb.org/
Généthon (maps of linked CA repeat markers for humans)	http://www.genethon.fr/
Cooperative Human Linkage Center (CHLC) (genotypes, marker data, linkage)	http://www.chlc.org/
Jackson Laboratory (data on mouse mapping)	http://www.jax.org/
Centre d'Étude du Polymorphisme Humain (CEPH) (YAC physical mapping data)	http://www.cephb.fr/bio/ceph_genethon_map.html/
Whitehead Institute (maps of linked CA repeats for mouse)	http://www.genome.wi.mit.edu/

Partially adapted from J. C. Murray et al., *Science* 265:2052 (1994).

Table 9.2 The Number of Genes Assigned to Human Chromosomes (June 15, 1997)*	
Chromosome	Gene Number
1	555
2	301
3	255
4	215
5	233
6	287
7	261
8	174
9	206
10	166
11	339
12	299
13	87
14	150
15	143
16	190
17	333
18	70
19	347
20	107
21	80
22	129
X	374
Y	25
Total	5,326

*Includes mapped and characterized genes (whose functions have been defined) stored in the Genome Database (GDB). Genes included in the GDB have been approved by the nomenclature and chromosome committees of the Human Genome Organization. Total genes unassigned to chromosomes: 1,555. The GDB is maintained by Johns Hopkins University School of Medicine, Baltimore, Maryland. The latest information on gene assignments to chromosomes can be obtained from the GDB web site: http://www.gdb.org/ (under Reports).

The Human Genome Organization (HUGO), Los Alamos National Laboratory, Los Alamos, New Mexico.

technologies are increasing the speed of sequencing and analysis, and advances are being made in automation of DNA purification and separation.

A major goal of automated sequencing has been to exceed 100,000 bases a day at less than 50 cents per base. Many rapid sequencing methodologies and sensitive, accurate detection methods are currently being explored (Figure 9.14 shows a schematic of a rapid sequencer). Methods that increase the rate of fragment separation require the use of high voltage capillary and ultrathin electrophoresis. Eventually gel-less sequencing methods will be perfected that will increase speed and accuracy several orders of magnitude.

ETHICAL, LEGAL, AND SOCIAL IMPLICATIONS

The Human Genome Project is providing valuable genetic information that will increase our understanding of human disease. However, social, legal, and ethical implications also must be addressed. Members of the scientific community have been in disagreement as to whether the benefits of the Human Genome Project outweigh the costs. Some researchers question the scientific merits of this monumental project, especially in an era of diminishing funding for basic scientific research. Some scientists argue that dwindling federal funds would be better spent on specific research projects, such as finding cures for heart disease, and cancer.

Other issues pertain to protection of personal information—in this case, genetic history. People have been denied insurance coverage because a family member has a genetically inherited disorder. Early in this century, compulsory sterilization programs were implemented in some states for people viewed as "not normal."

The NIH and DOE have established a working group for studying the Human Genome Project's ethical, legal, and social implications (which have been given the acronym ELSI). The DOE and the NIH have contributed 3 and 5% of their annual budget, respectively, to this initiative. The overall goal of the ELSI working group is to identify the social impacts of the project's discoveries and aid in genetics-related policy making. Here we examine only a few of the questions that must be addressed before completion of the Human Genome Project.

The Human Genome Project makes available genetic information that until only recently had not existed. Although much of this information will have a positive impact on society, other uses may increase discrimination, restrict reproduction, affect employment and insurance coverage decisions, and cause personal unhappiness to those who receive genetic information for which they are unprepared. Legislation may be necessary to protect individuals with known genetic predispositions to disease and with late-onset genetic disorders such as Huntington's disease. ELSI created an insurance task force to identify uses of genetic information by insurance companies and ways to prevent discrimination. The task force includes members of ELSI, the insurance industry, academia, and nonprofit health organizations.

New laws may be required to protect the confidentiality of genetic information. In 1990, the Americans with Disabilities Act (ADA), which prevents some kinds of genetic discrimination, was enacted. By July 1994, all companies with 15 or more employees had to comply with this federal law. Some issues remain unresolved.

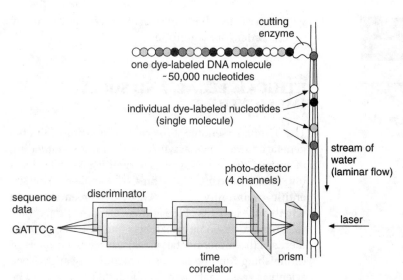

one dye-labeled DNA molecule
~ 50,000 nucleotides

individual dye-labeled nucleotides
(single molecule)

cutting
enzyme

photo-detector
(4 channels)

stream of
water
(laminar flow)

sequence
data

GATTCG

discriminator

laser

time
correlator

prism

Figure 9.14 Rapid DNA sequencing based on single molecule detection. In this method, each of the four bases is labeled with a different fluorescent tag. A DNA rapid sequencer detects single chromophores by laser-induced fluorescence in samples moving in a stream of water. After the bases in a fragment are fluorescently labeled, the fragment is bound to a support and moved through the stream of water. Individual bases are detected by laser-induced fluorescence as they are removed from the DNA molecule by an enzyme. As a molecule passes through the laser beam, a photon is emitted each time the molecule is cycled from the ground electronic state to an excited state. The emission spectra of the four fluorescent tags allow the four bases to be identified. The DNA rapid sequencer was developed by Los Alamos National Laboratory, Los Alamos, New Mexico.

For example, although people with a genetic disorder are protected, what about the family of a carrier of a genetic disorder, or the person who has discovered that he or she will get a debilitating or even fatal disease in the future?

Insurance companies worry that clients may withhold information about a genetic problem and obtain additional coverage, while others, after genetic testing, may find that their risks are low and purchase minimal insurance coverage. If insurance companies can obtain an individual's genetic information (in the form of test results) at the time of application, the applicant may avoid being tested so that risks are not disclosed before approval.

Often a person's medical information is released to employers and insurers, and sometimes to the Medical Information Bureau (an information resource for the insurance industry). Results from genetic testing could be disseminated to others without the person's knowledge or consent. Disclosure within families is also an issue. If a person tests positive for a late-onset genetic disorder such as Huntington's disease, perhaps other family members do not want to know whether they carry the gene. However, information about a person's genetic makeup can sometimes be inferred from that of close relatives. For example, if a person discovers that he or she has the gene for Huntington's disease, a parent must also have the gene. If the at-risk parent does not want to know, the child's test results provide unwanted information.

The ELSI working group is studying privacy issues from several perspectives. Guidelines for genetic databanks must be drafted to protect the right to privacy. State forensic laboratories store blood or other body fluids of convicted felons from which to obtain genetic "fingerprints" for later use. The Defense Department maintains a bank of tissue samples from members of the armed forces for use in identifying those killed in the line of duty. Here again, genetic information could be obtained without the person's knowledge or consent. No laws are in place to protect the individual's genetic privacy. How will genetic privacy be safeguarded?

How will everyone benefit from the knowledge generated by the Human Genome Project? That vast amount of information must somehow be assimilated and disseminated so that it can be applied. This knowledge will most likely influence how health-care workers, physicians, and genetic counselors are trained in the future. Future genetic breakthroughs will pose questions that must be answered, such as whom to screen, how to protect privacy, how to counsel clients, and how to train health-care personnel. Should genetic screening be conducted when there are no treatments for the genetic disorders discovered? Genetic analysis, counseling, and treatments will become much more complex as we learn more about the human genome. Perhaps current training programs for genetic counselors, physicians, and social scientists will no longer be adequate. We must also anticipate how new genetic information will affect the medical field. For example, once genes for genetic diseases are identified, we will have to determine which diseases receive priority funding for developing treatments.

Many patentable inventions will probably be an outcome of the Human Genome Project—from new technologies for sequencing and analyzing DNA, to new genetic screening test kits. Some investigators have aroused controversy by seeking patents for fragments of human DNA. In 1988, the Office of Technology Assessment suggested that Congress and federal granting agencies should encourage early filing of patent applications so that genome data can be disseminated rapidly. On the other hand, the National Research Council Committee on Mapping and Sequencing the Human Genome stated that DNA from the human genome should be a "public trust" and not "intellectual property." Nevertheless, in 1992, the NIH filed patent applications on

2,375 cDNA fragments as well as genes of unknown function and their products. What will be the impact on society if regions of the human genome are patented? The public may pay more than anticipated for the Human Genome Project—paying first, through taxes, to support the federally funded research, and paying a second time for the use of products and services that result from the research.

The information from the Human Genome Project has the potential to influence our lives more than any previous medical breakthrough. A major goal of ELSI is to help the general public understand issues related to the Human Genome Project such as the right to privacy, the potential for discrimination, and ethical considerations. Education is perhaps the most important way to prevent discrimination and maintain confidentiality. Medical professionals must have access to the latest information to aid in diagnosing diseases and prescribing new treatments. Scientists should be trained in the areas of ethics, the legal system, and sociology to address public concerns. At the same time, social scientists, legislators, and lawyers will have to become familiar with the science to more effectively evaluate the impact of the Human Genome Project.

General Readings

N.G. Cooper, ed. 1994. *The Human Genome Project: Deciphering the Blueprint of Heredity.* University Science Books, Mill Valley, California.

D.J. Kevles and L. Hood, eds. 1992. *The Code of Codes: Scientific and Social Issues In the Human Genome Project.* Harvard University Press, Cambridge, Massachusetts.

U.S. Department of Energy. 1992. *Human Genome: 1991-1992 Program Report.* Washington, D.C.

Additional Readings

T. Beardsley. 1996. Trends in human genetics. Vital Data. *Sci. Am.* 274:100-105.

C.J. Bult, et al. 1996. Complete genome sequence of the methanogenic Archaeon, *Methanococcus jannaschii. Science* 273:1058–1073.

D.T. Burke. 1991. The role of yeast artificial chromosomes in generating genome maps. *Curr. Opin. Genet. Dev.* 1:69-74.

M.J. Cinkosky, J.W. Fickett, P. Gilna, and C. Burks. 1994. Electronic publishing in GenBank. In N. G. Cooper, ed., *The Human Genome Project: Deciphering the Blueprint of Heredity.* University Science Books, Mill Valley, California, pp 270-272.

D. Cohen, I. Chumakov, and J. Weissenbach. 1993. A first-generation physical map of the human genome. *Nature* 366:698-701.

D.R. Cox, E.D. Green, E.S. Lander, D. Cohen, and R.M. Myers. 1994. Assessing mapping progress in the Human Genome Project. *Science* 265:2031-2032.

L.L. Deaven. 1994. Libraries from flow-sorted chromosomes. In N. G. Cooper, ed., *The Human Genome Project: Deciphering the Blueprint of Heredity.* University Science Books, Mill Valley, California, pp 236-246.

K.M. Devine. 1995. The *Bacillus subtilis* genome project: Aims and progress. *Trends Biotechnol.* 13:210-216.

W.F. Dietrich et al. 1994. A genetic map of the mouse with 4,006 simple sequence length polymorphisms. *Nat. Genet.* 7:220-225.

N.A. Doggett. 1994. The polymerase chain reaction and sequence-tagged sites. In: N.G. Cooper, ed., *The Human Genome Project: Deciphering the Blueprint of Heredity.* University Science Books, Mill Valley, California, pp 270-272.

C. Fields, M.D. Adams, O. White, and J.C. Venter. 1994. How many genes in the human genome? *Nat. Genet.* 7:345-346.

R.D. Fleischmann et al. 1994. Whole-genome random sequencing and assembly of *Haemophilus influenzae* Rd. *Science* 269:496-512.

C.M. Fraser et al. 1995. The minimal gene complement of *Mycoplasma genitalium. Science* 270:397-403.

K.L. Garver and B. Garver. 1994. The human genome project and eugenic concerns. *Am. J. Human Genet.* 54:148-158.

The Genome Directory. 1995. *Nature* 377 (supplement 28). Contains genome articles and detailed maps.

M.S. Guyer and F.S. Collins. 1993. The human genome project and the future of medicine. *Am. J. Dis. Child.* 147:1145-1152.

M.S. Guyer and F.S. Collins. 1995. How is the Human Genome Project doing, and what have we learned so far? *Proc. Natl. Acad. Sci. USA* 92:10841-10848.

B.M. Hauge et al. 1991. Mapping the *Arabidopsis* genome. *Symp. Soc. Exper. Biol.* 45:45-56.

J. Hodgkin, R.H.A. Plasterk, and R.H. Waterston. 1995. The nematode *Caenorhabditis elegans* and its genome. *Science* 270:410-414.

K.L. Hudson, K.H. Rothenberg, L.B. Andrews, M.J.E. Kahn, and F.S. Collins. 1995. Genetic discrimination and health insurance: An urgent need for reform. *Science* 270:391-393.

I.E. Järvelä et al. 1995. Physical map of the region containing the gene for Batten disease (*CLN3*). *Am. J. Med. Genet.* 57:316-319.

I.E. Järvelä et al. 1995. YAC and cosmid clones spanning the Batten disease (*CLN3*) region at 16p12.1-p11.2. *Genomics* 29:478-489.

B.M. Knoppers and R. Chadwick. 1994. The Human Genome Project: Under an international ethical microscope. *Science* 265:2035-2036.

E.S. Lander and N.J. Schork. 1994. Genetic dissection of complex traits. *Science* 265:2037-2048. (Published erratum in 1994. 266:353)

M.H. Meisler. 1996. The role of the laboratory mouse in the human genome project. *Am. J. Hum. Genet.* 59:764-771.

A.P. Monaco and Z. Larin. 1994. YACs, BACs, PACs, and MACs: Artificial chromosomes as research tools. *Trends Biotechnol.* 12:280-286.

V. Morell. 1996. Life's last domain. *Science* 273:1043–1045.

J.C. Murray et al. 1994. A comprehensive human linkage with centimorgan density. *Science* 265:2049-2054.

D.L. Nelson. 1991. Applications of polymerase chain reaction methods in genome mapping. *Curr. Opin. Genet. Dev.* 1:62-68.

M.V. Olson. 1993. The human genome project. *Proc. Natl. Acad. Sci. USA* 90:4338-4344.

S.D. Pena. 1996. Third World participation in genome projects. *Trends Biotechnol.* 14:74-77.

P.R. Rosteck, Jr. 1994. The human genome project: Genetic and physical mapping. *Trends Endocrin. Met* 5:359-364.

R. Schmidt, J. West, K. Love, Z. Lenehan, C. Lister, H. Thompson, D. Bouchez, and C. Dean. 1995. Physical and organization of *Arabidopsis thaliana* chromosome 4. *Science* 270:480-483.

M. Stoneking. 1997. The human genome project and molecular anthropology. *Genome Res.* 7:87-91.

T. Strachan, M. Abitbol, D. Davidson, and J.S. Beckmann. 1997. A new dimension for the human genome project: towards comprehensive expression maps. *Nat. Genet.* 16:126-132.

S.M. Tilghman. 1996. Lessons learned, promises kept: a biologist's eye view of the Genome Project. *Genome Res.* 6:773-780.

J. Weissenbach. 1993. A second generation linkage map of the human genome based on highly informative microsatellite loci. *Gene* 135:275-278.

N. Williams. 1995. Closing in on the complete yeast genome sequence. *Science* 268:1560-1561.

N. Williams. 1996. Yeast genome sequence ferments new research. *Science* 272:481.

N.D. Young. 1994. Plant gene mapping. In *Encyclopedia of Agricultural Science,* Vol. 3, Academic Press, Inc., San Diego, pp 275-282.

MEDICAL BIOTECHNOLOGY **10**

A new generation of **therapeutics** are targeting both genetic and acquired diseases, such as cystic fibrosis, hemophilia, familial hypercholesterolemia, heart disease, and cancer. Engineered vaccines and antibodies, "designer" drugs such as synthetic drugs and DNA drugs, and new methods of drug delivery are beginning to be used. Revolutionary therapeutics are in clinical trials, and many more are on the way. U.S. sales of 16 approved biotechnology-derived therapeutics and vaccines totaled approximately $2 billion in 1990. A recent survey by the Pharmaceutical Manufacturers Association indicated that over 100 products are in various stages of development. Although these new therapeutics will not completely replace traditional treatments, they will significantly enlarge the weapons arsenal against disease. In this chapter we look at a few of the many recent medical developments.

GENE THERAPY

Rapid progress is being made in what only a few years ago would have been a science fiction story—the use of genes in the treatment of patients. New technologies and a more friendly regulatory environment are encouraging investigations into human **gene therapy**: the transfer of a normal gene into cells to correct a specific disorder. One day in the not-too-distant future, genes will be regularly prescribed as "drugs" in the treatment of diseases.

Gene therapy has important implications in the treatment of acquired and genetic diseases, cancer, and even AIDS. Although the ethical implications of gene therapy continue to be debated, several hundred gene transfer trials have been approved by regulatory agencies and are being conducted. Much effort and funds have been devoted to the development of efficient gene transfer methods, and first-generation products are in preclinical or clinical trials.

Currently only gene therapy of somatic cells is feasible. Germ line gene therapy is not being conducted but is being hotly debated: The idea of manipulating the germ line cell to pass the genetic correction or modification on to succeeding generations has significant ethical and safety implications.

The first landmark gene therapy was performed at the NIH by Drs. W. French Anderson, Michael Blaese, and Kenneth Culver. In September 1990 they treated a four-year-old girl, Ashanthi DeSilva, for an inherited immunodeficiency called severe combined immunodeficiency (SCID). SCID is caused by a defective ADA gene that leads to an accumulation of adenosine deaminase, a toxic metabolic by-product. Ashanthi was treated with modified T-lymphocytes containing the adenosine deaminase (ADA) gene. Four months later a second patient, 11-year old Cynthia Cutshall, was treated. Both patients received gene-treated **stem cells** to permanently correct the genetic disorder. Each received a total of 11 to 12 gene infusions. Three years after treatment, more than 50% of Ashanthi's circulating T cells harbored the new gene. However, she still receives injections of PEG (polyethylenene glycol)-ADA, a synthetic form of the enzyme. Only 0.1 to 1% of Cynthia's circulating T cells contained the ADA gene, demonstrating that effectiveness varies. The only way to conclusively demonstrate that the gene therapy is effective is to withdraw the PEG-ADA over time. This would be the first demonstration that gene therapy completely cured a human disease.

GENE DELIVERY METHODS

The first step in gene therapy is to transform specific cells with a gene of interest. Major research has focused on finding methods for efficiently transferring genes to such cells as muscle, lymphocytes, hepatocytes, hematopoietic stem cells, and fibroblasts. Two developments would dramatically increase the efficacy of gene-based therapeutics: (1) efficient methods of introducing genes into specific cells, and (2) methods that effectively treat a disorder without aggravating the immune system or inactivating vital genes in the genome.

Two strategies for gene introduction are in use. In *ex vivo* gene therapy, cells are removed from the body, the gene of interest is inserted into them, the cells are cultured to multiply to a sufficient number, and they are then returned to the body by infusion or transplantation. If the patient's own cells (autologous cells) are used, rejection does not occur. The biggest technical hurdle for *ex vivo* gene therapy is the transplantation of transfected cells.

An example of *ex vivo* therapy is a nonviral approach called transkaryotic therapy. An individual's cells are obtained through a skin biopsy and the gene of

interest is inserted into them. Transformed cells are cultured *in vitro* to increase the number of cells expressing the desired protein, and are then injected under the person's skin, where the therapeutic protein is synthesized. The protein is circulated throughout the body.

In *in vivo* gene therapy, the gene is introduced directly into specific cells within the body. Vectors usually are required to target the DNA to specific cells. *In vivo* therapy presents some technical difficulties. Sometimes, the transferred gene is unstable and the product is expressed only transiently, instead of for the life of the cell. In addition, since these cells remain in the body and are not isolated from other cells during transfection, these methods are not as controlled and specific as targeted *ex vivo* gene therapy.

Both *in vivo* and *ex vivo* approaches are being evaluated. *Ex vivo* procedures offer one benefit: Since the vector is introduced only into the target cells and not into the body or circulatory system, the individual's immune system is not in direct contact with the vector. However, the method, in its current state of development, is time-consuming and expensive. Injectable *in vivo* treatments are more practical and will probably become the method of choice in the future, although, because a specific population of cells must be targeted, bone marrow stem cells and other types of circulating progenitor cells are an exception.

VIRAL DELIVERY

Viral delivery is the method used in most gene therapy experiments. Much research has focused on the development of safe and efficient viral vectors (Table 10.1). The two viruses most commonly used are the retrovirus and adenovirus. Herpes virus vectors offer another possibility, but they are not as well characterized, and may have risks not yet identified.

Retroviral Vectors

Most gene therapies today transfer genes to human cells by retroviral vectors (RNA viruses that infect human cells). Their popularity is due to their ability to target genes into many types of human cell, to package a wide variety of genes and promoters, and to integrate into the host genome. Unfortunately, however, integration into the genome occurs randomly, the site of insertion cannot be predicted, and only actively dividing cells are targets. Moreover, there is the possibility that viral insertion could inactivate indispensable genes (such as tumor suppressors) or activate oncogenes, thereby mutating the genome.

Adenovirus Vectors

Adenovirus vectors show much potential for *in vivo* gene therapy, since nondividing cells are readily infected, they have a large capacity for foreign genes,

Table 10.1 Viral Vector Characteristics

Vector	Capacity	Advantages	Disadvantages
Retrovirus	8 kb	Well-studied Few side effects Recombination unlikely Expression for long periods	Small insert size Transfection of actively dividing cells only Risk of insertional mutagenesis
Adenovirus	8–10 kb	Transfection of replicating and nonreplicating cells High expression of inserted gene Ideal for lung tissue Can be used in aerosols Rarely integrates into host genome	Risk of inflammatory response Only transient gene expression Vector lost during cell division Risk of recombination in host Other proteins expressed
Herpesvirus	~30 kb	Transfection of replicating and nonreplicating cells Especially good for neural tissue Rarely integrates into host genome	Risk of recombination in host Only transient gene expression Not as well characterized

and the foreign gene is highly expressed (increased protein production). These viruses also are not highly pathogenic. Adenovirus has become a desirable vector for the treatment of cystic fibrosis and other lung disorders, since nondividing respiratory tract cells are preferentially infected. Adenovirus does not integrate into the genome, so there is little risk of mutation. Although these viruses lyse the cells they infect, they have been engineered not to kill their host cells.

The use of adenovirus as a vector poses significant drawbacks: It can provoke an inflammatory immune response, and genes may function transiently because they do not integrate into the host chromosome. Another concern is the potential for adenovirus to replicate in infected cells. Some gene products are involved in the malignant transformation of cells, are toxic to cells, or may lead to an adverse immune response. Future adenoviral vectors will have minimal viral DNA to decrease immunogenicity while retaining transfection efficiency.

Nonviral Delivery Methods

Nonviral delivery methods may be a safer approach to gene therapy than viral methods. Some nonviral methods are discussed in chapters 4 and 7: electroporation, **liposome transfer**, microinjection, microinjectile bombardment. Current research focuses on gene targeting to specific cell types, gene delivery methods that do not trigger an immune response, and vectors that can survive the circulatory system in order to reach target cells.

GENE THERAPY MODELS

As effective gene therapies are developed for specific disorders (see Table 10.2 for examples), preclinical studies will be performed to develop appropriate protocols

for the targeting and expression of the gene, to assess the efficiency of the protocols, and to thoroughly evaluate the safety of the therapies. Many technical details are in the early stages of development; these inlude the construction of appropriate vectors and the design of efficient *ex vivo* and *in vivo* gene therapies. Gene therapies are being designed for many diseases; a few recent developments are described below.

Liver Diseases

Both *ex vivo* and *in vivo* gene therapies for liver diseases are being studied. Transfected hepatocytes (liver cells) have not been reintroduced into the liver or spleen with complete success. Only some 10% of the injected cells are incorporated into the liver. A more effective method of reintroducing transformed hepatocytes may be an hepatocyte graft.

Some liver therapy experiments already are being conducted. For example, the gene for the low-density lipoprotein **receptor** (LDLR) has been transferred into liver cells by retroviral and adenovirus vectors (by *ex vivo* and *in vivo* cell transduction, respectively). Although the expression of LDLR in cells has been transient, expression seems to be more prolonged if the gene therapy is preceded by a partial hepatectomy (excision of a piece of the liver). All future therapies probably will be *in vivo*, since *in vivo* transduction is efficient, and may present fewer technical problems that lower the efficiency of hepatocyte transplantation.

In June 1992, at the University of Michigan Medical Center, a 29-year-old patient with a rare, life-threatening form of hereditary coronary artery disease made medical history. Familial hypercholesterolemia (FH) is caused by a deficiency in LDLR, and leads to very high serum LDL-cholesterol levels—almost 10 times normal—which in turn leads to heart disease and often

Table 10.2 Examples of Genetic Disorders that are Candidates for Gene Therapy

Disease	Frequency	Gene Product	Target for Therapy
Cystic fibrosis	1/2,500 (Caucasians)	CFTR	Lung tissue
Duchenne's muscular dystrophy	1/10,000 (males)	Dystrophin (muscle)	Muscle tissue
Familial hypercholesterolemia	1/500	Liver receptor for low-density lipoprotein (LDL)	Hepatocytes
Hemoglobin defects (e.g., thalassemias)	1/600 for specific ethnic groups	Hemoglobin components	Bone marrow cells
Hemophilia	A 1/10,000 (males)	Blood clotting factor VIII	Fibroblasts or hepatocytes
	B 1/20,000 (males)	Blood clotting factor IX	Fibroblasts or hepatocytes
Severe combined immunodeficiency (SCID)	Very rare	Adenosine deaminase (ADA) in 25% of patients	Bone marrow or T lymphocytes

Adapted from I.M. Verma, *Scientific American*, 263:68–84 (1990).

sudden death. The patient was homozygous for a specific mutation in the LDLR gene that resulted in a defective LDL receptor. Consequently, cells could not take up cholesterol and metabolize it normally; the patient's LDL:HDL-cholesterol ratio was much higher than normal. *Ex vivo* gene therapy was conducted in the patient after the procedure was first tested in rabbits. Approximately 250 grams of the patient's liver was removed and placed in culture. Hepatocytes were harvested and mixed with a retrovirus containing a normal copy of the human LDLR gene. Approximately 25% of the cells were infected with the virus. The transfected cells were then transferred to the patient by a catheter inserted into the portal vein. Tests later confirmed that cells were expressing the normal LDLR gene, and the patient's LDL:HDL-cholesterol ratio, with the help of medication, was lowered. After this success, two children were subsequently treated with this procedure.

After preliminary data were submitted to the U.S. Recombinant DNA Advisory Committee (RAC) in December 1992, additional patients were approved for this gene treatment. The Institute for Human Gene Therapy was formed at the University of Pennsylvania, and in early 1994, two "gene transplants" were conducted successfully.

Lung Diseases

In vivo methods of gene therapy are being studied intensively for the treatment of diseases that affect the airways. Genes have been efficiently transferred by adenovirus vectors to airway epithelial cells in the cotton rat. In these experiments, genes have been expressed only transiently. A method that may be useful is liposome-mediated gene transfer: A gene is encapsulated into an artificial phospholipid vesicle so that it diffuses into cell membranes and releases the gene into cells.

Cystic fibrosis affects the airways and other organs, such as the pancreas and intestines. In 1989, the cystic fibrosis (CF) gene was found on chromosome 7. Since then, researchers have elucidated the function of the encoded protein: the cystic fibrosis transmembrane conductance regulator (CFTR); Figure 10.1 shows a schematic of the CFTR protein. This protein is being used to develop effective treatments for both alleviating the symptoms of CF and permanently curing the genetic disorder. CFTR is a cyclic AMP (cAMP)-sensitive, low-conductance chloride channel that is normally present in the membranes of secretory epithelial cells in the pancreas, lung, intestine, and sweat glands. In CF patients, the epithelial cells do not properly transport chloride ions in response to the presence of cAMP.

The relationship between gene mutation and expression of phenotypic characteristics is being intensely studied. CF patients express many physical characteristics such as increased mucus production, bacterial infections in the lungs, and altered epithelial cell transport that affects epithelial tissues such as the intestines. One important question is whether these problems are the direct effects of faulty CFTR protein and lack of cAMP-sensitive chloride conductance in membranes. More than 200 mutations that lead to CF have been identified—examples include improper protein processing, faulty regulation, overexpression of the protein, nonsense, frameshift, and splice mutations; 70% of CF patients harbor the mutation ΔF508, which results in a

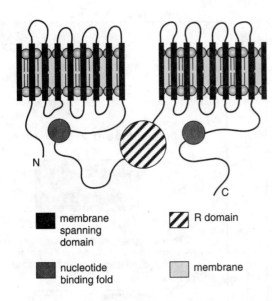

■ membrane spanning domain	▨ R domain
■ nucleotide binding fold	▨ membrane

Figure 10.1 A schematic of the CFTR protein inserted into the cell membrane. A channel is formed that allows the movement of chloride ions. The R domain blocks the channel opening, possibly to regulate the flow of ions through the channel.

three-nucleotide deletion and loss of the amino acid phenylalanine in the CFTR. This defective CFTR protein is not processed properly and so cannot be inserted into the cell membrane.

Gene therapy has been successful in both epithelial cell culture and transgenic CF mice. The CFTR gene was experimentally transferred by liposomes into lung epithelia, alveoli, and tracheas of various CF model mouse strains with a disrupted CFTR gene. Although the CF **genotype** was still present, the CFTR gene was expressed, and the sodium-absorbing and chloride-secretion activities of epithelial cells were identical to those of normal cells. These studies also show only transient expression. An effective gene therapy will require longer-term expression of the CFTR; otherwise, repeated treatments will be required. Clinical trials have yielded mixed results.

Questions remain as to how effective this type of therapy will be in humans, whether the other symptoms of CF will disappear with gene therapy, how long the correction will last after *in vivo* introduction of the normal gene, and whether a patient will develop an adverse immune reaction to repeated treatment with the therapeutic DNA.

Commercialization most likely will focus on *in vivo* transfection targeted to the lung or nasal area by inhalation of aerosols (such as DNA-liposome). Aerosols will be less costly and time-consuming and much simpler to use, especially when repeated treatments are required. Treatment of sites other than the lungs, such as the pancreas, will require a more complicated *in vivo* method because the tissue is not directly accessible.

Hematopoietic Diseases

Cells of the circulatory system are generated by pluripotent, undifferentiated hematopoietic stem cells located in the bone marrow. These cells differentiate into erythrocytes (red blood cells), leukocytes (white blood cells), and platelets. Erythrocytes develop in response to erythropoietin, a hormone produced by the kidneys. The hormone, colony-stimulating factor, induces stem cells to differentiate into five types of leukocytes—the lymphocytes, B-cells (which produce antibodies), and T-cells (which destroy antibody-bound antigens)—and four types of phagocytes, which digest foreign debris in the bloodstream. The B lymphocytes are involved in cell-mediated immune responses and the T lymphocytes in antibody-mediated immune responses, as Figure 10.2 shows. (Appendix C presents a basic review of the immune defense system.) Platelets, found in bone marrow, are formed from large cells called megakaryocytes. These cells, which can circulate throughout the body, protect the body from injury and foreign invaders.

The transduction and reintroduction of hematopoietic stem cells with a normal gene would ensure that a normal gene product is made continuously. Thus, genetically modified cells would be made for the life of the patient. Stem cells, present in low numbers, are found only in bone marrow, and unfortunately most seem to be inactive. Methods must be developed for identifying stem cells that are active in contributing to hematopoietic cells. The *in vitro* assays currently used (for example, 14-day colony-forming unit spleen assay) for quantifying the different hematopoietic progenitors do not show which stem cells will directly lead to the formation of hematopoietic cells (that is, those with reconstituting activity).

Results from studies in mice indicate that treatment of cells with various compounds increases the transduction efficiency of hematopoietic cells. For the successful transduction of bone marrow cells, mice are pretreated with the drug 5-fluorouracil (5-FU). If, during or just before transduction, growth factors such as **interleukin**-3 or interleukin-6 are added to bone marrow cells being cultured, the proportion of active stem cells seems to increase, and possibly the number of cells in the population susceptible to viral infection does as well (and may even increase the potential for successful transplantation).

Circulated Gene Products

Some gene products must be targeted to the circulatory system for distribution. Success depends on several factors: the stability of the gene, the type of product, the half-life of the target cell, the type and age of target cells that will synthesize and deliver the product to the circulatory system, cell culture methods, the mode of gene delivery to cells, the region of cell transplantation (that

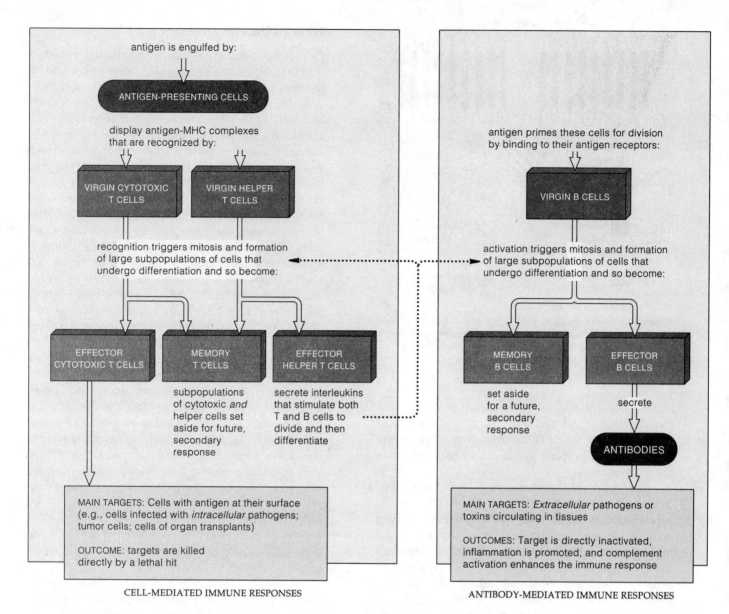

Figure 10.2 Overview of the major interactions of white blood cells, the B and T lymphocytes, during an immune response. When an antigen is detected, both cell types are activated. Memory cells are formed during the primary response, the first encounter with a particular antigen. A secondary response, a faster and stronger immune reaction that includes memory cells, occurs when the antigen is detected a second time.

is, where the cells are implanted), and any adverse immune response that might develop. Fibroblasts, keratinocytes, and myoblasts are among the cells that have been used in targeting experiments. In experiments, both myoblasts and fibroblasts have expressed a transferred gene successfully. For example, human factor IX (a blood-clotting factor) has been synthesized in cells and circulated for more than six months after injection of transduced myoblasts into muscle tissue. Growth hormone also can be expressed in myoblasts. In other experiments, fibroblasts have expressed β-glucuronidase for more than five months, thereby correcting the lysosomal storage disorder in β-glucuronidase-deficient

mice. Retrovirus vector transduction of keratinocytes and subsequent transplantation have been only partially successful, since only low levels of expression are obtained and for only a short time. *In vivo* methods of gene introduction and product delivery to the circulatory system have yet to be developed. These methods would be less time-consuming than the transduction and reimplantation of isolated cells.

Encapsulated cell implants that release their products show much promise (Figure 10.3). These cells can be placed in immunoprotective microcapsules made of hydrogels such as alginate while they receive both nutrients and oxygen from the bloodstream. Cell products are

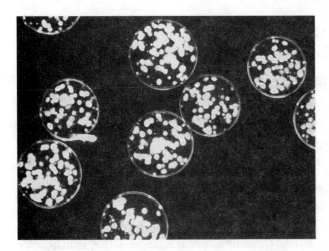

Figure 10.3 Photomicrograph of alginate-encapsulated recombinant mouse fibroblasts at day 8 post-encapsulation. Fibroblasts were transfected by calcium phosphate precipitation with a plasmid encoding the human growth hormone and the mouse metallothionein promoter. The encapsulated cells continued to proliferate optimally and secrete human growth hormone (through the microcapsule membrane) at the same rate as nonencapsulated cells.

free to move out of their compartment and into the bloodstream for circulation (see the discussion of the pancreas in the Tissue Engineering section below).

Cancer and Autoimmune Diseases

Many therapies for the treatment of a variety of cancers are being explored. One promising therapy for brain tumors is the transfer of the *Herpes simplex* virus (HSV) thymidine kinase gene into tumor cells by retrovirus vectors, making transformed cells sensitive to antiviral drugs. The virus infects only the dividing tumor cells, leaving nondividing normal tissue uninfected. When ganciclovir is administered to the patient, tumor cells harboring the HSV thymidine kinase gene are destroyed.

Other potential therapies involve the genetic manipulation of the immune system. Potential cancer therapies have involved the genetic modification of T-lymphocytes, which are extremely important in preventing and destroying tumors. Tumor-infiltrating lymphocytes (TIL) may be used to synthesize large amounts of **cytokines** (for example, interleukin-2) and other antitumor gene products directly at the tumor site. Although the efficacy of TIL therapy has not been well established, this type of therapy could alleviate the toxic effects of administering cytokines systemically.

Researchers are attempting to transduce tumor cells with genes that encode products that stimulate an antitumor response (that is, tumor "vaccines"). This approach is used to make a tumor more immunogenic. For example, an effective cancer treatment may be one that provokes an immune response by manipulating tumor cells to express cytokines such as granulocyte-macrophage colony-stimulating factor (GM-CSF), tumor necrosis factor, or interleukin-2. Cytokines and other antitumor products have resulted in tumors being "attacked." Studies have demonstrated that transduced tumor cells also can promote the rejection of nontransduced tumor cells.

Other cancer gene therapies involving tumor cell transduction include the introduction of toxic genes, the inhibition of oncogenes, and the introduction of tumor-suppressor genes. The effective treatment of cancer is likely to require a multifaceted approach. For example, the use of lymphocytes, genetically modified tumor cells, and cells that secrete tumor antigens may be used in combination with chemotherapy, radiation, and the systemic use of cytokines.

Gene therapies also are being developed for the treatment of acquired immunodeficiency syndrome (AIDS). These therapies include the genetic modification of cells, such as T-lymphocytes or hematopoietic cells, that are resistant to HIV infection. "Vaccination" would prevent HIV infection or prevent the release of virus from cells.

COMMERCIALIZATION OF GENE THERAPY

Before a gene therapy can be commercialized, it must be extensively tested on both animals and humans to determine its safety and efficacy. Before gene therapy tests can be conducted on humans, approval must be obtained by the appropriate regulatory agencies—FDA, the Recombinant DNA Molecule Program Advisory Committee (RAC) within NIH, etc. The NIH has spent over $200 million on funding the development and testing of gene therapy methods. Private companies have also invested heavily in gene therapy, and have conducted 60% of all gene therapy trials. To minimize risk to the public, all NIH-funded clinical research protocols for gene therapy have been reviewed for safety and risk assessment by the RAC panel of 20 scientists and nonscientists. Nonscientists typically have focused on safety and ethics, while scientists have examined the technical details of protocols.

A gene therapy protocol must be tested at several levels before it can be established for widespread use. First, in preclinical trials it is tested in *in vitro* experiments and on laboratory animal models. Clinical trials follow: In phase I trials, the procedure is tested on a small number of human volunteers (usually six to 10), primarily to determine dose and route of delivery and to assess the procedure's toxicity and safety. In phase II trials, to determine efficacy and collect additional toxicity and safety information, the number of human test subjects is increased. If the therapy is shown to be effective and safe, phase III trials are conducted on an even larger group of people. Information obtained from phase I and II clinical trials is incorporated, and a comprehensive investigation of the therapeutic role of the drug is conducted. When the treatment is approved, any remaining questions regarding safety and efficacy are addressed in phase IV clinical tests.

A large number of gene therapy protocols are in preclinical development and phase I clinical trials, the first important steps toward commercialization. According to a recent review of gene therapy progress (*Science*, August 25, 1995, vol. 269), the RAC has approved more than 100 therapies for clinical trials, most of them for the treatment of cancer. Through June 1995, the RAC had approved 106 clinical trials involving a total of 567 patients. In 85% of these trials, the vector of choice for transferring therapeutic genes has been retrovirus or adenovirus; 76 of the 106 used retrovirus.

Of the 106 trials, almost half (51) are for cancers, because these afflict such large numbers of people—which also means large numbers are available for experimental therapy. In many of these protocols, cancer cells or HIV-infected cells are stimulated to produce proteins that in turn stimulate the immune system to attack. In 30 of the cancer trials, genes, such as interleukin-2, which encode molecules that signal the immune system, are inserted into tumor cells to stimulate the immune response so that tumor cells are recognized and attacked by the immune system. In 11 studies the herpes virus thymidine kinase gene is inserted into tumor cells to make them sensitive to the drug ganciclovir.

Of the many genetic diseases that have been well studied, only cystic fibrosis, Gaucher, severe combined immunodeficiency or SCID, hemophilia, and familial hypercholesterolemia have gene therapy protocols that the RAC has approved for use. Gene therapy protocols for many other inherited and acquired diseases are being studied and tested (examples include various cancers and AIDS).

In AIDS gene therapy trials, HIV proteins have been expressed in patients' cells so that they become antigenic and the immune system attacks the infected cells expressing these antigens. Other trials have involved using molecules that compete with natural viral proteins required for virus function, thereby disrupting HIV function.

A small number of trials have involved actual gene therapy; that is, transferring a gene to correct a faulty gene. Twenty trials have focused on single-gene deficiencies such as adenosine deaminase (ADA) or cystic fibrosis (CF). Eleven of the 20 have involved replacing the defective chloride transport gene (CFTR) by transferring to the lung tissue a normal CFTR gene via an adenovirus vector. (Unfortunately, some CF patients have an immune reaction to the vector.) Three trials have focused on the metabolic disorder, Gaucher's disease.

As recently as 1991, gene therapy was considered to be entirely experimental with little prospect of becoming a major medical advance in the near future. In fact, many viewed gene therapy as science fiction—something that might only be possible sometime in the next century. Today, however, it is becoming a reality, and steady progress in the area of gene therapeutics is being made. Scientists are working diligently to overcome the biggest problems in successful gene therapy: low gene transfer efficiencies (virus-based vectors seem to be the best method to date) and low levels of expression.

Gene therapy has been shown to be a relatively safe technology. In more than five years of clinical experimentation, there has been virtually no risk to the general public. Now the question has been raised as to how the NIH-RAC should review gene therapy protocols in the future. The FDA already monitors clinical trials and examines therapeutic agents for safety and efficacy. To streamline the review process and prevent delays, researchers in the future may be able to submit one application for review by both the NIH and FDA.

Many companies are actively developing gene therapy protocols and methods for delivering genes and other therapeutics; examples are liposomes, microspheres, nasal delivery, transdermal therapy. A few of the recent developments are highlighted here. (For an extensive list of drugs and treatments under development, including gene and nongene therapies, see *Genetic Engineering News*, August 1995, pp. 12–16; for examples of developments being pursued by biotechnology companies, see *Genetic Engineering News*, April 14, 1994, p. 8, and June 15, 1997, p. 35). Transkaryotic Therapies, Inc., in Cambridge, Mass., is developing safe, nonviral methods of transferring genes into patients. Skin fibroblasts obtained from biopsies are being transfected with specific genes by electroporation. Cells are injected subcutaneously and remain at the site of injection while expressing the desired protein product. Genes are expressed constitutively, although second-generation gene products will be regulated by specific control sequences. One project entering clinical trials is an erythropoietin gene-activation therapy to treat chronic anemia caused by kidney disease. A "genetic switch" (to turn on the gene) is inserted into cells by homologous recombination.

Targeted Genetics Corp. (TGC), in Seattle, has developed an *ex vivo* protocol for HIV gene therapy that is now in phase I clinical trials. HIV-specific CD8+ cytotoxic T lymphocytes (CTLs) are isolated from HIV-positive patients and the thymidine kinase gene is inserted into these lymphocytes via a retrovirus. After their numbers have been increased in culture, the transformed lymphocytes are returned to the patient. (Recall that the thymidine kinase gene makes transfected cells sensitive to the drug ganciclovir.) TGC is working on a second-generation HIV therapy in which cytokine and cytokine-receptor genes are incorporated into the HIV-specific CTLs. CTLs are activated in response to cytokines, and, when these lymphocytes contact HIV-infected cells, they begin proliferating. Normally CTLs require CD4+ helper T cells (which produce cytokines) for activation, but HIV infects and destroys CD4+ cells.

Viagene, in San Diego, is focusing on immunotherapy with retroviral vectors to transfer genes encoding antigens that stimulate the production of disease-specific CTLs (against HIV and cancers, for example). Both *in vivo* and *ex vivo* immunotherapies are in phase I trials. GenVec in Rockville, Maryland, is using *in vivo* radiation-controlled gene therapy in treating cancer. In preclinical studies on a variety of cancers, a radiation-sensitive regulatory sequence is transferred to tumor cells by an adenovirus vector. When cells are irradiated, genes causing cell death are switched on. This radiation-sensitive regulatory sequence is essentially a "gene switch" that sensitizes tumor cells to low levels of radiation.

VACCINES

Vaccines are essential to the ongoing struggle to eradicate infectious diseases. Through increased understanding of the process of infection, the immune response, and the factors that contribute to virulence, recombinant DNA technology has made it possible to develop a new generation of vaccines. New vaccine vectors and new delivery approaches, immunoenhancers, and even nucleic acids will be used. Vaccines are being developed against pneumococcal pneumonia, malaria, herpes virus (Cytomegalovirus, Varicella-Zoster), diarrhea (rotavirus, prevalent in children), respiratory syncytial virus (respiratory infections), measles, and cholera.

Vaccine Vectors

Often the best vaccine is a live, attenuated derivative of the disease-causing organism because it stimulates a strong, long-lasting immune response. Developing such a vaccine does present problems; among them are difficulties in propagating the pathogenic organism while maintaining immunogenicity, and difficulties in finding methods for attenuating an organism's virulence. New-generation vaccines will involve cloning one to several genes from the pathogen into a nonpathogenic organism that can be administered orally. This nonpathogenic vector expresses the virulent organism's genes and, thus, can invoke an immune response against the antigens of the host. A variety of vectors are being tested, including *Salmonella*, adenoviruses, herpes viruses, avipoxviruses (avian host-specific), vaccinia virus, and polio viruses that can express proteins from a variety of pathogens. These vectors, engineered to express high levels of the appropriate foreign immunogen, will serve as an efficient and potent antigen delivery system. A recombinant vaccine against *Borrelia burgdorferi*, which causes Lyme disease, soon will be tested in clinical trials.

Nucleic Acid Vaccines

Nucleic acids show much promise as new-generation vaccines. For example, when a plasmid vector that harbors a gene encoding an antigen is injected into muscle tissue, expression of the antigen produces an immune response. This method offers a new approach to vaccines whereby only selected components of a virus are expressed. Studies have demonstrated that nucleic acid immunization is long-lasting and induces antigen-specific humoral and cell-mediated immune responses even though the plasmid DNA is not incorporated into the chromosomes. A particle gun can deliver the nucleic acid vaccine directly to immunologically competent skin cells. This type of vaccine precludes the use of a virus (attenuated, live), since DNA is used alone. Intravenous and mucosal routes for nucleic acid vaccine delivery also are appropriate, since the adjacent lymphoid-associated tissues are activated in response to the presence of antigens. Some important safety questions remain, however. For example, what are the implications of an immune response to the introduced nucleic acid? What problems might arise if the DNA integrates into the patient's genome?

Immunoenhancing Technology

Although safer, more efficient vaccines are being developed by recombinant DNA technology, many of these are not highly immunogenic, and researchers are looking for ways to boost their immunogenicity. Methods include: (1) modifying the conformation of an antigen; (2) inducing the synthesis of cytokines, which act directly on T-lymphocytes to stimulate various immune responses; (3) activating macrophages; and (4) encapsulating antigens in microspheres or liposomes so that, rather than being proteolyzed in the stomach, they can be targeted to the gut where they activate the lymphoid system and stimulate mucosal immune responses. These immunoenhancers are being studied extensively in the hope that newly engineered vaccines (such as new subunit vaccines) will be more potent.

SYNTHETIC DNAS

Synthetic DNAs are being designed to serve as drugs in the future. **Antisense** DNA and triplex DNAs may someday serve as drugs against viruses and cancers. Recall that messenger RNA encoded by antisense DNA blocks translation of its complementary messenger RNA by binding to it, and triplex DNAs are formed when a nucleic acid binds to double-stranded DNA to form a three-stranded molecule. These two DNAs, then, disrupt the function of viruses and cancer cells by blocking their production of deleterious proteins. Antisense oligonucleotides are the first DNA therapies to reach the clinical trial stage.

Antisense oligonucleotides (short nucleotide sequences), designed to recognize specific sequences, enter cells, either by diffusion across the cell membrane or by endocytosis (the membrane engulfs the molecule). Oligonucleotides must be at least 15 nucleotides to bind tightly and specifically to nucleic acid sites.

In the early 1980s, scientific investigators learned that some microbes made antisense RNA to regulate gene expression; these microbes base paired with complementary "sense" RNA to prevent translation. It was later discovered that plants and animals also use antisense to regulate gene expression. In the course of studying this phenomenon, researchers realized that this cellular process could have important implications in medicine. They reasoned that antisense molecules could be designed as drugs by introducing them into the body, where they would decrease expression of a particular gene by selectively inhibiting translation. An antisense gene also can be introduced into the chromosomal DNA. The antisense mRNA that is synthesized during transcription is complementary to the endogenous mRNA that is made. Antisense molecules seem to interfere with translation either by preventing ribosomes from "reading" the mRNA or by causing the enzyme ribonuclease H to degrade the message (that is, these two sequences hybridize with one another so that a dramatically reduced amount of endogenous mRNA is present for translation into protein). The antisense RNA lacks signals for translation, so the "anti message" cannot be translated.

Antisense therapy is an important treatment for diseases in which either there is a loss of control over gene regulation or a gene is overexpressed. In the first case, mRNA (and therefore protein) is present all the time; in the second case, there is too much mRNA. The first antisense drugs are likely to be for viral infections. An oligonucleotide that inhibits a viral gene involved in replication of papillomavirus (the cause of genital warts) is currently in clinical trials. Cancer-specific antisense therapies are also being developed. For example, oligonucleotides against acute myelogenous leukemia, a progressive blood cancer, are being studied. One of the targets is the RNA encoded by the *p53* gene (normally a tumor suppressor gene), which seems to be overexpressed in patients with this type of leukemia. Another target is the RNA transcribed by the *c-myb* gene, which is usually involved in the normal proliferation of blood cells. Abnormal regulation contributes to leukemia. Antisense technology also may be useful in the battle against HIV. Oligonucleotides are being targeted to the HIV *gag* gene RNA product, which is necessary for viral replication.

One of the newest of the DNA therapies is triplex technology, in which oligonucleotides bind to DNA instead of RNA. Instead of disrupting the hydrogen bonding of the double helix, the oligonucleotides establish hydrogen bonds with purines in the target site of the duplex, resulting in a three-stranded molecule, or triplex. Oligonucleotides that are rich in cytosine and thymine orient themselves parallel to the purine-rich strand of the target duplex. The oligonucleotides that have guanine and thymine align themselves antiparallel to the purine-rich strand.

In *in vitro* studies, synthetic triplex-forming oligonucleotides bind to target control regions or to coding regions of genes, where they selectively block transcription into messenger RNA. The oligonucleotides being studied usually bind only one strand of the double helix, the one that contains most of the purines (guanine and adenine). A drawback of triplex therapy, at least in its current stage of development, is that more than half of the targets contain purines on both strands of the duplex. To circumvent this problem and to increase the range of DNA targets, researchers are designing new types of oligonucleotides with different chemical groups. Triplex DNA therapy is not yet in human clinical trials, although testing may begin within the next several years.

THERAPEUTIC RIBOZYMES

Methods for destroying specific RNAs by catalytic RNAs exploit the self-cleavage properties of these **ribozymes**. Two oligonucleotides make up the self-cleavage domain: One oligonucleotide serves as the catalytic surface, and the other, called the substrate, contains the site of cleavage. The complementarity of these two oligonucleotides allows hydrogen bonding so that cleavage can take place. Ribozymes now are synthesized to cleave mRNA, a process with important implications for medicine (Figure 10.4).

One goal of researchers in RNA-targeted therapeutics is to use synthesized ribozymes to inhibit mammalian virus production. Ribozymes that have potential for this application come from plant pathogens. Examples are the hairpin, shown in Figure 10.5a, which is derived from tobacco ringspot virus satellite RNA and attacks the 5' leader regulatory sequence encoded by the HIV genome; the hammerhead (named for its shape), shown in Figure 10.5b; the axehead from hepatitis δRNA; the group I intron; and RNase P found in all cells. Other investigations have used the hammerhead motif found in numerous viroid RNAs to cleave mRNA encoded by *rex* and *tax* regulatory gene products of bovine leukemia virus (BLV), which is similar to human

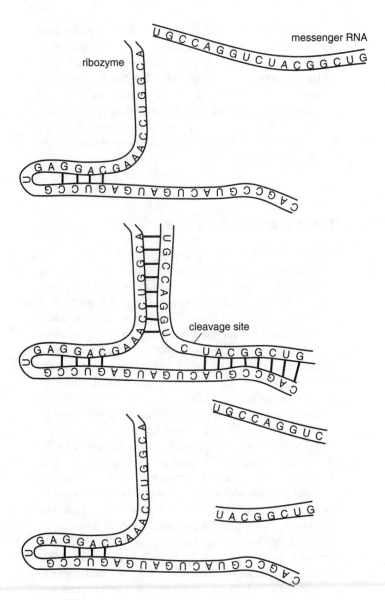

Figure 10.4 A ribozyme cuts an RNA sequence at a specific site by binding to it, thereby inactivating the target RNA.

T-cell leukemia viruses. Multiple hammerheads can attack several targets in HIV RNA.

In preliminary experiments with ribozymes, a custom-designed ribozyme must demonstrate cleavage *in vitro*. Various viral markers have been used to determine the success of ribozyme cleavage; for example, HIV and BLV coat protein antigen p24 and their RNAs, HIV *tat* gene product, *tax* gene product, and BLV reverse transcriptase activity are indicators of viral infection. The lack of expression of the gene product or viral marker (that is, the absence of RNA or protein) indicates that ribozyme cleavage of the target RNA sequence to be cleaved exceeds 95%. All studies showed at least a 70% reduction of RNA target sequence.

Ribozyme probably is more stable in cells than single-stranded antisense RNA because the ribozyme's complex, highly folded structure makes it resistant to intracellular nuclease digestion. One potential disadvantage is that ribozymes, which are much larger than antisense RNAs, may be difficult to transfer into cells. Transfer agents such as liposomes may be necessary.

Ribozymes also have been engineered to cleave the RNA of the HIV virus in a sequence-specific way, a first step toward the use of ribozymes as a treatment against HIV. Ribozymes are being studied that have the potential to cleave the HIV virus at specific sites after they are placed inside cells of HIV patients. In *in vitro* studies, several ribozymes, each directed against a different target in the HIV genome and each acting independently, have increased efficiency of HIV-RNA cleavage. Thus, a combination of ribozymes, each with different specificities, may be an effective therapy in the future. Regulatory regions, rather than structural, may someday be the targets of choice. A ribozyme that attacks a regulatory sequence common to all viral strains will overcome the

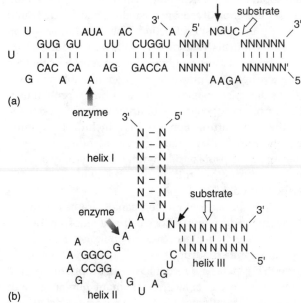

Figure 10.5 Secondary structures of two different types of ribozymes that have both enzymatic and substrate properties. Together the enzyme and substrate form (a) a hairpin derived from tobacco ringspot virus satellite RNA and (b) hammerhead derived from plant viroid RNA. In the hammerhead, helix I and III are involved in binding the target sequence (substrate) and the nucleotides that are not base paired that flank helix II on the 5' and 3' sides are part of the catalytic core. Helix II and its tetranucleotide loop can vary in sequence and length. Small arrows show sites of cleavage in the substrate sequence; large solid arrows show catalytic RNA; large open arrows show substrate (target sequence). N = nucleotides.

problem of high viral mutation rates, because the regulatory regions should be conserved among viral strains.

In 1993, the NIH granted researchers permission to test ribozyme therapy in HIV-infected patients (approval by the FDA is also required). A number of gene therapy trials are in progress. In one preliminary study, T-cells infected by HIV received the gene for an HIV-specific ribozyme. These cells were cultured to increase the number of engineered cells prior to placing them in the patient. T-cells with incorporated ribozyme were monitored to determine how long they remained intact in the body. Ribozymes reduce viral replication either by preventing the infecting virus from incorporating its genome into the host cell's DNA or by preventing expression of viral genes. For this therapy to be effective, a patient must periodically receive engineered cells or, if a permanent supply of engineered cells is desired, a bone marrow transplant with these cells.

SYNTHETIC DRUGS

Engineered proteins and antibodies were the first-generation therapeutics used to treat acute diseases by recombinant DNA technology. Unfortunately, these molecules are short-lived and usually must be delivered directly to their site of action. The second-generation pharmaceuticals that rapidly followed include small stable **peptides** and synthetically derived drugs that are effective against chronic diseases. Small stable synthetic drugs designed to either block or mimic a protein may be part of a new generation of drugs we can expect to see in the near future. Benefits may include fewer side effects than with protein, greater stability, and easy delivery by capsule or pill. Synthetic drugs may one day play an important role in the treatment of diabetes, arthritis, and AIDS.

Many biotechnology companies are beginning to invest in the development of small synthetic compounds. For example, Genentech, Inc., in San Francisco, is developing a synthetic drug that will substitute for human growth hormone and recognize the appropriate receptor. This drug could replace recombinant human growth hormone currently used to treat dwarfism. A small synthetic compound might not produce the adverse reactions that sometimes occur with recombinant protein.

To chemically synthesize a drug, researchers must determine the structures of the protein and its receptor and, if possible, how they interact. A small compound is designed to have the appropriate three-dimensional structure with a receptor-recognition site and amino acids specific for receptor recognition and binding. Sometimes the natural protein can have a very complex structure. For example, some proteins may bind two different receptors (and require two binding sites) with amino acids specific for these receptors separated by some distance.

Synthetic drugs may be used to block cell surface receptors, thus preventing, for example, a virus from infecting cells. These antiviral compounds would have a huge market, since effective antiviral drugs are in high demand. Other applications may include the use of small synthetic compounds as antagonists. For example, a synthetic antagonist could block the action of tumor necrosis factor (TNF), a cytokine that contributes to septic shock and arthritis.

TISSUE ENGINEERING

The new field of tissue engineering focuses on the development of substitutes for damaged tissues and organs. Millions of people lose the function of their tissues or organs, and maintain their health by resorting either to mechanical devices (kidney dialysis machines, pacemakers, and artificial hearts are examples), organ transplants, or reconstructive surgeries (Table 10.3 presents a partial list). Tissue or organ damage often leads to death. The costs of treating patients are astronomical, with estimates exceeding $400 billion per year. More than eight million surgical procedures are performed each year in the United States to correct or alleviate problems associated with tissue or organ failure. The scarcity of suitable organ donors is a serious problem, and most transplant candidates die while waiting for an organ.

Tissue engineering provides a viable alternative to the problem; it combines the principles of engineering and biology to aid in the formation of new tissue or in the implantation of functional cells. Tissue-inducing compounds such as growth factors can serve as signals to stimulate the growth and development of different tissues. Isolated cells, engineered before transplantation if necessary, can replace nonfunctional cells. The cells can reside within a matrix, such as collagen or a synthetic **polymer**, that is sometimes incorporated into a patient's tissue. To reduce the chance of rejection, transplanted cells can remain isolated from host tissues. Isolated cells often re-form the correct tissue structure if the appropriate conditions are provided. Scaffolds or networks of polymers may serve as a substrate for cells (Figure 10.6) to help induce the formation of functional and morphologically correct tissues. Polymers may either be nondegradable and remain in the patient permanently or degrade over time and disappear from the tissue. A negative side effect of the use of scaffolds, however, may be rejection of the implant by the immune system.

Since transplanted cells require an adequate supply of oxygen and nutrients, they cannot be located far from a blood supply. Vascularization might be induced using various growth factors to enable transplanted cells to reside at a distance from a pre-existing supply.

A variety of tissues derived from ectoderm, mesoderm, and endoderm are being engineered. For example,

Table 10.3 The Incidence of Organ and Tissue Damage in the United States and the Number of Patients or Treatments	
Disease/Injury	Number of Procedures or Patients/Year
Skin	
Burns*	2,150,000
Diabetic ulcers	600,000
Neuromuscular disorders	200,000
Spinal cord and nerves	40,000
Bone	
Joint replacement	558,200
Bone graft	275,000
Facial reconstruction	30,000
Cartilage	
Patella resurfacing	216,000
Meniscal repair	250,000
Arthritis (knee)	149,900
Arthritis (hip)	219,300
Fingers and small joints	179,000
Tendon repair	33,000
Ligament repair	90,000
Blood vessels	
Heart	754,000
Large and small vessels	606,000
Liver	
Liver cirrhosis	175,000
Liver cancer	25,000
Pancreas	
Diabetes	728,000
Intestine	100,000
Kidney	600,000
Bladder	57,200
Hernia	290,000
Breast	261,000
Blood transfusions	18,000,000
Dental	10,000,000

*Approximately 150,000 individuals are hospitalized and 10,000 die each year.

Adapted from R. Langer and J. P. Vacanti, *Science* 260:920-925. Copyright 1993 American Association for the Advancement of Science. Used by permission of the publisher and the author. See this paper for original references.

ongoing studies are exploring tissue replacements for the nervous system, skin, cornea of the eye, liver, pancreas, bones and cartilage, muscles, and even blood vessels.

Skin

Skin transplantations often are required after a severe burn and other skin injuries. Researchers are exploring ways to simulate skin. For example, patients have received an upper layer of silicone to prevent dehydration (a serious problem in burn patients), and a lower layer of collagen/chondroitin-sulfate, which stimulates the formation of vascular and connective tissue. By acting like the dermis that is lost through injury, these materials decrease scarring and facilitate healing. To eliminate the need to later graft an epidermal layer, epidermal cells from the patient are added to the collagen/chondroitin-sulfate layer before it is placed on the patient.

Other methods of replacing skin include the *in vitro* culturing of human neonatal dermal fibroblasts on a matrix that can be cryopreserved for later use. These cells are placed in the wound and a skin graft placed over the top. After vascular tissue develops under the epidermal graft, a dermis-like tissue forms. Epidermal cells from the patient also can be cultured, although three or four weeks are usually required to obtain the number of cells required to cover a large area.

Liver

Hepatocyte transplantation may one day replace traditional liver dialysis, thus eliminating the need for hookup to a mechanical device to remove toxins from the body. Hepatocytes may be placed within hydrogel microcapsules and injected into the body or placed on substrates, such as a polymer network, that are implanted. Transplanted hepatocytes have been shown to function in animal experiments, where they have produced albumin and removed by-products of metabolism. Such systems may one day serve as permanent liver replacements if they can be stably maintained within the patient.

Before being transplanted, hepatocytes must be cultured *in vitro*. The cells are placed between two hydrated collagen layers, and their physiological state is determined by assaying the functional markers that are secreted. Hepatocytes are then attached to a matrix-coated polymer network or microcarrier so that they remain active and in a differentiated state. After transplantation, the area must become vascularized so that hepatocytes receive oxygen and nutrients. In animal experiments, not all liver cell types have been represented; thus, the transplant cannot function completely like a liver. Progress is being made, however, since hepatocytes can form tissues that resemble the liver, form bile ducts, and remove bilirubin.

Pancreas

Many Americans die from diabetes or its complications each year. When the islet cells of the pancreas are destroyed, a loss of blood glucose control occurs. The

a

b

Figure 10.6 Tissue engineering offers hope to the hundreds of thousands of patients who need functional tissues and organs. Scaffolding allows cells to multiply in the appropriate location during tissue regeneration. (a) Tubular biodegradable scaffolds for tubular tissue engineering of veins, arteries, and intestines. From left to right: poly(L-lactic acid) foam tube, poly(glycolic acid) nonwoven tube, and two poly(glycolic acid-co-lactic acid) composite tubes of different sizes. Tubes are tailored to different dimensions, mechanical properties, and degradation rates to meet the different types of tubular tissue engineering. (b) Scanning electron micrograph of poly(glycolic acid) nonwoven scaffolds for tissue engineering.

diabetic's glucose levels are normally controlled by diet and/or injections of **insulin**, depending on the severity of the disease. Unfortunately, the symptoms often worsen with age. Tissue engineering research has focused on the transplantation of encapsulated functional pancreatic islet cells. Islet of Langerhans cells (insulin-producing pancreatic cells) that have been alginate-encapsulated are being used in experiments to develop a safe, efficient method of providing insulin to diabetics. The semiporous microspheres protect insulin-producing cells from attack by the immune system while allowing glucose to reach the cells and insulin to be secreted into the bloodstream. Usually porcine islet cells are used as donor cells, but one day genetically engineered human islets that produce insulin may be used. In experiments using this method, rats have shown normal glucose and insulin levels for more than two years.

Various methods have been used to achieve functional transplants in animal models. Islet cells have been immobilized by a polysaccharide alginate within hollow semiporous fibers that are then implanted intraperitoneally in experimental animals. In other experiments, islet cells have been placed in a tubular membrane that enables glucose and insulin to pass freely through the pores; the membrane has been bound to a polymer that is grafted onto blood vessels. Antibodies and lymphocytes that would have generated an immune response were too large to pass into the membrane tube containing the islet cells.

XENOTRANSPLANTATION

One way to alleviate the severe shortage of available organs for transplantations may be animal-to-human transplants, or **xenotransplants**. There may one day be organ farms, on which livestock are raised to provide organs for medical use. Although many immunological problems exist—not to mention ethical and legal problems—researchers continue to look for ways to prevent rejection of animal organs, such as by identifying human "shield" proteins that protect organs from attack by the immune system. If genes encoding these proteins can be transferred to donor animals, their organs will express human proteins, thereby reducing the chance of rejection. Experiments with primates and mice seem promising. Shield proteins such as membrane cofactor protein (MCP) and decay accelerating factor (DAF) are bound to the surface of human cells. These proteins inactivate complement proteins in order to prevent them from marking cells for attack by the immune system. Pigs may be the most suitable organ donor because the major organs of pigs and humans are of similar size and shape, and because few diseases are transmitted from pigs to humans. In experiments, pig eggs have been transformed with MCP and DAF genes, as well as CD46 and CD59, and have produced pigs that produce the shield proteins. Astrid, born December 23, 1992, was the first transgenic pig to produce shield proteins (DAF and MCP). By 1994, two transgenic lines of pigs had been produced, each producing a variety of shield proteins. In 1996, a new generation of descendants from the original Astrid line were produced, and may produce enough shield proteins to become a source of donor organs for clinical trials.

The success of xenotransplants can be tested by perfusing a transgenic organ, such as a pig's heart or kidney, with human blood and looking for signs of immune rejection. Before human clinical trials can be approved, pig-to-primate transplants most likely will be conducted to test for organ rejection and safety.

Although progress has proceeded rapidly, ethical issues remain, such as whether it is appropriate to genetically manipulate animals for the use of their body parts. Although the question will continue to be debated, patients in need of organ transplants will benefit greatly from having an alternative source.

ANTIBODY ENGINEERING

An antibody molecule or immunoglobulin binds to a specific target protein or antigen. The antigen–antibody complex mobilizes phagocytes and other immune system components. Ultimately, the foreign cell is lysed and phagocytized. Antibody engineering has become an exciting area in recent years. While antibodies normally work at the cell surface, antibody engineering is a form of gene therapy that directs antibodies to a specific region *inside* cells. Antibodies may one day be used in cells to block virus replication or the action of proteins that adversely affect health, such as oncoprotein activity, which results in uncontrolled growth of cells.

Recently, antibodies placed in cultured cells reduced cell infectivity of HIV. An antibody called F105 attached itself to a key HIV envelope protein, gp120, that is required for the virus to bind to and infect host cells. In experiments, the F105 gene was streamlined to consist primarily of nucleotides encoding the amino acids that bind to gp120. The gene was engineered to encode a modified F105 that is targeted to the endoplasmic reticulum (ER), where the HIV envelope protein is made from a large precursor protein, gp160. Thus, the addition of DNA that encodes a short leader sequence for targeting the F105 protein to the ER would allow the antibody to bind to the gp120 portion of the precursor gp160. The antibody is held within the ER by attachment to an ER protein called BiP, and can bind to gp160 and actually reduce the amount of gp120 generated. Since gp160 cannot be processed, it is most likely degraded within the ER. With a reduced amount of gp120 on cell surfaces, HIV-infected cells may no longer be able to fuse with other cells to form the cell-killing network that is thought to promote the loss of cell immune response. The virus released from engineered antibody-producing cells were more than 1,000 times less infectious than virus released from HIV-infected cells that do not produce antibodies.

The next logical step is to develop a method for producing engineered F105 antibody in all HIV-infected cells. Technical difficulties may prevent this step from being achieved in the near future. The engineered antibody only weakens HIV; it does not kill it. Second-generation antibodies that prevent replication may be a better way to prevent HIV infection. Engineered antibodies against replication proteins, Rev or Tat may be effective, although these proteins are made in the cytoplasm, and researchers do not yet know how to target and keep antibodies in areas other than the ER.

In the future, antibodies may be targeted to any desired location within cells to inhibit or knock out the activity of a protein, virus, or even cancer cells. Antibodies against oncoproteins would be a very valuable therapy.

CELL ADHESION-BASED THERAPY

Cell-cell interaction, or adhesion, is essential for the development and function of multicellular organisms. These intercellular interactions are mediated by cell-adhesion molecules (CAMs) that are expressed on cell surfaces. CAMs are extremely important to tissue and organ formation during embryonic development, and to other processes such as inflammation, wound healing, and blood clotting. Cell adhesion is a highly complex process, involving multimeric association with a large number of CAMs on each cell surface. This association ensures that cell-cell adhesion does not occur spontaneously or that the wrong cell types adhere to one another. CAMs are grouped into five classes according to structure, and they differ in the strength and duration of the cell-cell interactions they mediate. The cadherins are involved in the stable interactions between cells in epithelia and organs, and therefore mediate strong cell-cell interactions and contribute to the integrity of organ structure. In contrast, selectins facilitate the transient cell-cell interactions of leukocytes and endothelial cells. Leukocytes leave the main circulatory system to enter tissues and "roll" along endothelial surfaces. Since they are mobile cells of the immune system, forming only transient associations with other cells, they usually do not express cadherins on their surfaces. In transient interactions, bonds form and break between the selectin adhesion molecules on platelets, endothelial cells, and leukocytes. Diseases are caused by both the failure to stick and the abnormal sticking of CAMs to other cells.

Cell adhesion technology is revolutionizing the medical field, and will be important in the treatment of many diseases. CAMs are being modified for therapeutic use, and have attracted the interest of biotechnology companies (see *Genetic Engineering News*, November 15, 1996, p. 36). Companies are developing CAM-based therapies for multiple sclerosis, atherosclerosis, inflammation, cancer, osteoporosis, and a multitude of other medical disorders. Cell adhesion is being manipulated as a form of treatment for diseases that involve CAMs. For example, researchers at the small Boston biotechnology company Leukon are developing treatments for autoimmune diseases such as lupus and rheumatoid arthritis. Glycotech, in Rockville, Maryland, is focusing on the development of anti-inflammatory agents that function by preventing cells of the immune system from attaching to various adhesion molecules.

Integrins

Among the most promising adhesion molecules are the **integrins**, a family of receptor proteins found on the surfaces of most cells. They facilitate cell-cell attachment (such as in the inflammatory response) and cell–extracellular matrix binding. A subclass of integrins—B-3 integrins—are of particular interest because only three amino acids (arginine, glycine, and aspartate) are required for binding to proteins. Recently a gel called Telioderm was developed by Telios Pharmaceuticals (San Diego) to heal wounds more quickly. A synthetic peptide present in the gel contains the recognition site for the three amino acids; the site attracts the integrins of desirable cells, thereby promoting healing. In clinical trials, Telioderm significantly increased the rate at which chronic dermal ulcers healed. This gel may be the first commercial product based on adhesion molecule technology.

The inhibition of cell-cell interactions has important medical implications. Normally, platelets adhere to an injury site in a blood vessel, producing a clot that stops the bleeding. However, life-threatening clots sometimes form. An integrin receptor on platelets can bind to the three amino acids on the blood protein, fibrinogen. One fibrinogen can bind to several platelets simultaneously, causing many cells to aggregate. If the clot becomes large enough, the vessel may become blocked.

Drugs are being developed to prevent blood clots, which often cause strokes and heart attacks. The market for such drugs in the United States could exceed $550 million a year. Drugs containing the three-amino-acid peptide (that is, receptor blockers) may be able to decrease the likelihood of a dangerous clot by competing with fibrinogen to reduce the extent of crosslinking. Other future treatments to prevent extensive crosslinking may use a monoclonal antibody fragment to block the platelets' fibrinogen-binding receptors.

Integrin technology may also help prevent osteoporosis, a debilitating disease in which the body's absorption of bone tissue results in brittle bones. In osteoporosis, specific cells bind to the integrins of bone cells and destroy those cells. Loss could be slowed by blocking the integrins on bone cells to prevent the binding of cells that cause the absorption of bone. The three amino acids that bind to integrins could be incorporated into an agent that could compete with these bone-absorbing cells.

Inflammation

The **inflammation** response may be amenable to cell adhesion technology. Inflammation normally helps protect tissues from infection and damage. Normally, leukocytes (white blood cells) adhere to the endothelial cells of blood vessels and migrate into the injured tissue by way of capillaries. This invasion requires several types of CAMs, including selectins and integrins. As these white blood cells move along the inner vessel lining, signals activate them and increase their binding strength through integrins expressed on their cell surface. White blood cell integrins bind tightly to endothelial adhesion molecules. These adhesion molecules called selectins increase along the endothelium of blood vessels that attract leukocytes. The white blood cells change shape to move through blood vessel walls and into tissue to destroy foreign materials (bacteria, for example) at the injury site. However, the inflammatory response sometimes results in a reaction that leads to problems such as asthma or autoimmune disease. In injuries that significantly reduce or cut off blood flow, the damaged, oxygen-deprived tissue sends a signal that, through a complex series of events, activates the inflammatory response. Unless this response is reduced, it leads to tissue damage when blood flow resumes. These injuries require the use of anti-inflammatory drugs, as do all inflammatory diseases such as multiple sclerosis, asthma, rheumatoid arthritis, lupus, and other autoimmune diseases.

A new generation of cell adhesion technologies will reduce the inflammatory response. Drugs that block selectins on endothelial cells or leukocyte integrins may reduce inflammation to injured tissues. Traditional therapies for reducing chronic and acute inflammation have targeted the enzymes on the surfaces of activated leukocytes at the site of inflammation. These enzymes destroy pathogens, but can damage host tissues and organs during injury or when there is an autoimmune disease. Since selectins mediate cell-cell binding by way of their carbohydrate moieties, a new type of therapeutic intervention using small carbohydrate analogs that compete with selectin binding may be possible.

Anti-inflammatory drugs based on cell adhesion technology—that is, CAM antagonists—could have a market value of $10 billion a year. These CAM antagonists may decrease the inflammatory response as needed in the treatment of heart attacks and stroke, as well as frostbite and other injuries that damage tissue. They could become important drugs for a large number of inflammatory diseases.

Cancer and Metastasis

Adhesion molecule biotechnology may also play a role in cancer therapy. Adhesion molecules on metastatic cancer cells differ from those on nonmetastatic cancer cells, and may play a role in spreading cancer from the primary site. To spread, cancer cells must evade the immune system. They may do so by binding to other cells (such as platelets) that circulate throughout the body or by expressing surface adhesion molecules found on normal cells. Metastatic pancreatic tumor cells carrying the adhesion molecule CD44, normally found on lymphocytes, are an example. In the future, cells may be identified as metastatic through assays for specific adhesion molecules on the cell surface, or tumors may be treated with anti-adhesion molecules so that they are detected by the immune system.

The future seems very promising for adhesion molecule biotechnology. A variety of diseases may be amenable to adhesion molecule-based therapy once the specific cell adhesion molecules involved in each disease are identified. Treatment may require either inhibiting or promoting cell-cell adhesion.

DRUG DELIVERY

An effective drug is useful only if a reliable method of delivery is available. How could a large protein or peptide be administered—other than by a painful injection—and protected from digestion or degradation? Drugs such as proteins, DNA, and even synthetic molecules could be inhaled as aerosols through the mouth or nose. Nebulized deoxyribonuclease (DNase), recently tested in clinical trials, may become a new therapy for CF patients. DNase breaks down the mucus that accumulates in the lungs. A drug for osteoporosis, calcitonin, has been tested as a nasal spray. The lungs are a good route of entry for peptides and other molecules. They can readily reach the bloodstream through capillary-rich lung tissue, since only a thin epithelial layer separates the drug from the circulatory system. Oral delivery of peptides and larger proteins encapsulated by tiny microspheres ranging in size from 50 nanometers to 20 micrometers may protect these molecules so that they can be absorbed by the intestine.

The skin may also be an effective route for administering small molecules. A transdermal patch used in combination with an electric current can transfer peptides into the skin and circulatory system. Patches have been used effectively to administer, for example, medications for motion sickness, nitroglycerin for angina, and nicotine to help people quit smoking. However, the molecules of many drugs are too large and insoluble to be moved through skin by current methods. A process called iontophoresis or electrotransport can force molecules through the skin. Since like electric charges repel, a small electric current applied to the skin essentially "pushes" charged proteins through openings such as hair follicles and sweat glands.

Alza Corporation of Athlone, Ireland, is testing a skin patch that applies a current of 3 to 12 volts for periods of minutes to hours. Another method being tested,

transdermal electroporation, uses pulses of several hundred volts lasting only a few milliseconds. This method is thought to open pores so the skin becomes more permeable to molecules such as proteins.

Medical biotechnology has come a long way since recombinant human insulin was commercialized in 1982. Gene therapy is no longer a treatment of the future; designer drugs soon will be commonplace. As scientists identify more genes and the mutations that lead to genetic diseases, unravel the complexities of their regulation, elucidate protein-protein interactions, and find the causes of acquired diseases, powerful biotechnological treatments will be developed.

General Readings

W. French Anderson. 1995. Gene Therapy. *Sci. Am.* 273:124-128.

A.M.L. Lever and P. Goodfellow, eds. 1995. *Gene Therapy*. Churchill Livingstone, Edinburgh.

Additional Readings

S. Abramowitz. 1996. Towards inexpensive DNA diagnostics. *Trends Biotechnol.* 14:397-401.

S. Agrawal. 1996. Antisense oligonucleotides: towards clinical trials. *Trends Biotechnol.* 14:376-87.

S. Agrawal and S. Akhtar. 1995. Advances in antisense efficacy and delivery. *Trends Biotechnol.* 13:197-199.

W.H. Allen. 1995. Farming for spare body parts. *Bioscience* 45:73-75.

E.W.F.W. Alton and D.M. Geddes. 1995. Gene therapy for cystic fibrosis: A clinical perspective. *Gene Ther.* 2:88-95.

M. Barinaga. 1993. Ribozymes: Killing the messenger. *Science* 262:1512-1514.

R.M. Blaese. 1997. Gene therapy for cancer. *Sci. Am.* 276:111-115.

T.L. Blundell. 1994. Problems and solutions in protein engineering—towards rational design. *Trends Biotechnol.* 12:145-148.

J. Bossart and B. Pearson. 1995. Commercialization of gene therapy—the evolution of licensing and rights issues. *Trends Biotechnol.* 13:290-294.

M.K. Brenner. 1995. Human somatic gene therapy: Progress and problems. *J. Intern. Med.* 237:229-239.

J.S. Brugge. 1993. New intracellular targets for therapeutic drug design. *Science* 260:918-919.

J.M. Burke. 1997. Clearing the way for ribozymes. *Nature Biotechnol.* 15:414-415.

S.M. Chamow and A. Ashkenazi. 1996. Immunoadhesins: principles and applications. *Trends Biotechnol.* 14:52-60.

J.S. Cohen and M.E. Hogan. 1994. The New Genetic Medicines. *Sci. Am.* 271:74-82.

Collection of articles. 1994. Frontiers in Medicine: Vaccines. *Science* 265:1371-1400.

A. Cuthbert. 1994. Cystic fibrosis gene update. *J. Royal Soc. Med.* 87(Suppl. 21):2-4.

B. Dodet. 1993. Commercial prospects for gene therapy- a company survey. *Trends Biotechnol.* 11:182-189.

V.J. Dzau, R. Morishita, and G.H. Gibbons. 1993. Gene therapy for cardiovascular disease. *Trends Biotechnol.* 11:205-210.

S. Edington. 1994. A new force in biotech: tissue engineering. *Bio/Technology* 12:361-365.

C. Eng. 1997. Genetic testing: the problems and the promise. *Nat. Biotechnol.* 15:422-6.

P.L. Felgner. 1997. Nonviral strategies for gene therapy. *Sci. Am.* 276(6):102-106.

T. Friedmann. 1997. Overcoming the obstacles to gene therapy. *Sci. Am.* 276(6):96-101.

C.K. Goldman et al. 1997. *In vitro* and *in vivo* gene delivery mediated by a synthetic polycationic amino polymer. *Nat. Biotechnol.* 15:462-6.

S.S. Hall. 1995. IL-12 at the crossroads. *Science* 268:1432-1434.

J. Hubbell. 1995. Biomaterials in tissue engineering. *Bio/Technology* 13:565-575.

F.M. Huennekens. 1994. Tumor targeting: activation of prodrugs by enzyme-monoclonal antibody conjugates. *Trends Biotechnol.* 12:234-239.

M. Kiehntopf, E.L. Esquivel, M.A. Branch, and F. Herrmann. 1995. Ribozymes: Biology, biochemistry, and implications for clinical medicine. *J. Mol. Med.* 73:65-71.

E.J. Kremer and M. Perricaudet. 1995. Adenovirus and adeno-associated virus mediated gene transfer. *Brit. Med. Bull.* 51:31-44.

P.E. Lacy. 1995. Treating diabetes with transplanted cells. *Sci. Am.* 273:50-58.

R. Langer and J.P. Vacanti. 1993. Tissue engineering. *Science* 260:920-925.

R. Langer and J.P. Vacanti. 1995. Artificial organs. *Sci. Am.* 273:130(3)-133.

R.P. Lanza. J.L. Hayes and W.L. Chick. 1996. Encapsulated cell technology. *Nat. Biotechnol.* 14:1107-11.

R. Malouin. 1994. Surgeon's quest for life: The history and the future of xenotransplantation. *Perspect. Biol. Med.* 37:416-428.

E. Marshall. 1995. Gene therapy's growing pains. *Science* 269:1050-1055.

A. Mire-Sluis. 1993. Cytokines and disease. *Trends Biotechnol.* 11:74-77.

K. Mitani and C. Thomas Caskey. 1993. Delivering therapeutic genes-matching approach and application. *Trends Biotechnol.* 11:162-166.

A.S. Moffat. 1993. Going back to the future with small synthetic compounds. *Science* 260:910-912.

R.C. Mulligan. 1993. The basic science of gene therapy. *Science* 260:926-931.

M.R. Natowicz, J.K. Alper, and J.S. Alper. 1992. Genetic discrimination and the law. *Am. J. Hum. Genet.* 50:465-475.

R. Nowak. 1994. Xenotransplants set to resume. *Nature* 266:1148-1151.

R.B. Parekh and C.J. Edge. 1994. Selectins- glycoprotein targets for therapeutic intervention in inflammation. *Trends Biotechnol.* 12:339-345.

L.S. Parker and E. Gettig. 1995. Ethical issues in genetic screening and testing, gene therapy, and scientific conduct. In F.E. Bloom and D.J. Kupfer, eds., *Psychopharnacology: The Fourth Generation of Progress*, Raven Press, Ltd., New York, pp 1875-1881.

N.R. Rabinovich, P. Mcinnes, D.L. Klein, and B.F. Hall. 1994. Vaccine technologies: View to the future. *Science* 265:1401-1404.

J. Rennie. 1994. Grading the Gene Tests. *Sci. Am.* 270(6):88-97.

J.H. Richardson and W.A. Marasco. 1995. Intracellular antibodies: development and therapeutic potential. *Trends Biotechnol.* 13:306-310.

H.J. Rothig. 1994. Clinical trials of gene/biotechnology products. *Methods Find. Exp. Clin. Pharmacol.* 16:539-544.

R.M. Sade. 1994. Issues of social policy and ethics in gene technology. *Methods Find. Clin. Pharmacol.* 16:477-489.

S.T. Sigurdsson and F. Eckstein. 1995. Structure-function relationships of hammerhead ribozymes: from understanding to applications. *Trends Biotechnol.* 13:286-289.

J.A. Smythe and G. Symonds. 1995. Gene therapeutic agents: The use of ribozymes, antisense, and RNA decoys for HIV-1 infection. *Inflamm. Res.* 44:11-15.

E.J. Sorscher and J.J. Logan. 1994. Gene therapy for cystic fibrosis using cationic liposome mediated gene transfer: A phase I trial of safety and efficacy in the nasal airway. *Hum. Gene Ther.* 5:1259-1277.

J. Travis. 1993. Biotech gets a grip on cell adhesion. *Science* 260:906-908.

D.L. Vaux. 1995. Ways around rejection. *Nature* 377:576-577.

R.G. Vile and S.J. Russel. 1995. Retroviruses as vectors. *Brit. Med. Bull.* 51:12-30.

G.J. Waine and D.P. McManus. 1995. Nucleic acids: Vaccines of the future. *Parasitol. Today* 11:113-116.

B.M. Wallace and J.S. Lasker. 1993. Drug delivery—Stand and deliver—getting peptide drugs into the body. *Science* 260:912-913.

L. Walters. 1991. Human gene therapy: ethics and public policy. *Hum. Gene Ther.* 2:115-122.

T. Wang. et. al. 1997. An encapsulation system for the immunoisolation of pancreatic islets. *Nat. Biotechnol.* 15:358-62.

A.B. Wedel. 1996. Fishing the best pool for novel ribozymes. *Trends Biotechnol.* 14:459-65.

H.M. Weintraub. 1990. Antisense RNA and DNA. *Sci. Am* 262(1):40-46.

K.A. Whartenby, C.N. Abboud, A.J. Marrogi, R. Ramesh, and S.M. Freeman. 1995. The biology of cancer gene therapy. *Lab. Invest.* 72:131-145.

D. White. 1996. Alteration of complement activity: a strategy for xenotransplantation. *Trends Biotechnol.* 14:3-5.

N.A. Wivel and R. Walters. 1993. Germ-line gene modification and disease prevention: Some medical and ethical perspectives. *Science* 262:533-538.

FORENSICS AND DNA PROFILING 11

DNA as a means of identifying individuals is finding important use in criminal and civil proceedings, in searching for missing people, and in determining close genetic relationships (such as establishing paternity). Most recently, the criminal trial of O.J. Simpson increased public awareness of forensic genetics as no previous event has. Although human DNA is 99 to 99.9% identical from one individual to the next, DNA identification methods use the 1 to 0.1% of DNA that is unique to generate a unique "fingerprint" or identification tag for an individual. DNA "fingerprinting" also is called DNA profiling and DNA typing.

Protein polymorphisms have been used to detect genetic differences among individuals since the late 1960s. Among the protein markers used were the ABO blood groups, histocompatibility antigens (MHC) or human leukocyte antigen (HLA), and red blood cell enzymes. The major problem with using protein polymorphisms is their limited variability. Thus, when they are used for forensic analysis, the results really exclude an individual rather than identifying an individual, since the probability is high that a match between two samples could represent a chance event.

DNA fingerprinting was first described in England in early 1985 by Dr. Alec Jeffreys at the University of Leicestershire. Jeffreys developed a method for identifying individuals by their DNA banding patterns, and suggested that these DNAs would be useful in paternity disputes, immigration investigations, and forensics. At the time, fears were expressed that legal problems might preclude the use of this method. However, in April 1985, DNA fingerprinting was used to resolve an immigration dispute in England: A boy living outside the United Kingdom was granted entry because DNA profiling demonstrated that he was the son of a U.K. resident.

DNA was subsequently used as evidence in murder and rape cases. In October 1986, a prime suspect in a murder case was vindicated by DNA fingerprinting. When two schoolgirls were raped and murdered in Leicestershire, England, Jeffreys' DNA probes were used in Southern hybridizations (described in Chapter 4) of DNA from a collected semen sample. The tests demonstrated that the suspect was not the rapist. (A conventional criminal investigation eventually identified the murderer.)

Shortly after this crime, DNA testing was used in the United States, in the conviction of an individual for assault and rape. By 1987, DNA fingerprinting was admitted as evidence in both the United States and the United Kingdom. In 1988, the United Kingdom Home Office and Foreign and Commonwealth Office ratified the use of DNA fingerprinting to resolve family relationships in immigration disputes. However, by 1989, DNA fingerprinting was under attack in the United States. Most debate has centered on its use in criminal identification. Its use in paternity testing and determining familial relationships has generated less controversy.

During the late 1980s and early 1990s, U.S. courts of law questioned whether DNA profiles should be admissible in criminal trials. Questions of scientific validity and procedural consistency were raised. In response, the United States Office of Technology Assessment (OTA) issued a review finding that, used with the proper procedures and controls, DNA fingerprinting was a scientifically valid way of determining individuality. However, criticism and debate focused on the statistical validity of DNA fingerprints in the identification of criminals.

In May of 1992, the United States National Research Council published a report that confirmed the scientific validity of DNA identity testing and issued appropriate guidelines. DNA evidence now is admissible, and reliability and accuracy have been continually improved. Many commercial laboratories have been created to meet the growing demand for forensic and parentage analysis (Cellmark Diagnostics and Lifecodes Corporation are examples). More than 25 countries use DNA fingerprinting for forensics and paternity testing.

SATELLITE DNA

Repetitive DNA

The human genome comprises three billion base pairs of DNA. Much of the human genome is composed of repetitive DNA for which functions have not been completely established. Two major classes of repetitive DNA have been identified: (1) tandemly repetitive sequences of satellite DNA make up approximately 10% of the genome, and (2) interspersed repetitive DNA makes up 5 to 20% of the genome. Interspersed (so called because it is scattered throughout the genome) repetitive DNA is further subdivided into short and long sequences; sequences of fewer than 500 base pairs are called SINES, of 500 base pairs or more are called LINES.

Satellite DNA was first defined as a fraction of DNA that banded out at a different place from the rest of the DNA during ultracentrifugation of genomic DNA in a CsCl (cesium chloride) gradient. The GC content of genomic DNA influences the position of the DNA in the tube during buoyant density gradient centrifugation. Eukaryotic DNA often forms more than one band during centrifugation, with the minor bands called "satellites."

Tandemly repeated DNA comprises long macrosatellite regions near the centromere and shorter microsatellite and minisatellite DNAs that are scattered within the genome (Table 11.1). Some repetitive DNA is composed of a short sequence or motif, sometimes called a core or unit sequence, that is repeated many times in tandem (that is, end to end). These tandem repeats form bands of a different buoyant density from the main genomic DNA because the GC content deviates significantly from other genomic DNA (although not all satellite DNA has a different GC content). The repeat units vary from very short ones, two to six base pairs in length, to longer repeats of 300 base pairs and more. In this context, satellite DNA refers to the short head-to-tail tandem repeats of specific sequence motifs. The number of times the core sequences are repeated in tandem varies; thus these sequences are called variable number of tandem repeats (VNTRs). A VNTR refers to a single locus that has alternative alleles characterized by the differences in the number of times the core sequence is repeated in tandem. A VNTR locus is hypervariable when many alternative alleles exist in the population.

Many hypervariable loci have been identified in the human genome and are found on every chromosome—sometimes close to genes but more often in stretches of noncoding regions. In fact, much of our DNA comprises noncoding, short tandemly repeated sequences. For example, a 20–base sequence (a unit or core sequence) may be tandemly repeated 20 to 30 times or more. Often slight sequence variations may occur in some of the unit sequences. For example, the main unit may be

-AAGGGCACCAGAGACCCGGA-

However, a single base change could yield

-AAGGGCACCAG__G__GACCCGGA-.

Other variations include changes not only in sequence but also in the number of times a unit is repeated in the locus. Thousands of VNTR loci are present throughout the human genome. A VNTR found only at one locus is a single-locus repeat, and sequences that are found at many loci along the chromosomes are called multilocus repeats.

Alec Jeffreys and his colleagues determined that when a region of the highly polymorphic sequences is homologous, probes can be used in hybridization. To detect polymorphisms, the DNA is digested with an appropriate restriction enzyme that cuts just outside the VNTR region, the DNA fragments are separated by gel elctrophoresis, blotted onto a nylon membrane, and hybridized with a complementary probe that yields the banding pattern or fingerprint (Southern blot hybridizations are discussed in Chapter 4). Digesting the DNA with a specific restriction endonuclease allows the difference in length of a VNTR locus between two individuals to be detected. The differences in the length of the fragments depend on the number of repeating sequence units and

Table 11.1   The Two Classes of Repetitive DNA in Humans: Tandemly Repetitive Satellite DNA and Interspersed DNA. n = the number of repeats; x = the base of a nucleotide

Designation	Size Range	Examples	Features
Satellite DNA			
Microsatellites	<1 kb	$(AC)_n$ repeats	$(XXX)_n$, $(XXXX)_n$ are more useful
Minisatellites	1–30 kb	Probes called 33.6, 33.15 3′α HVR	Multilocus (common core)$_n$ Single locus Multiple repeats in tandem: VNTRs
Macrosatellites	Megabases in size	Alpha satellites	Primarily in centromere/telomeres
Interspersed repeats	≈300 bp >500 bp to 10 kb	*Alu* repeats *Kpn* or L1 repeats	Interspersed repeats are not always internally repetitive or tandemly repeated

Adapted from R. J. Trent, *Molecular Medicine: An Introductory Text for Students*, Table 8.1. Churchill Livingstone, 1993.

not on the length of the DNA flanking the repeats (which is invariable) where the enzyme cut. Hypervariable loci, reflected as length polymorphisms, are useful in distinguishing individuals and in determining parentage and other close relationships. When the variabilities from different loci of an individual are compiled, the composite profile becomes specific to an individual. DNA isolated from blood, tissues, and hair roots can be used to produce individual-specific DNA "profiles."

Interestingly, satellite DNA is found only in eukaryotic organisms and not in prokaryotic organisms. Biological functions have been suggested for satellite DNA; one suggestion is that they provide sequences necessary for the pairing of homologous chromosomes during meiosis and sites of recombination.

Microsatellites

Microsatellites are simple repeated sequences or short tandem repeats (STR) less than 1 kb in total length. STRs are composed of mono-, di-, tri-, or even tetranucleotide units tandemly repeated so that the overall length usually varies from 70 to 200 base pairs. The most common motif observed in the human genome is $(CA)_n$–$(GT)_n$ (n equals the number of repeats). The motif is dispersed throughout the genome of all eukaryotes and is present in high copy number. Some 50,000 to 100,000 copies are found every 30 kilobases in human DNA.

The best described repeat is $(AC)_n$ where n can vary from 10 to 60. Approximately 50,000 $(AC)_n$ repeats exist in the human genome. Since they are widely distributed, these polymorphisms serve as important markers in mapping the human genome. STRs are also important in familial studies of genetic diseases and for use in paternity disputes. Because of their hypervariability (that is, their many allelic forms), they are more valuable than traditional RFLPs, which have only two allelic forms—present or absent (RFLPs are discussed in more detail later in the chapter).

Since the length of the repeats varies only slightly (such as when the repeat is based on only two nucleotides), only one microsatellite locus at a time should be used in an analysis. Other microsatellites such as $(AGC)_n$ and $(AATG)_n$ have been identified, and yield greater differences in alleles (that is, fragment lengths), since the repeats are based on three or four nucleotides.

Minisatellites

Minisatellites were first characterized between 1986 to 1987; they were discovered in close association with the α-globin gene locus on chromosome 16 and the immunoglobin heavy chain gene locus on chromosome 14. Minisatellites are usually located near the ends of chromosomes in the telomere region. Loci may number in the thousands, although each locus has a distinct repeat unit. These sequences are used to detect length polymorphisms (that is, length alleles), and are useful in generating DNA fingerprints. Many minisatellites share a core sequence that is a part of the repeat unit of each minisatellite locus, although some minisatellites appear to lack a common core sequence.

Minisatellites seem to be of greater value than microsatellites in the laboratory since the repeat unit is longer. DNA fragments are usually 1 to 30 kb; thus, the variation in repeat length is greater than in microsatellites. Since the repeats are hypervariable, the chance of finding polymorphic alleles is extremely high.

Macrosatellites

Macrosatellite DNA is located near the centromeres and telomeres (the chromosome ends). These DNA sequences are very large—megabases in length—and usually require a special type of electrophoresis for separation (pulse field gel electrophoresis). Their length makes these DNAs subject to breakage (thus resulting in short fragments), and they are therefore not used in forensic analysis, especially since most tissue used for forensic work is at least somewhat degraded.

POPULATION GENETICS AND ALLELES

A locus refers to the physical location of a segment of DNA encoding an RNA or a protein (that is, a gene), or to a specific sequence that is not a gene. The ability to detect alternative forms of a locus or sequence (that is, alleles) depends on the methods used. For forensic analysis, VNTRs are used. Since humans are diploid organisms, each individual has two alleles per locus. Alleles from both the mother and father contribute to an individual's genotype or genetic makeup. A homozygote has two copies of a VNTR of the same overall length, while a heterozygote has two copies of different lengths. In a homozygote, the VNTR alleles may be composed of different sequences, and can be detected if the DNA is sequenced. Many alleles can exist in a population (although each individual would have only two alleles at each locus), with the maximum number of different alleles possible being 2 × the number of people in the population. Some DNA regions are hypervariable—that is, large numbers of alleles exist—while some are invariable and have only one allele for a gene or locus. Typically, polymorphic genes encoding proteins are represented by two to six alleles. The VNTRs used in forensics and individual typing are hypervariable and may have 100 or more alleles.

Allelic polymorphisms in a population are maintained by random genetic drift and / or natural selection. When the frequency with which a combination of banding patterns occurs is estimated to exceed the frequency found in the population, the evidence is overwhelming that the pattern is individualistic. Therefore, when conducted with appropriate controls, DNA fingerprinting is an important identification tool.

MULTILOCUS MINISATELLITE VNTRS

Alec Jeffreys developed the first minisatellite probes that were used in courts of law. These DNA probes, 33.6 and 33.15, comprised core units, (AGGGCTGGAGG), repeated 18 times, and (AGAGGTGGGCAGGTGG), repeated 29 times, respectively. These probes detected minisatellites at multiple loci in the genome. In Jef-

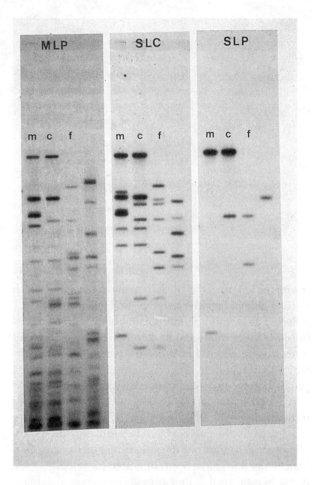

Figure 11.1 Two types of probes are used for identification testing: multilocus and single-locus. Shown here are three autoradiograms for a paternity test (m = mother's DNA, c = child's DNA, f = true father's DNA, unlabeled lane is excluded alleged father's DNA). Far left panel: Multilocus probes allow the detection of multiple repetitive DNA loci that are located on more than one chromosome. Since up to 20 to 30 bands often are obtained, the chances of two randomly selected people having a match at all band positions is extremely low; thus, the DNA profile constitutes a banding pattern that is unique for each person. Far right panel: Single-locus probes allow the detection of a single repetitive DNA locus on one chromosome. At most, two bands are observed, one for each of the two alternate alleles (that is, of different sizes) present at the locus on each member of the chromosome pair. Only one band is observed if the alleles are the same size. Middle panel: Many single-locus probes are available; to increase the sensitivity of discrimination, several are used to examine different loci. In this autoradiogram a single-locus cocktail comprised of four different single-locus probes was used in the hybridization.

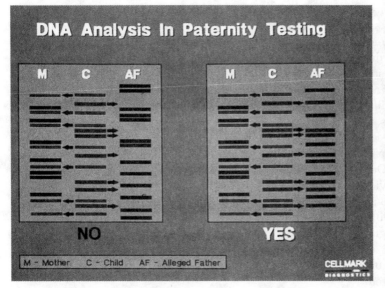

DNA Analysis In Paternity Testing

M   C   AF   M   C   AF

NO   YES

M - Mother   C - Child   AF - Alleged Father

CELLMARK
DIAGNOSTICS

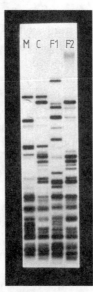

M C F1 F2

Figure 11.2   To establish paternity, DNA fingerprints of the mother, child, and alleged father are compared. The bands from the mother and child that co-migrate are identified, and the bands remaining in the child's fingerprint pattern are compared to those of the alleged father. The bands that do not match the mother's must come from the biological father. Left: schematic example: NO means alleged father is excluded and YES means alleged father is identified as the biological father. Right: autoradiogram (M = mother, C = child, F1 = alleged father #1, excluded, F2 = alleged father #2 identified as the biological father).

freys' multilocus probes, approximately 17 variable DNA fragments for each individual are observed, ranging in size from 3.5 to greater than 20 kilobases (Figure 11.1). Many smaller DNA fragments are also present but are ignored in the analysis, since they are not well resolved by gel electrophoresis. These probes have also been useful in producing individualized fingerprints in a wide variety of animal, plant, and bird species. Thus, fingerprinting also has applications in plant and animal breeding, conservation biology, and population genetics. Since the 33.6 and 33.15 tandemly repeated probes were developed, many other useful probes have been isolated that allow the detection of variable DNA fragments.

Multilocus human DNA fingerprinting is supported by genetic and population data. Jeffreys' multilocus probes have been statistically evaluated to determine the proportion of bands that are shared by unrelated individuals. The data are sufficiently reliable that DNA fingerprinting has been routinely used in paternity disputes (Figure 11.2) and immigration disputes (Figure 11.3). Multilocus fingerprints are more difficult to interpret than single-locus banding patterns for several reasons: the large number of bonds usually generated, incomplete cutting of the DNA, DNA degradation, low DNA recovery, and problems of sorting out DNAs in mixed DNA samples from more than one individual. Despite the technical problems, DNA fingerprinting has been used successfully in criminal cases.

SINGLE-LOCUS MINISATELLITE VNTRS

Single-locus VNTR patterns for microsatellites generate only one DNA fragment (one allele) or two DNA frag-

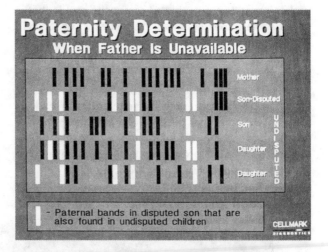

Paternity Determination
When Father Is Unavailable

Mother
Son-Disputed
Son
Daughter
Daughter

UNDISPUTED

▯ - Paternal bands in disputed son that are also found in undisputed children

CELLMARK
DIAGNOSTICS

Figure 11.3   DNA fingerprinting can be used when a claim of relationship is in doubt, as, for example, in immigration disputes in which people seek entry into the United States on the grounds that they are blood relatives of U.S. citizens. Paternity can be established even when genetic information from only one of the parents is available. The DNA band pattern of the person in question can be reconstructed by analyzing the DNA of other blood relatives such as siblings. In this example, the mother's DNA fingerprint is compared to her children's. The remaining bands come from the biological father; thus, the paternal bands of the undisputed children can be compared to the child in question.

ments (two different alleles), as Figure 11.1 shows. One band is generated for each allele. Thus, two bands are observed for heterozygous loci and one band for homozygous loci. Simple banding patterns (one or two different bands) are much easier to compare than is a pattern composed of many.

Minisatellite probes are useful for generating locus-specific DNA fingerprints. Many of these minisatellites have been used as genome landmarks for the human genome. A large number of cloned minisatellites are available and are useful for forensic analysis. For forensic work, a suite of unlinked probes selected from different loci generate enough information for determining identity. The more probes (that is, loci) used, the more informative the analysis (Figure 11.4). Thus, the banding pattern becomes a unique fingerprint that serves as an individual's "signature." Multilocus probes have been used often in the past, since one probe reveals the pattern of VNTRs located throughout the genome. One probe yields much more information about the individual, providing that the bands can be resolved. However, the use of single-locus probes is technically less difficult—for example, it eliminates problems with co-migration of alleles and provides better resolution of bands—and is less likely to produce artifacts (Figure 11.5). To reach the degree of variability that can be generated with one multilocus probe, at least four or five single-locus probes must be used. Several single-locus probes generate a meaningful composite profile that is used to determine the statistical significance of a match. When several single-locus VNTR probes are used in an analysis (one filter is stripped of the probe and rehybridized using the second probe), the composite DNA profile that is generated is usually unique to a particular person. By determining the frequencies of various alleles at different loci in specific populations, one can calculate a frequency for combinations of alleles (Table 11.2). The probability for separate alleles to have come from the same individual then is calculated.

RESTRICTION FRAGMENT LENGTH POLYMORPHISMS (RFLPS)

Simple RFLP analysis (as opposed to the VNTR analysis just discussed) allows one to detect a single base change in a DNA sequence. A restriction endonuclease recognizes a specific sequence and cuts the DNA in that region. For example, EcoRI recognizes (the EcoRI recognition site is underlined):

-TGAATTCG-

-ACTTAAGC-

and cuts both DNA strands to give

-TG   AATTCG-

-ACTTAA   GC-

If one of those six bases in the recognition sequence is changed, the DNA is no longer cut by EcoRI, and the resulting DNA fragment now is longer than if it had been cut with the enzyme. If the restriction enzyme recognition sequence borders, or is within, a gene or sequence of interest, changes in that sequence (that is, a change in the size of that DNA fragment) can be detected by cutting DNA with that enzyme and using Southern blot hybridization. Variant alleles are detected in this manner.

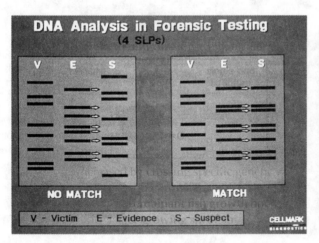

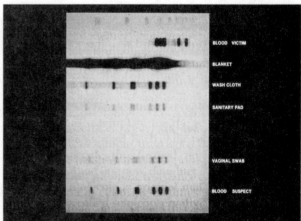

Figure 11.4 The use of four single-locus probes increases the discriminatory ability of DNA fingerprinting. Single-locus probes are approximately 10 times more sensitive than multilocus probes and can generate a fingerprint from as little as 20–50 nanograms of sample DNA. In this example from a criminal case, the DNA band patterns from a specimen retrieved at the crime scene and a sample from a suspect are compared. This information and additional circumstantial evidence provide compelling evidence that DNA band patterns from the two samples are not likely to be identical solely by chance. Left panel: schematic of DNA analysis showing exclusion of suspect as the source of evidence DNA (that is, no match) and a match indicating the suspect as the possible source of evidence DNA; Right panel: autoradiogram.

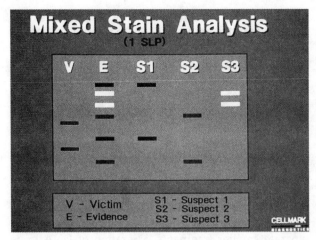
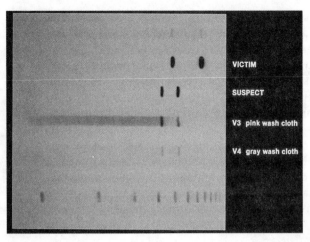

Figure 11.5 A major benefit of single-locus probes is their ability to resolve mixed stains, especially important in criminal evidence when blood samples may be mixed. The mixture can be analyzed with one single-locus probe to determine how many individuals were involved. For example, if six bands were observed, the mixture contained samples from at least three individuals. Several single-locus probes then could be used sequentially to increase the accuracy of identifications. (Left, schematic example; right, autoradiogram.)

Screening for RFLPs is extremely powerful, and is used in disease detection such as for the "sickle" allele (the cause of sickle cell anemia) of the β-globin genes. RFLP analysis is beneficial when alleles (markers that do not have an observable effect on an individual) are linked to mutated genes that encode genetic diseases. Individuals can be screened to determine whether they harbor a particular RFLP and thus the genetic defect that can be passed on to progeny.

Table 11.2 Frequency Calculations of a Common Genotype, D2S44-D14S13-D17S79-D18S27, in Four Populations	
Locus	DNA Fragment Size (kb)
D2S44	11.5, 10.5
D14S13	6.0, 5.0
D17S79	3.5, 3.4
D18S27	6.2, 4.7
Race	Likelihood of Genotype
Black	1:4,000,000
Caucasian	1:700,000
Hispanic	1:1,200,000
Oriental	1:400,000

Adapted from K. C. McElfresh, D. V.-F., and I. Balazs. *Bioscience* 43:149–157 (1993).

METHOD OF DNA PROFILING

The methods used to obtain a DNA fingerprint are discussed in Chapter 4: restriction enzyme digestion, gel electrophoresis of DNA fragments, hybridization with a VNTR probe, and autoradiography to visualize the bands on x-ray film. To generate a DNA profile, DNA is isolated from body fluids such as urine, semen, or saliva and from tissue such as blood, bone, or skin. The DNA is cut at specific sites with restriction enzymes, most commonly *Hinf*I, *Hae*III, *Pst*I, and *Taq*I.

The DNA profile generated on x-ray film must be analyzed to determine the positions (that is, the sizes) of the DNA fragments or bands. DNA fragments of known size are run on the same gel as the samples. The distance migrated is inversely correlated with the size or length of the fragment. A standard curve determines the relationship of the distance migrated and the length of the fragment, and is used to calculate the length of unknown band sizes.

The goal of DNA profiling is to exclude particular suspects or to determine the probability of a match between samples collected from the crime scene and samples collected from a suspect. For the profile to be valid, the frequency of the selected sequences in the general population must be used for comparison.

TECHNICAL CONSIDERATIONS

The need to standardize methods to ensure reproducibility of DNA fingerprints has generated much discussion. When stringent methods are not followed,

DNA profiles might wrongly exclude a suspect rather than falsely incriminate. Standardization of methods ensures reproducibility of results.

To prevent artifacts and maintain consistency, care should be taken to preserve the integrity of DNA, to completely digest the DNA with restriction enzymes, to standardize hybridizations, and to select appropriate probes that have been demonstrated to be stably inherited. Errors can occur through contamination of the sample, DNA degradation, difficulties in interpreting the bands on the x-ray film, and statistical misinterpretation of a match. Artifacts (for example, extra bands, missing bands, band shifting) could provide false information.

DNA

Not all DNA from samples collected is of the same quality or quantity. The quantity of the DNA may be limited and the DNA may be partially or almost completely degraded. To further complicate the situation, samples from one person may be contaminated with samples from another, or even with bacterial DNA. To prevent degradation, tissue samples should be collected and packaged as soon as possible and placed on ice. DNA should be extracted upon reaching the laboratory. If immediate DNA isolation is not possible, the tissue can be frozen. Experts recommend that samples be frozen either in an ultracold freezer at –70°C, on dry ice at –120°C, or in liquid nitrogen at –196°C. Although DNA degrades readily in warm, wet conditions (DNA from dried fluids such as blood and from cells is quite stable), it can survive for thousands of years in some types of environment. Examples are DNA isolated from fossilized plants, insects in amber, mummies, and even fossilized bones. The small amount of DNA obtained is usually very fragmented.

DNA Digestion If DNA is incompletely digested by restriction enzymes, the DNA profile will be inaccurate. The DNA fragments will be larger than they would have been if all sites had been cut. Reasons for incomplete digestion of DNAs include contaminants from tissues or reagents in the digestion solution, low enzyme activity, or suboptimal digestion conditions. Methylated DNA also can influence the digestions; if the DNA is methylated at the sites recognized by the selected restriction enzyme, the sites will not be cut. Enzymes that are sensitive to methylation (*Hpa*II, for example) may have to be avoided. The enzymes used in DNA fingerprinting are not affected by methylation (an exception is *Hinf*I, which cuts methylated DNA inefficiently).

Sometimes, if optimal conditions are not maintained (if, for example, the wrong salt concentration is in the buffer), enzymes cut at a site different from the normal recognition sequence, generating artifacts in the DNA profile. Under low salt conditions, the enzyme *Eco*RI recognizes another sequence, –AATT-, in addition to –GAATTC-. This recognition of the wrong sequence is typically called "star" activity. A profile generated with *Eco*RI "star" would show additional, smaller bands because of excessive digestion.

Gel Electrophoresis The DNA banding pattern is strongly influenced by gel electrophoresis conditions. Several parameters influence the band patterns of DNA profiles. The concentration of agarose or polyacrylamide affects the migration of DNA fragments. Large DNA fragments require a lower concentration for better resolution, while a higher percentage is used to separate smaller fragments. Migration of DNA fragments in gels is influenced by the voltage gradient applied across the gel. A high voltage does not resolve large fragments well; a very low voltage decreases the separation of small fragments. The buffer type and concentration in the gel and buffer chamber also affect the banding pattern in the gel. A high salt concentration decreases band separation and raises temperature. The depth of the overlay buffer also affects the migration rate of DNA. If low voltage is used for a long period of time during electrophoresis, small fragments can diffuse from the gel into a deep overlay buffer. The temperature also affects the quality of the DNA profile, since migration rates increase as temperature increases. The thickness of the gels can also vary, but typically thin gels allow better resolution. Even ethidium bromide staining significantly affects DNA fragment migration.

Finally, the position of the wells of the gel can affect DNA migration rate. Sometimes DNA loaded in the middle of the gel migrates more rapidly than DNA loaded closer to the edges. The FBI monitors migration differences by loading standard DNA molecular weight markers in every third lane (the first, fourth, seventh, tenth, and so on). During sample loading, care must be taken not to mix samples in adjacent lanes. Overflow from wells will mix samples and give erroneous DNA profiles (that is, similar band patterns). Overloading the well with too much DNA causes artifacts, such as broad bands, that decrease resolution and the accuracy of measurements.

Probe Selection When the exact sequence of the VNTR repeats is known, synthetic oligonucleotide probes usually are used. The probe must include all known sequence variants (between alleles) of the VNTR repeating unit. The DNA also should be hybridized with a bacterial probe (usually a bacterial ribosomal RNA gene) to detect bacterial sequences. Often, forensic tissue is contaminated with bacteria, and VNTR probes are known to hybridize to bacterial DNA, thereby giving inaccurate DNA profiles.

POLYMERASE CHAIN REACTION

PCR amplifications of VNTR DNA regions also detect polymorphisms. PCR (described in Chapter 4) is easy to conduct, and results are obtained relatively quickly. Very small amounts of DNA—as little as the DNA from a single cell—from tissue or body fluids can be collected at a site and used as a template for amplification (that is, *in vitro* replication). DNA that has been degraded can also be amplified, since primers used in PCR amplification can anneal even to a small fragment of DNA (such as from skeletal remains and from the saliva on a postage stamp, as was obtained in the Unabomber investigation). DNA can be amplified in large enough quantity so that another laboratory can confirm results.

The extreme sensitivity of PCR makes cross-contamination a very serious problem. Trace amounts of contaminating DNA from tissues, fluids, bacteria, and laboratory equipment such as pipettes can be amplified along with the target DNA. Unfortunately, the conditions under which forensic samples are usually collected subject them to cross-contamination. Additional bands will be visible and confound results, possibly with devastating consequences. Careful selection and use of controls and adherence to appropriate protocol will ensure that cross-contamination is identified.

Usually PCR products (or sometimes the oligonucleotides) are spotted onto membrane and hybridized with the labeled probe (or the PCR product). To detect genetic variation, sequence-specific probes are used to detect the different sequences of alleles. Short sequence-specific oligonucleotides (15 to 30 bases) are hybridized against PCR products of the locus in question by dot blot hybridization. After exposure to x-ray film, or color development if a nonradioactive method of labeling is used, spots show where hybridization occurred. The sensitivity of this method allows PCR products differing by single bases to be detected.

Commercial forensic kits can distinguish the alleles that define different genotypes. The Cetus Corporation, a prominent biotechnology company, markets DQα typing kits. Its probe strip allows subtyping of samples and easy comparison of results. DQα locus variability is detected by dot blot hybridization with sequence-specific oligonucleotides. Degraded samples can be assayed by PCR. DQα is one of three families of genes on chromosome six that encode HLA Class II proteins. The DQα locus encodes the subunit of the DQ protein, a highly polymorphic protein of the major histocompatibility gene complex (MHC) that is involved in the immune response. DNA analysis involves the detection of six alleles that define 21 genotypes. Sequence differences include at least five base pairs over a 30 base pair length of the hypervariable region. Two other sequence differences in two of the alleles flank the hypervariable region. Much work has been conducted on polymor-

phisms and population frequencies. Population frequency data have been obtained for the Caucasian population (much less information is available for other ethnic populations) and demonstrate that there is approximately a 7% chance that two random individuals share the same DQα genotype.

Several marker systems have been established for forensic analysis. The ideal marker allele for efficient amplification by PCR should have 100 to 500 base pairs— small enough that fragments can be amplified from degraded DNA. Also, all alleles should be resolved by electrophoresis and hybridization. Ideally a large number of alleles should be available for use, and no single allele should be shared among individuals.

National standards and controls on PCR testing as well as on methods of collecting samples will increase accuracy and reliability. DNA standards would also increase consistency. The method of choice for forensic laboratories is VNTR analysis, despite some of the advantages of PCR, although increased testing of PCR methodology may change the order of preference. Most state crime laboratories use VNTR analysis instead of PCR because its use is well established and it has been promoted by the Federal Bureau of Investigation.

DIGITAL DNA TYPING

Minisatellite repeat coding is a digital method of DNA profiling. DNA profiling typically involves the detection of the banding pattern. Digital typing, developed by Jeffreys in 1991, does not have such problems as band shifting and the need to size specific DNA bands. Minisatellite repeat coding or maps are not based on the number of repeat units in specific loci, but instead on the variation within one locus that is demonstrated by restriction enzyme digestion. After PCR amplification of the locus, the DNA is treated with the restriction enzyme *Hae*III. Two classes of repeats are obtained, one that is cut with *Hae*III (5′GGCC3′) and the other that is not cut because a *Hae*III recognition site is not present (for example, 5′GCCC3′ instead). Numbers are assigned to each repeat. If *Hae*III cuts the PCR-amplified fragment, a "1" is assigned. An uncut fragment is assigned a "2." When one allele at a locus is cut by *Hae*III and the other allele remains uncut, a "3" is assigned (meaning both "1" and "2" are present). Several minisatellite loci are combined to generate a complex numerical pattern, thus providing more information. In other words, if seven loci were used, the barcode would have seven numbers—for example, "2133212." Each locus is assigned one number depending on whether the PCR-amplified fragments are cut, uncut, or both cut and uncut by *Hae*III. A barcode for one sample (say from blood or tissue) or one individual can readily be compared to others. The primary benefit of this method is that by eliminating the need for sizing DNA bands it also eliminates the ambiguity due to band shifting and co-migrating bands.

POPULATION COMPARISONS

The biggest challenges in forensic analysis are, one, to calculate the probability of coincidental matches by comparison with a reference population and, two, to settle on the criteria to be used for determining what constitutes a relevant reference population. What constitutes appropriate reference populations is being debated.

Individual DNA polymorphisms must be used in the context of population data to determine the probability that a match occurs by chance alone. For comparisons to be valid, they must be made against an ethnic group relevant to the individual; the normal distribution of the different alleles in the population must be known. To demonstrate a match, identical patterns between two individuals or an individual and a sample must be established. The patterns then must be shown to be unique (that is, the probability of the bands being present in another person should be extremely low). What is the probability that a DNA match is random? The answer requires a thorough and careful study of different populations to determine the distribution of the particular RFLP. Unfortunately, data for only a limited number of samples have been collected.

Determining the probability of a DNA profile match requires that several assumptions be made. One assumption is that the alleles for loci of the VNTR probes used in the laboratory segregate independently, and thus are not linked. Detecting linkage requires that a large population be sampled, which may not be possible in smaller ethnic groups. Population geneticists disagree about the significance of population comparisons. Many laboratories are establishing local population databases (for example, for African American, Caucasian, Hispanic, and Asian populations). These databases assume that random mating is occurring, which may or may not be the case in a particular group within a community.

THE FRYE TEST

Pretrial hearings are usually conducted to determine the admissibility of scientific evidence. At these hearings, the judge evaluates all evidence before presenting it to the jury. Expert witnesses for both the defense and the prosecution evaluate the evidence. Potential problems with the methods and the statistical significance are presented. Although DNA profiling is widely accepted in the courtroom, parties to a case may have to submit to a Frye hearing to demonstrate that a scientific principle is well established in the field. The defense often requests

such a hearing to determine the validity of a company's DNA testing. Since the technology has been tested numerous times and the scientific basis is no longer challenged, rejection of DNA evidence is based primarily on such factors as methods used to collect DNA and possible contamination of samples. Probably the greatest area of controversy resides with the statistical probabilities of a match.

The Frye test arose out of a judgment rejecting the admissibility of evidence based on polygraph testing because it was deemed insufficiently scientific to fall within the realm of "expert opinion evidence." The following is often cited from the 1923 *Frye v. United States* case:

"While the courts will go a long way in admitting expert testimony deducted from a well-recognized scientific principle or discovery, the thing from which the deduction is made must be sufficiently established to have gained general acceptance in the particular field in which it belongs. We think the systolic blood pressure deception test has not yet gained such standing and scientific recognition among physiological and psychological authorities as would justify the courts in admitting expert testimony deducted from the discovery, development, and experiments thus far made."

Thus, the method or technique that is used to generate the evidence must be well established, reliable, accepted by the scientific community, and supported by scientific principles. The Frye test is thus intended to prevent the presentation to a jury of invalid opinions based on unreliable experimental procedures or evidence. To demonstrate that the method is foolproof is not required, however. Also, differing opinions as to the appropriate conclusions from the method or technique can be expressed.

Even though a court may accept the reliability of DNA profiling methods for forensic analysis, the reliability of the evidence generated by DNA profiling may be subject to doubt in a particular case. The court could rule the evidence is inadmissible on the grounds that the scientific consensus on the validity of the DNA analysis method did not extend to the material itself if the DNA was degraded or even contaminated.

Witnesses used by the proponents of a particular method must satisfy certain criteria: They must be experts in the field and possess the academic and professional credentials that allow them to understand the scientific principles involved and any controversy involving the reliability of the method. These expert witnesses are often scientific investigators or technicians who perform these methods as part of their work.

General acceptance of a method, however, does not indicate reliability; therefore, proof of reliability still must be demonstrated. Studies are necessary to determine the prevalence of sequences in populations. Sample collection, RFLP methodology, and statistical interpretation must be standardized and rigorous methods applied. Published guidelines, entitled *Technical Working Group on DNA Analysis Methods*, for commercial laboratories have become the standard in the industry. Other guidelines have been published, for example, by the California Association of Criminologists. Standardization will ensure that the methods are reliable and generate accurate results.

DNA DATABASES

A variety of information about individuals (including DNA and other means of identification) can be stored in databases for future retrieval. Genetic information about populations can be gathered from DNA databases, or data banks as they are often called (since tissue and blood also can be stored). Information databases store detailed personal information about individuals from medical records and from insurance and financial reports. These data are available to insurance companies (to prevent insurance fraud and often to set insurance premiums), hospitals, and employers for obtaining additional information about an individual.

Identification databases include descriptive information about a person such as physical characteristics (eye color, height, weight), fingerprints, dental records, and genetic traits such as blood type and histocompatibility antigens. The government has used this information to identify missing individuals or even criminals.

The information provided to data banks comes from forensic collections, tissue samples taken during surgery or biopsies, blood samples, neonatal screenings, and samples taken from members of the armed forces. Databases and banks are being established across the country. The FBI, insurance companies, the military, and state facilities are beginning to collect and store genetic information. These include the FBI's shared national database of convicted felons (called CODIS or Combined Data Index System), the Department of Defense's armed services pool of biological samples, and Virginia's saliva and blood bank of convicted felons. A national database has been suggested. Large-scale sample collection would provide information for large databases, and if collected information was anonymous, such information on the population genetics of different ethnic groups would be invaluable for calculating statistical probabilities of DNA profile matches.

Issues of privacy, however, must first be addressed. Information from DNA databases could be misused and privacy infringed upon. Genetic information from sample analysis and data about genetic disease, current medical problems, health risks (such as genetic susceptibilities and risky behaviors), and other personal information could be readily obtained. Stored materials are susceptible to unauthorized use if safeguards are not in place. Specific regulations must be implemented to prevent discrimination and breach of privacy. Donors must be told exactly what data are stored, and must consent to the use of biological samples. In addition, current or potential employers or insurers must not base any decisions on genetic information and biological samples stored in banks.

General Readings

P.R. Billings, ed. 1992. *DNA On Trial: Genetic Identification and Criminal Justice.* Cold Spring Harbor Laboratory Press, Cold Spring Harbor, New York.

T. Burke, G. Dolf, A.J. Jeffreys, and R. Wolfe, eds. 1991. *DNA Fingerprinting: Approaches and Applications.* Birkhauser Verlag, Basel.

M. Farley and J.J. Harrington, eds. 1991. *Forensic DNA Technology,* Lewis Publishers, Inc., Chelsea, Michgan.

B. Herrmann and S. Hummel, eds. 1994. *Ancient DNA,* Springer-Verlag, New York.

K.M. Hui and J.L. Bidwell, eds. 1993. *Handbook of HLA Typing Techniques.* CRC Press, Boca Raton, Florida.

H.C. Lee and R.E. Gaensslen, eds. 1990. *DNA and Other Polymorphisms in Forensic Science,* Year Book Medical Publishers, Inc. Chicago.

B.S. Weir, ed. 1995. *Human Identification: The Use of DNA Markers.* Kluwer Academic Publishers, Boston.

Additional Readings

J. Baird. 1992. Forensic DNA in the trial court 1990-1992: A brief history. In P.R. Billings, ed., *DNA On Trial: Genetic Identification and Criminal Justice.* Cold Spring Harbor Laboratory Press, Cold Spring Harbor, New York, pp 61-77.

P.L. Bereano. 1992. The impact of DNA-based identification systems on civil liberties. In P.R. Billings, ed., *DNA On Trial: Genetic Identification and Criminal Justice.* Cold Spring Harbor Laboratory Press, Cold Spring Harbor, New York, pp 119-128.

D.A. Berry. 1992. Statistical issues in DNA identification. In P.R. Billings, ed., *DNA On Trial: Genetic Identification and Criminal Justice.* Cold Spring Harbor Laboratory Press, Cold Spring Harbor, New York, pp 91-108.

B. Budowle, K.L. Monson, and J.R. Wooley. 1992. Reliability of statistical estimates in forensic DNA typing. In P.R. Billings, ed., *DNA On Trial: Genetic Identification and Criminal Justice.* Cold Spring Harbor Laboratory Press, Cold Spring Harbor, New York, pp 79-90.

T. Burke and M.W. Bruford. 1987. DNA fingerprinting in birds. *Nature* 327:149-152.

R. Chakraborty and K.K. Kidd. 1991. The utility of DNA typing in forensic work. *Science* 254:1735-1739.

L.C. Chen. 1994. O. J. Simpson case sparks a flurry of interest in DNA fingerprinting methods. *GEN (Genet. Eng. News)* 14:1 and 30.

B. Devlin, N. Risch, and K. Roeder. 1991. Estimation of allele frequencies for VNTR loci. *Am J. Hum. Genet.* 48:662-667.

J.T. Epplen. 1994. DNA fingerprinting: simple repeat loci as tools for genetic identification. In B. Herrmann and S. Hummel, eds., *Ancient DNA*, Springer-Verlag, New York, pp 13-30.

P. Helminen, C. Enholm, M.L. Lokki, A.J. Jeffreys and L. Peltonen. 1988. Application of DNA "fingerprints" to paternity determinations. *Lancet*, March 12:574-576.

G. Herrin, Jr., L. Forman, and D.D. Garner. 1990. The use of Jeffreys' multilocus and single locus DNA probes in forensic analysis. In H. C. Lee and R. E. Gaensslen, eds., *DNA and Other Polymorphisms in Forensic Science*, Year Book Medical Publishers, Inc. Chicago, pp 45-60.

W.G. Hill, A.J. Jeffreys, J.F.Y. Brookfield, and R. Semeonoff. 1986. DNA fingerprint analysis in immigration test-cases. *Nature* 322:290-291.

A.R. Hoelzel and W. Amos. 1988. DNA fingerprinting and 'scientific' whaling. *Nature* 333:305.

S. Hummel and B. Herrmann. 1994. General aspects of sample preparation. In B. Herrmann and S. Hummel, eds., *Ancient DNA*, Springer-Verlag, New York, pp 159-68.

A.J. Jeffreys et al. 1987. Highly variable mnisatellites and DNA fingerprints. *Biochem Soc. Trans.* 53:165-180.

A.J. Jeffreys and D.B. Morton. 1987. DNA fingerprinting in dogs and cats. *Anim. Genet.* 18:1-15.

A.J. Jeffreys and S.D.J. Pena. 1993. Brief introduction to human DNA fingerprinting. In S.D.J. Pena, R. Chakraborty, J.T. Epplen, and A.J. Jeffreys, eds., *DNA Fingerprinting: State of the Science*. Birkhauser Verlag, Basel, pp 1-19.

A.J. Jeffreys, V. Wilson, and S.L. Thein. 1985. Hypervariable "minisatellite" regions in human DNA. *Nature* 314:67-73.

K.C. McElfresh, D. Vining-Forde, and I. Balazs. 1993. DNA-based identity testing in forensic science. *Bioscience* 43:149-157.

L. Mueller. 1991. Populations genetics of hypervariable human DNA. In M. Farley and J.J. Harrington, eds., *Forensic DNA Technology*, Lewis Publishers, Inc., Chelsea, Michgan, pp 51-62.

P.J. Neufeld and N. Colman. 1990. When science takes the witness stand. *Sci. Am.* 262(5):46-53.

S.J. Odelberg and R. White. 1990. Repetitive DNA: molecular structure, polymorphisms, and forensic applications. In H. C. Lee and R. E. Gaensslen, eds., *DNA and Other Polymorphisms in Forensic Science*, Year Book Medical Publishers, Inc. Chicago, pp 26-44.

N. Risch and B. Devlin. 1992. On the probability of matching DNA fingerprints. *Science* 255:717-720.

G. Sensabaugh, D. Crim, and C. Von Beroldingen. 1991. The polymerase chain reaction: Application to the analysis of biological evidence. In M. Farley and J.J. Harrington, eds., *Forensic DNA Technology*, Lewis Publishers, Inc., Chelsea, Michigan, pp 63-82.

W. Thompson and S. Ford. 1991. The meaning of a match: Sources of ambiguity in the interpretation of DNA prints. In M. Farley and J.J. Harrington, eds., *Forensic DNA Technology*, Lewis Publishers, Inc., Chelsea, Michigan, pp 83-152.

K. Weising, J. Ramser, D. Kaemmer, G. Kahl, and J. T. Epplen. 1991. Oligonucleotide fingerprinting in plants and fungi. In T. Burke, G. Dolf, A.J. Jeffreys, and R. Wolfe, eds., *DNA Fingerprinting: Approaches and Applications*. Birkhauser Verlag, Basel, Switzerland, pp 312-329.

REGULATION, PATENTS, AND SOCIETY 12

Rapid progress in biotechnology and the development of new products have raised legal, safety, and public policy issues. The legal and safety issues raised by the deliberate release of transgenic organisms into the environment—as, for example, when microorganisms are used in bioremediation and when transgenic food crops are grown—are beginning to be addressed. Only a coordinated effort among federal regulatory agencies will ensure that new products are evaluated and approved in a cost-effective and efficient manner. A coordinated regulatory framework must allow for a thorough evaluation of product safety and benefit to society, and must foster a continuing open discussion of issues concerning safety and levels of tolerable public risk.

Many of the new products of biotechnology are either living organisms or the products of living organisms. Companies investing in research and development are seeking intellectual property rights to many of their living investments. The patenting of living organisms and genomes has been steeped in controversy and has not been completely resolved, especially in Europe.

Although powerful new biotechnologies have the potential to alleviate medical, agricultural, and environmental problems, they can also pose problems if we do not reflect on the implications of these developments and evaluate their risks.

REGULATION OF BIOTECHNOLOGY

The safety of modern biotechnology has been debated since the first recombinant DNA experiments in 1973 (described in Chapter 2). Such statements as "we are playing God" and "we are changing the course of evolution" have been made many times over the years. Although the focus of concern has changed through the years, from fear of recombinant molecules, to release of genetically engineered organisms into the environment, to genetically engineered foods, the public continues to express fears. Nevertheless, many concerns have been assuaged as the years have passed and predicted disasters did not occur.

As scientific knowledge increased and experience with recombinant DNA molecules and methods accumulated, the National Institutes of Health (NIH) Recombinant DNA Advisory Committee (RAC) relaxed laboratory guidelines pertaining to recombinant DNA research. Relaxed regulations and positive public perceptions have encouraged the development of products from recombinant DNA technology. Regulatory agencies treat pharmaceuticals produced by this technology exactly the same as other pharmaceuticals. The process of production is no longer a regulatory consideration; only the product itself is evaluated for safety.

THE COORDINATED
BIOTECHNOLOGY FRAMEWORK

As biotechnology became more commercialized and applications proliferated, most experiments fell outside the biomedical focus of the NIH and within the jurisdiction of other federal agencies. In the April 1982 NIH-RAC revisions, experiments in which genetically engineered organisms (GEOs) would be deliberately released into the environment, were subject to review by the RAC and approval by the Institutional Biosafety Committee (IBC) of the federally funded or voluntarily complying organization. Having established the first criteria for reviewing field test proposals for GEOs, the NIH gradually reduced its role, inviting other regulatory agencies to review proposals and eventually transferring authority to the Environmental Protection Agency (EPA) and U.S. Department of Agriculture (USDA). However, as the number of proposals for field tests increased, it was seen that a new coordinated regulatory structure was needed.

In November 1985, the Office of Science and Technology Policy (OSTP) established the Biotechnology Science Coordinating Committee (BSCC), a statutory interagency coordinating committee within the OSTP. Made up of representatives of the USDA, NIH, EPA, the National Science Foundation (NSF), and the Food and Drug Administration (FDA), the BSCC was to address scientific problems raised in regulatory and research applications, promote consistency in review and assessment, identify gaps in scientific knowledge, and facilitate cooperation among federal agencies.

The first tasks of the BSCC were to develop a federal policy on biotechnology and to define a multi-agency regulatory structure that would assign jurisdictions to the various federal agencies. This policy was formulated by representatives from 18 agencies that made up the Administration's Domestic Policy Council. The OSTP published the final version of the Federal Policy on Biotechnology in the *Federal Register* on June 26, 1986. This policy, known as the Coordinated Framework for the Regulation of Biotechnology, or simply the Coordinated Framework, was meant to regulate products in all stages from research and development to marketing, shipment, and use. The framework incorporated policy statements from the three primary federal agencies (USDA, FDA, and EPA); established a systematic, coordinated review process for risk analysis and assessment; and defined the levels of review and regulation. The framework was meant to encourage a common language in reviewing and assessing proposals, thereby ensuring consistency in regulation, with the ultimate goal of streamlining and simplifying the regulatory procedures. In addition, agency jurisdiction was defined for different products and consistency in regulatory review was encouraged to prevent unnecessary delays in the review process. (For example, if an organism was both a pest control agent and a plant pathogen, the framework established which agency had primary jurisdiction.)

In setting policy for deciding whether a particular GEO should be granted a field trial, the BSCC determined how much scientific information was required and the level of risk. The policy goal was to minimize the risks while still encouraging innovation and product development. In this regulatory framework, agencies were directed to focus on the product and not on the mode of production. Products of recombinant DNA technology were treated exactly like organisms that were not genetically manipulated. Indeed, the primary feature of the Coordinated Framework was that it required no new legislation or agencies for regulating genetically engineered organisms; existing regulatory statutes and agencies were determined to be adequate. Among such statutes that now apply to many areas of biotechnology research, testing, and commercialization are the Toxic Substances Control Act, the Federal Insecticide Fungicide and Rodenticide Act, the Federal Plant Pest Act, the National Environmental Policy Act, the Plant Quarantine Act, and the Virus-Serum-Toxin Act. Although appropriate in some situations, existing statutes have shown increasingly apparent inadequacies.

Four federal agencies were charged with overseeing product regulation (Table 12.1). For example, the EPA oversaw the deliberate environmental release of genetically modified organisms for pest and pollution control; the USDA oversaw the use of genetically engineered organisms with agricultural plants and animals; and the NIH oversaw genetically engineered organisms that might affect public health. Interagency coordination was not entirely smooth, and conflicts often resulted in deadlocks and lack of progress.

The BSCC proved unable to resolve conflict among regulatory agencies, and, to facilitate cooperation, the Federal Coordinating Council on Science, Engineering, and Technology (FCCSET) replaced it in late 1990 with a new interagency committee: the Biotechnology Research Subcommittee (BRS) of the Committee on Health and Life Sciences. Although the BRS functions similarly to the BSCC, its membership is larger; in addition to the agencies listed for the BSCC, it includes representatives from the DOE (Department of Energy), AID (Agency for International Development), NASA (National Aeronautics and Space Administration), DOD (Department of Defense), OMB (Office of Management and Budget), DOC (Department of Commerce), and the Department of Interior.

IS THE FRAMEWORK EFFECTIVE?

The Coordinated Framework has been criticized for several reasons. A major problem is inadequate legal authority for oversight of all recombinant products and

Table 12.1 Regulatory Agencies and Related Statutes*

National Institutes of Health (NIH)

    Recombinant DNA Advisory Committee (RAC)**

Environmental Protection Agency (EPA)

    Toxic Substances Control Act (TSCA)

    Federal Insecticide, Fungicide, and Rodenticide Act (FIFRA)

    National Environmental Policy Act (NEPA)

Food and Drug Administration (FDA)

    Public Health Service Act (PHSA)

    Federal Food, Drug, and Cosmetic Act (FFDCA)

United States Department of Agriculture (USDA)

    Animal and Plant Health Inspection Service (APHIS)

    Agriculture Recombinant DNA Research Committee (ARRC)

    Food Safety Inspection Service (FSIS)

        Federal Plant Pest Act (FPPA)

        Plant Quarantine Act (PQA)

        Virus-Serum-Toxin Act

*For a discussion of federal agencies and their regulatory statutes see Appendix E.

**Now primarily oversees gene therapy.

genetic engineering research activities. Implementation of policies and regulations also presents problems. Important areas of biotechnology are not covered by any statutes, and there seem to be many areas of unclear jurisdiction. The framework may not cover many animals that could potentially be released into the environment; for example, some critics fear that laboratory-produced genetically engineered fish could be released without violating any regulations. The framework also is not completely clear on the commercialization of transgenic crops. The Federal Plant Pest Act (FPPA) requires that the USDA determine the impact of crops on the environment, but does not clearly cover the commercial development of most crop plants.

No statutes slow the increased use of herbicides that has come about partially as a result of engineered herbicide-tolerant crops. With the increased use of pesticides and the advent of pesticide-producing plants, insects are becoming resistant to chemical pesticides. Neither the Federal Insecticide Fungicide and Rodenticide Act (FIFRA) nor the FPPA has clear application to these critical areas. How do we ensure that these issues are addressed appropriately? Another area of concern is the limited authority of the Toxic Substances Control Act (TSCA): This chemical control statute may not apply directly to living microorganisms.

Overall, the regulatory framework does not appear to meet current needs. It does not provide biotechnology companies with a clear idea of what they are allowed or not allowed to do in the course of developing genetically engineered organisms and their products for commercialization, especially those of agricultural and environmental significance. Confusion over which statute applies to particular activities, and whether certain activities even are covered by statutes, has given rise to fear that the environment may suffer.

Experience has revealed gaps, redundancies, and inconsistencies in the regulatory process. Coordination among the three main agencies (FDA, USDA, EPA) has sometimes been difficult. For example, if the same microorganism is both a plant pest and an animal pathogen, it is subject to both USDA and EPA review and regulation, as is a microorganism that is both a pesticide and a plant pathogen. A complex regulatory process could create delays in approval, as could a regulatory dispute arising from the question whether a substance is, for example, considered a veterinary biologic (USDA) or a new animal drug (FDA). A recombinant bovine interferon might raise this question. Genetically modified animals to which no statutes specifically apply require that the framework be re-examined.

Many of the problems described lie in the fact that the legislation used to regulate genetically engineered organisms was designed primarily for the chemical industry, such as the TSCA. No one could have predicted such a use: Organisms and chemicals present very different risks. Unfortunately, statutes within the Coordinated Framework have not been able to efficiently regulate the products of biotechnology. Many regulatory procedures are extensive and costly, and yet fail to cover certain engineered organisms.

Many questions remain to be answered. Unique genetically modified microorganisms, plants, and animals will be used to produce pharmaceuticals, foods, and chemicals. How will these organisms and products be regulated? Which agencies will have oversight? Are we prepared for such rapid advances in biotechnology? The framework is being reassessed, and steps are being taken to modify the regulation of genetically engineered plants and animals in order to ensure that the regulatory framework is consistent and comprehensive, reduces confusion, and protects the environment and the public.

THE DELIBERATE RELEASE OF GENETICALLY ENGINEERED ORGANISMS

The first requests for approval to deliberately release genetically engineered organisms in field trials were submitted to the NIH-RAC in 1982. Despite critics' concerns about the environmental consequences, no specific guidelines and regulations had been drafted by federal agencies in anticipation of such applications. Among the first applications for field trials was one for testing the genetically modified soil microbe *Pseudomonas syringae* to determine its effect on frost damage to plants (as was described in Chapter 5). This landmark case stirred a great deal of controversy.

The *P. syringae* bacterium normally resides on plant surfaces and secretes a protein that facilitates the formation of ice crystals at low temperatures. Ice crystals cause tissue damage that costs U.S. citrus growers over one billion dollars a year in losses. Dr. Steven Lindblow and his colleagues at the University of California at Berkeley reasoned that if an ice-minus strain were sprayed onto leaves before the natural bacteria could colonize the surfaces, plants could survive lower temperatures without significant damage. Ice-minus deletion strains of *P. syringae* were constructed by deleting the gene that encodes the ice-nucleation protein.

To obtain NIH-RAC permission, a proposal was submitted by the applicants, announced in the *Federal Register*, and reviewed by a panel of experts. In 1983, after carefully examining the proposal, the NIH-RAC granted permission to proceed with field tests. Concerned citizens immediately challenged the review process. The Foundation on Economic Trends, headed by Jeremy Rifkin, filed a suit to block the tests. The suit was successful, and a judge blocked the tests on the grounds that the NIH-RAC had not requested an environmental impact assessment or held a hearing as required by federal law.

In late 1984, the EPA claimed jurisdiction to regulate recombinant bacteria used as pesticides. Ice-minus *P. syringae* was considered a pesticide that controlled naturally occurring wild-type *P. syringae*. The EPA received two separate applications to apply ice-minus *Pseudomonas* species to potato plants as frost protectants. In November, Advanced Genetics Sciences, Inc. (now DNA Plant Technology Corp.) applied for clearance to spray ice-minus *P. syringae* and *P. fluorescens* on the flowers of the plants to determine whether the growth of ice-nucleation bacteria would be prevented or reduced, thus reducing frost injury between 0 and −5°C. In December, Lindow applied for clearance to treat potato seed segments with ice-minus *P. syringae* just before planting them, and then to spray the plants with the bacteria once foliage appeared, to assess the effect on frost damage.

The EPA conducted an extensive assessment of the ice-minus strain by evaluating its ecological impact, human health implications, and environmental fate. The EPA held public meetings, and commissioned evaluations by the NIH and FDA. The EPA's General Council Office, Office of Pesticide Program Review, and its committees on Toxic Substances, Research and Development Policy Planning, and Evaluation each assessed the proposed field trial plans. The California Department of Agriculture also was involved in the evaluation, since the field tests would be conducted in California. In addition, a science advisory panel composed of a plant pathologist, a microbiologist, and an ecologist reviewed the proposals. In short, this review process was extensive, costly, and even redundant. (There was consensus, however, that in time the review process could be streamlined without sacrificing quality.)

Opponents of the tests expressed fears that deliberately released GEOs might replace natural organisms in the environment, spreading into important niches and devastating "the balance of nature," and might transfer introduced genes to other organisms (such as by conferring herbicide resistance to weed species). Both applications anticipated these objections. The vicinities of the test sites were to be monitored to detect dissemination by wind or insects, and after each experiment, the foliage was to be incinerated and plant debris was to be disked into the soil.

The Hazard Evaluation Division of the EPA concluded that neither of the proposed set of ice-minus field trials would pose a significant risk to the environment or to human health. Permission was granted (although fearful residents near the test site temporarily blocked the tests and further delayed experimentation through court orders). Field trials were finally conducted in 1987, and data indicated that the engineered bacteria were effective and neither persisted after application nor were dispersed into the environment. Nevertheless, engineered ice-minus microorganisms are not likely to be marketed. Natural deletions of the ice-nucleation gene have been isolated, and will be marketed under the name Frostban. A nonengineered product will have an easier time making it through the regulatory maze.

The ice-minus experiments paved the way for the many field trials conducted since. The outdoor application of recombinant microorganisms seems to be safe. No transfer of genes to indigenous organisms has been detected, and the GEOs seem to remain at the site of application. Genetically engineered crop plants have been introduced into the environment with little objection, although some fear that genes could be transferred to weedy species through cross-pollination. Some proponents argue that recombinant crop plants differ little from those generated by classical plant breeding methods.

RISK ASSESSMENT

The public and environmental scientists alike have voiced numerous fears—for example, that genetically modifying plants and animals will turn benign organisms into economically destructive pests or will enhance existing pests through hybridization (sexual out-crossing common in plants); that new biotechnology products (plant pesticides, for example) may cause harm to nontarget species such as insects; that a recombinant virus used in the production of recombinant organisms might detrimentally infect other organisms; that recombinant organisms may eliminate indigenous species through competition, depletion of valuable resources and nutrients, and by incomplete degradation of toxic wastes to even more toxic by-products during bioremediation.

Those assessing risk must answer many questions. For example, when a transgenic organism is released into the environment, what is the probability of gene transfer, and how can the movement of genes be monitored? What information is needed to assess risk adequately? How much baseline data is required? Risk assessment must take into account numerous factors, both intrinsic and extrinsic. Intrinsic factors include the genetic modification, the molecular biology, natural history, and genetic background of the organism to be released. Characteristics that can be expressed include the conversion of a nonpathogen to a pathogen, the ability to resist host defense systems, and the ability to invade cells, increase host range, and survive adverse conditions. Extrinsic factors are the characteristics of the immediate environment such as selection pressure, habitat quality, nontarget organisms that might become recipients, and the density of organisms that directly or indirectly affect the organisms to be released. Introductions can be modified forms of resident organisms, forms that exist elsewhere, or resident organisms disabled in such a way as to require a supplement for survival.

HOW MUCH RISK?

Risk is defined as an estimate of the probability that an event will have an adverse effect and an estimate of the magnitude of that effect. In other words, what is the probability that something bad will happen, and, if it does happen, how serious are the consequences?

Those charged with developing regulatory plans to ensure safety must balance the degree of oversight, or management, with the level of risk. Comprehensive risk analysis thus requires both risk assessment and risk management. Risk assessment uses scientific data to identify and estimate potential adverse effects; risk management is the process of weighing alternatives to select the most efficient and effective regulatory plan. Technical, economic, political, and social factors affect the manner in which assessment and management are carried out by regulatory agencies under legislative mandates. When weighing potential risks versus benefits versus regulatory costs, decision makers must make value judgments.

Risk can be broken down into several components. The first, *quantitative risk assessment*, has already been described; it is an empirical or experimental concept that provides an estimate of the probability that something negative will happen. *Acceptable level of risk* states a trade-off between risk and benefit. Do the costs, or risks, outweigh the benefits to society? The risk in question is weighed against both the benefit and the cost of eliminating the risk (for example, either by increasing safeguards or by forgoing the benefits—that is, not adopting the technology in question). The *cultural context of risk* encompasses the psychological and social aspects of risk. How do members of a community or society perceive the level of risk and how do they perceive the benefit and where do they set the balance between the two? Actual risk (that is, the experts' probabilistic risk assessment) does not usually parallel perceived risk. *Risk communication* is an essential component; the public perception of risk is influenced by the way that risk assessment is communicated by the media in a variety of contexts.

All of the components of risk just enumerated influence policy makers. The public is still grappling with the question of what constitutes an acceptable level of risk. How much risk will be tolerated, and, more important, how much are we willing to pay to make a genetically engineered organism completely safe?

For an engineered organism to pose a risk to the environment (such as by displacing indigenous organisms) or to human and animal health (such as by the spread of a pathogen), it must be able to survive, reproduce, and disseminate beyond the area where it was released and intended to function, or it must be able to transfer genetic material to other organisms. Humans, plants, animals, and ecological systems may be put at risk of injury and environmental displacement or disruption.

Risks to the environment also may arise from market forces. Suppose, for example, that a single company produced an herbicide and engineered herbicide-resistant crop plants that must be used with it. If farmers became completely dependent on the combination, the environmental risks could include the increased use of agricultural chemicals (pesticides and herbicides) or decreased genetic diversity among organisms.

PATENTS AND BIOTECHNOLOGY

The ethics of patenting plants, animals, and their genes have been hotly debated for several years. Proponents of such patents have argued that in the absence of patents, which demonstrate legal ownership of a product, there might not be incentive to invest large amounts of money for research and development. Some have also argued that the incentive of such patents may accelerate progress in biotechnology. The commercialization of biotechnology has benefited from the potential for patenting products and organisms resulting from recombinant DNA technology; companies are willing to take greater risks and invest more funds into research and development if they stand to benefit by receiving more profits on legally protected new products.

In addition to patents, legal protection of intellectual property rights can take the form of trade secrets (to protect technical details and formulas from unauthorized disclosure and use), trademarks (names for products) and service marks, trade dress (ornaments, shapes, esthetic objects), or copyrights. Patents are the most valuable type of intellectual property protection; by giving the owner exclusive rights to market a product or invention and thereby potentially earn substantial profits, they encourage technological innovation, investment, and the development of beneficial new products. In addition, any products that are derived from the original product also are protected. A patent is meant to exclude others from using, producing, or selling the legally protected product or invention without purchasing the rights to do so. In the United States, patent protection begins on the date the patent is granted and ends 20 years from the date the application was filed, after which others can use the invention freely.

The Patent Process

Biotechnology innovations fall within all three categories of patentable inventions: products or composition of matter, methods of use, and manufacturing processes. Of the three, products are most easily protected because patent infringement can be determined by examination of a product.

Several conditions must be met for patentability. The invention must be new (i.e., not previously published or presented anywhere), useful, and nonobvious to one skilled in the field. A patent application must include several elements.

- The technical field to which the invention applies must be described.

- A background search must identify the problems to be solved as well as describing the prior "art" (including information that is publicly available); this search aids in determining the scope of the patent claim for the invention.

- How the invention improves upon the prior art must be described.

- A summary must enumerate the fundamental components of the invention.

- A description of the invention and the indispensable steps for constructing the invention must be sufficiently detailed for someone skilled in the field to reconstruct it.

- Finally, the application must include claims that outline the elements protected by law. (The claim must be clearly stated, must define the aspects of the invention to be protected, and must describe the scope, prior art, and infringement. A claim cannot be so broad that it infringes on prior "art." On the other hand, if the claim is too narrowly defined, the applicant may risk losing property claims; an attorney familiar with both patent law and science can be invaluable in helping to define the claims.)

In the United States, a grace period allows scientists to publish their information within one year of applying for a patent without forfeiting their patent rights (Japan and Canada also have grace periods). Thus, information about the invention or patentable discovery cannot be disclosed in any manner—be it a scientific meeting or publication—more than one year before the application is submitted. In Europe, the application must be filed before any information can be disclosed by publication or other means.

Companies in the United States that are granted U.S. patents often seek a patent in another country, but patent protection is inconsistent from country to country; definitions of what constitutes a patentable invention vary, as do the patent laws themselves. A patent application that is acceptable in the United States might be rejected in some countries and in others might be too narrowly defined to give the filer adequate protection. Rapid biotechnological advancements have led to suggestions that international patent laws be established.

History of Patenting Plants and Animals

In 1930, the United States Congress passed the Plant Patent Act, which allows asexually propagated plant varieties to be patented. Plant varieties produced by cuttings, budding, and grafting could be patented (but not seeds). With the Plant Variety Protection Act of 1970, Congress extended protection to sexually propagated plants (excluding first-generation hybrids). Plant breeders now could have exclusive rights over the propagation and sale of sexually propagated plant varieties. Important crops produced by the classical breeding methods of crossing, screening for desirable characteristics, and selection through seeds could be patented.

Patents for living organisms had been filed unsuccessfully until 1980, when, after a landmark Supreme Court case, the United States Patent and Trademark Office determined that life forms were patentable. In *Diamond v. Chakrabarty,* the United States Supreme Court ruled that there is no distinction between living and nonliving matter and that living matter is patentable. Ananda Chakrabarty had filed an application for a patent on a genetically modified bacterium from the genus *Pseudomonas,* which could break down crude oil (see Chapter 5, Microbial Biotechnology). The bacterium harbored at least two plasmids that provided the genes for hydrocarbon degradation. In his application, Chakrabarty sought to patent the bacterium, the process of producing its characteristics, and the method of dispersal. The Patent and Trademark Office rejected the organism claim, stating that the genetic manipulations that resulted in the oil-degrading product did not constitute a product of manufacture or a new composition of matter. The examiner operated on the supposition that living things are not patentable subject matter and that microorganisms are products of nature.

Since the inventor claimed the bacterium, the Supreme Court had to decide whether a living organism that had been modified by a person was patentable subject matter. The Court determined that "anything under the sun that is made by man" is patentable and ruled in favor of the patent, stating that the issue is not whether something is living or nonliving but whether it is a product of human modification or of nature.

In the past, microorganisms could be patented if they were a part of a process and the direct use was identified. Many such patents have been granted in the fermentation and pharmaceutical industries dating back to 1873, when Pasteur was issued a patent for purified yeast. In 1987, the Patent and Trademark Office determined that all nonhuman multicellular living organisms, including animals, can be patented if certain criteria are met. Shortly thereafter, scientists were granted a patent for a genetically altered triploid Pacific oyster that was produced by hydrostatic pressure, a process that did not use recombinant DNA methods. In 1988, the first genetically engineered animal was patented (both the process and product): the Harvard mouse (also known as the oncomouse), which has a breast cancer susceptibility gene that made the animal extremely sensitive to carcinogens. This mouse was to serve as a cancer model on which to test new anticancer drugs and to determine the role of environmental factors in cancer development. The patent covers all animals that have a variety of cancer genes inserted into the genome. In Europe, the road to patenting living organisms has not been as smooth; the European Patent Office has not readily granted authority to patent genetically modified plants or animals.

The patenting of animals has generated more heated debate than perhaps any other area of patent law. Again, proponents argue that animal patents stimulate research and development by protecting investment during the development of new technologies. They argue that genetically engineered animals will be more resistant to disease and will go to market faster, thus reducing farmers' costs.

Opponents of animal patenting argue that private ownership of animals (i.e., monopoly ownership) should not be allowed; that genetic manipulation may be cruel and cause undue harm to animals, may increase the use of animals in research and commercial endeavors, and may even decrease the genetic diversity of commercial animals. There have been claims that family farmers cannot afford to pay for the use of patented animals, and that the higher volume of animal production will drive prices lower and mean the eventual demise of the small family farm. Farmers often pay a heavy price economically when surplus agricultural production drives prices down.

As a cautionary example, opponents of patents on living organisms cite the commercialization of hybrid crops that resulted from patent protection of seeds. Large, multinational companies invested heavily in plant breeding and the development of new varieties with desirable traits. Hybrid crop plants were patented and commercialized. Companies concentrated on a few high-yielding varieties and gained control of world markets. Farmers worldwide have become dependent on a small variety of high-yielding crop plants that require costly inputs of fertilizers and pesticides. One concern is that patents may encourage the development of too many pesticides and herbicides that are linked to the patented crops produced by the same companies. Thus, chemical usage is perpetuated by the patent system and the farmer becomes even more dependent on the use of agricultural chemicals to maintain high crop productivity.

By embracing high-technology crops, farmers in the developing world often discard ancient crops that harbor resistance to disease and pests and have adapted to harsh (e.g., nutrient-poor) environments. The world increasingly is relying on a small number of food crops that have lost much genetic diversity. Whether patents are the impetus is difficult to determine, but some critics warn that we should proceed with caution. The Patent and Trademark Office is not required to examine the environmental or social implications of a patent application, nor has Congress limited patents because a patented item may have adverse consequences. The public ultimately will be charged with determining which patented products are of social, medical, environmental, and agricultural benefit.

WHAT ABOUT THE FUTURE?

In the recent past, some constituencies have requested a five-year moratorium on granting patents for genetically modified animals so that Congress can completely assess the economic, ethical, and environmental issues raised by the patenting of animals. Although the animal patent controversy has diminished somewhat, the issue probably will continue to be debated.

We are witnessing the beginning of a new era of patented products that will significantly affect our lives. We must determine how to proceed with new knowledge and technologies and how to use them within the confines of the law. If they are used wisely we can at the very least expect an increase in the quality of food, pharmaceuticals, and chemicals that can be used in a multitude of ways to enhance the quality of life; progress in human medicine and environmental cleanup; and increased agricultural productivity. The legal, ethical, economic, and social implications of modern biotechnology must be thoroughly examined if we are to ensure that biotechnology products contribute in positive ways.

General Readings

P. Barker, ed. 1995. *Genetics and Society.* The H.W. Wilson Company, New York.

L. Busch, W.B. Lacy, J. Burkhardt, and L.R. Lacy. 1992. *Plants, Power, and Profit: Social, Economic, and Ethical Consequences of the New Biotechnologies.* Blackwell Scientific Publications, Cambridge, Massachusetts.

R.S. Crespi. 1988. *Patents: A Basic Guide to Patenting in Biotechnology.* Cambridge University Press, Cambridge.

B.D. Davis, ed. 1991. *The Genetic Revolution: Scientific Prospects and Public Perceptions.* The Johns Hopkins University Press, Baltimore, Maryland.

K.A. Drlica. 1994. *Double-Edged Sword: The Promises and Risks of the Genetic Revolution.* Addison-Wesley Publishing Company, Reading, Massachusetts.

L.R. Ginzburg. 1991. *Assessing Ecological Risks of Biotechnology.* Butterworth-Heinemann, Boston, Massachusetts.

W.H. Lesser, ed. 1989. *Animal Patents: The Legal, Economic, and Social Issues.* Stockton Press, New York, N.Y.

National Research Council. 1989. *Field Testing Genetically Modified Organisms: Framework for Decisions.* National Academy Press, Washington, D.C.

G.T. Tzotzos, ed. 1995. *Genetically Modified Organisms: A Guide to Biosafety.* CAB International, United Kingdom.

Additional Readings

R.G. Adler. 1984. Biotechnology as an intellectual property. *Science* 224:357-363.

J.H. Barton. 1991. Patenting life. *Sci. Am.* 264(3):40-46.

D.B. Berkowitz. 1990. The food safety of transgenic animals. *Bio/Technology* 8:819-825.

A.H. Berks. 1994. Patent information in biotechnology. *Trends Biotechnol.* 12:352-364.

G. Bogosian and J.F. Kane. 1991. Fate of recombinant *Escherichia coli* K-12 strains in the environment. *Adv. Appl. Microbiol.* 36:87-131.

R.S. Crespi. 1992. What's new in patent law? *Trends Biotechnol.* 10:108-110.

D.J. Drahos. 1991. Field testing genetically engineered microorganisms. *Biotechnol. Adv.* 9:157-171.

P.D. Kelly. 1991. Are isolated genes "useful"? *Bio/Technology* 10:52-55.

L. Kim. 1993. *Advanced Engineered Pesticides.* Marcel Dekker, New York. (See section III, Regulatory, pp 321-419.)

E. Marshall. 1991. The patent game: Raising the ante. *Science* 253:20-24.

M. Ratner. 1990. Survey and opinions: barriers to field-testing genetically modified organisms. *Bio/Technology* 8:196-198.

J.M. Tiedje, R.K. Colwell, Y.L. Grossman, R.E. Hodson, R.E. Lenski, R.N. Mack, and P.J. Regal. 1989. The planned introduction of genetically engineered organisms: Ecological considerations and recommendations. *Ecology* 70:298-315.

Appendix A

Summary of NIH Guidelines, July 1976

This summary of National Institutes of Health (NIH) guidelines of July 1976 defines four levels of physical containment, P1 through P4, and three levels of biological containment, EK1 through EK3, in order of increasing stringency. Experiments are assigned levels according to their potential risk.

a. Shotgun experiments using *E. coli* as the host	
Non-embryonic primate tissue	P3+EK3 or P4+EK2
Embryonic primate tissue or germ line cells	P3+EK2
Other mammals	P3+EK2
Birds	P3+EK2
Cold blooded vertebrates, non-embryonic	P2+EK2
embryonic or germ line	P2+EK1
If vertebrate produces a toxin	P3+EK2
Other cold blooded animals and lower eukaryotes	P2+EK1
If Class 2 pathogen*, produces a toxin, or carries a pathogen	P3+EK2
Plants	P2+EK1
Prokaryotes that exchange genes with *E. coli*	
Class 1 agents (non-pathogens)	P1+EK1
Low risk pathogens (for example, enterobacteria)	P2+EK1
Moderate risk pathogens (for example, *S. typhi*)	P2+EK2
Higher risk pathogens	banned
Prokaryotes that do not exchange genes with *E. coli*	
Class 1 agents	P2+EK2 or P3+EK1
Class 2 agents (moderate risk pathogens)	P3+EK2
Higher pathogens	banned

In all above cases, if DNA is at least 99% pure before cloning and contains no harmful genes, either physical or biological containment levels can be reduced one step.

b. Cloning plasmid, bacteriophage and other virus genes in *E. coli*

Animal viruses**	P4+EK2 or P3+EK3
If clones free from harmful regions	P3+EK2
Plant viruses	P3+EK1 or P2+EK2
99% pure organelle DNA, Primates	P3+EK1 or P2+EK2
other eukaryotes	P2+EK1

  Impure organelle DNA: shotgun conditions apply.

Plasmid or phage DNA from hosts that exchange
genes with *E. coli*

  If plasmid or phage genome does not contain harmful
  genes or if DNA segment 99% pure and characterised   P1+EK1

  Otherwise, shotgun conditions apply.

Plasmids and phage from hosts which do not exchange
genes with *E. coli*

  Shotgun conditions apply, unless minimal risk that
  recombinant will increase pathogenicity or ecological
  potential of the host, then   P2+EK2 or P3+EK1

c. Animal virus vectors

Defective polyoma virus+DNA from non-pathogen	P3
Defective polyoma virus+DNA from Class 2 agent	P4
If cloned recombinant contains no harmful genes and host range of polyoma unaltered, reduce to	P3
Defective SV40+DNA from non-pathogens	P4

  If inserted DNA is 99% pure segment of prokaryotic
  DNA lacking toxigenic genes, or a segment of eukaryotic
  DNA whose function has been established and which
  has previously been cloned in a prokaryotic host-vector
  system, and if infectivity of SV40 in human cells unaltered   P3

Defective SV40 lacking substantial section of the late
  region+DNA from non-pathogens, if no helper used and no
  virus particles produced   P3

Defective SV40+DNA from non-pathogen can be used to
  transform established lines of non-permissive cells under
  P3 provided no infectious particles produced. Rescue of
  SV40 from such cells requires   P4

d. Plant host-vector systems

P2 conditions can be approximated by insect-free greenhouses, sterilization of
  plant, pots, soil and runoff water, and use of standard microbiological practice.

P3 conditions require use of growth chambers under negative pressure and routine
  fumigation for insect control.

Otherwise, similar conditions to those prescribed for animal systems apply.

*Classes for pathogenic agents as defined by the Center for Disease Control.

**cDNAs synthesised *in vitro* from cellular or viral RNAs are included in above categories.

Date	Containment Requirements (selected experiments)	Oversight for Government-Funded Work	Oversight for Work in the Private Sector	Prohibitions
December 1978	Containment levels work with *E. coli* K12 ranges from P1EK1 to P3EK2	Prior review and approval by IBCs; registration with NIH	Voluntary registration	Six classes; cloning of DNA from class 3, 4, or 5 from P1 to P3 pathogens or from oncogenic viruses classified as moderate risk
				Cloning of genes encoding toxins
				Creation of plant pathogens likely to have enhanced virulence or host range.
				Deliberate release into the environment
				Transfer of drug resistance trait not known to occur naturally
				Large-scale experiments (greater than 10 liters)
January 1980 (response to Rowe-Campbell proposal)	Experiments using *E. coli* K12 that are not prohibited and not exempt: P1	*E. coli* K12 experiments that are not prohibited and not exempt: registration with IBCs; no prior review, except for experiments involving expression of genes Other experiments: prior review and approval by IBCs; registration with NIH	Voluntary compliance scheme for private sector, with procedures for protection of trade secrets and commercial and financial information	Six classes retained
November 1980 (response to Singer proposal)	Experiments using *E. coli* K12 and *S. cerevisiae* systems that are not prohibited and not exempt: P1	Elimination of NIH oversight for all experiments assigned containment levels in the guidelines; IBCs responsible for reviewing research for compliance with the guidelines	RAC responsible only for setting containment requirements for L-S processes	Five classes of prohibition; prohibition of the creation of plant pathogens with increased host range and virulence deleted

Date	Containment Requirements (selected experiments)	Oversight for Government-Funded Work	Oversight for Work in the Private Sector	Prohibitions
July 1981 (response to IBC chairs' proposal)	Experiments using approved *E. coli, S. cerevisiae*, and *B. subtilis* systems that are not prohibited: exempt; P1 recommended	Nonexempt experiments assigned containment levels in guidelines; prior review and approval by IBCs		Prohibition on the formation of rDNA containing the genes for biosynthesis of toxins limited to a few lethal toxins; other four classes retained
October 1981 (response to Lilly proposal)				Prohibition lifted on large-scale processes using approved *E. coli, S. cerevisiae,* and *B. subtilis* systems; such processes require only IBC prior review and approval
April 1982 (response to Baltimore-Campbell proposal)	For nonpathogens: P1	Elimination of NIH oversight for all experiments and processes except for three classes of previously prohibited experiments; IBCs and/or principal investigators responsible for all other experiments and processes		All prohibitions removed; cloning of genes encoding highly lethal toxins, deliberate release into the environment, and transfer of drug resistance trait not known to occur naturally require RAC review and NIH and IBC approval before initiation

Sources: Department of Health, Education, and Welfare, National Institutes of Health, "Guidelines for Research Involving Recombinant DNA Molecules," *Federal Register* 43 (22 December 1978): 60108-31; Department of Health, Education and Welfare, National Institutes of Health, "Guidelines for Research Involving Recombinant DNA Molecules," *Federal Register* 45 (29 January 1980): 6724-49; Department of Health, Education, and Welfare, National Institutes of Health, "Recombinant DNA Research: Actions under Guidelines," *Federal Register* 45 (21 November 1980): 77372-81; Department of Health and Human Services, National Institutes of Health, "Guidelines for Research Involving Recombinant DNA Molecules," *Federal Register* 46 (1 July 1981): 34462-87; Department of Health and Human Services, National Institutes of Health, "Recombinant DNA Research: Actions under Guidelines," *Federal Register* 46 (30 October 1981): 53980-85; Department of Health and Human Services, National Institutes of Health, "Guidelines for Research Recombinant Involving Recombinant DNA Molecules," *Federal Register* 47(21 April 1982):17180–98.

Appendix C

The Immune Response: An Overview

The body has several lines of defense against foreign invaders (Table C.1): various barriers at the surface, non-specific responses such as inflammation, and specific immune responses that involve white blood cells (Table C.2) or defense compounds such as antibodies and complement proteins. Plasma proteins that together comprise the complement system circulate in the blood in an inactive form. They are readily activated when they bind to antibodies that have attached to the antigens on a foreign invader such as a pathogen. The activated complement proteins in turn activate many other types of complement proteins that protect the body and tissues

Table C.1 Three Lines of Defense Against Pathogens

Barriers at Body Surfaces (*nonspecific* targets):

1. Intact skin; mucous membranes at other body surfaces

2. Infection-fighting substances in tears, saliva, etc.

3. Normally harmless bacterial inhabitants of body surfaces that outcompete pathogenic visitors

4. Flushing effect of tears, urination, and diarrhea

Nonspecific Responses (*nonspecific* targets):

1. Inflammation

   a. Fast-acting white blood cells (neutrophils, eosinophils, basophils)

   b. Macrophages (also take part in immune responses)

   c. Complement proteins, blood-clotting proteins, other infection-fighting substances

2. Organs with phagocytic functions (e.g., lymph nodes)

Immune Responses (*specific* targets):

1. White blood cells (macrophages, T cells, B cells)

2. Communication signals (e.g., interleukins) and chemical weapons (e.g., antibodies, complement proteins)

Table C.2 Major White Blood Cells and Their Roles in Defense

Cell Type	Main Characteristics
Macrophage	Phagocyte; takes part in nonspecific defense responses; presents antigen to T cells; cleans up and helps repair tissue damage
Neutrophil	Fast-acting phagocyte; takes part in inflammation but not in sustained responses
Eosinophil	Secretes enzymes that attack parasitic worms
Basophil and mast cell	Secrete histamines, prostaglandins, other substances that act on small blood vessels, producing inflammation; also have roles in allergies
Lymphocytes:	(All take part in most immune responses; following antigen recognition, all form clonal populations of effector cells and memory cells.)
1. B cell	Effectors secreté four classes of antibodies (IgA, IgE, IgG, and IgM) that protect the host in specialized ways
2. Helper T cell	Effectors secrete interleukins that stimulate rapid divisions and differentiation of both B cells and T cells
3. Cytotoxic T cell	Effectors kill infected cells, tumor cells, and foreign cells by way of a lethal hit
Natural killer (NK) cell	Cytotoxic cell of undetermined affiliation (may be a kind of lymphocyte); kills infected cells and tumor cells by way of a lethal hit

Sources: (Table C.1 and Table C.2): C. Starr and R. Taggart, *Biology: The Unity and Diversity of Life,* 7th ed. Copyright © 1995 Wadsworth Publishing Co., Inc.

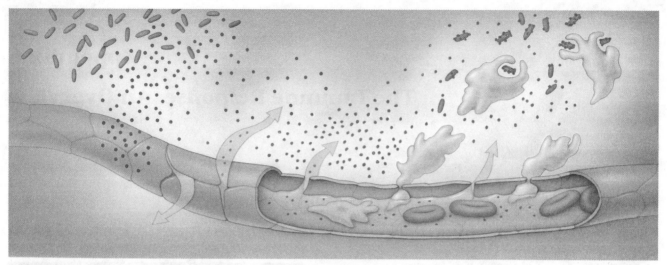

a Bacteria invade a tissue. They kill cells or release harmful metabolic by-products.

b The substances released by bacteria and by damaged or killed body cells accumulate in the tissue.

c The substances make the tissue's small blood vessels more permeable. Plasma fluid and various plasma proteins escape into the tissue.

d Some plasma proteins attack bacteria. Others create chemical gradients that facilitate migration of phagocytes to the issue. Still others repair tissue damage (as by clotting mechanisms).

e Phagocytic white blood cells engulf bacteria.

Figure C.1 Acute inflammation from a bacterial infection. Blood flow to the area is increased and white blood cells enter the tissue to kill the invading bacterial cells.

from damage by invaders; they protect through lysis of target cells or by nonspecific inflammation that draws phagocyctic white blood cells to the damaged tissue (Figure C.1). Four types of white blood cells are involved in defense and repair of tissue: neutrophils, basophils, eosinophils, and macrophages.

The body's immune system initiates a variety of defenses against an antigen that is recognized as foreign. Membrane proteins collectively called the major histocompatibility complex (MHC) mark cells of the body as "self"; MHC markers are also involved in the defense against foreign invaders. Some MHC markers are common to all cells and others are found only in lymphocytes and macrophages. Macrophages engulf invading cells and express the foreign cell surface antigens as antigen-MHC complexes on their own cell surface. These antigen-presenting cells activate cell-mediated (Figure C.2) and antibody-mediated immune responses (Figure C.3). Once activated, B and T lymphocytes rapidly divide and differentiate into effector cells that attack the enemy and memory cells that remember the antigen that activated the immune response. When the invader (i.e., antigen) reappears, a more rapid and stronger immune response is initiated.

Figure C.2 A cell-mediated response begins after an antigen-MHC complex activates a cytotoxic T cell.

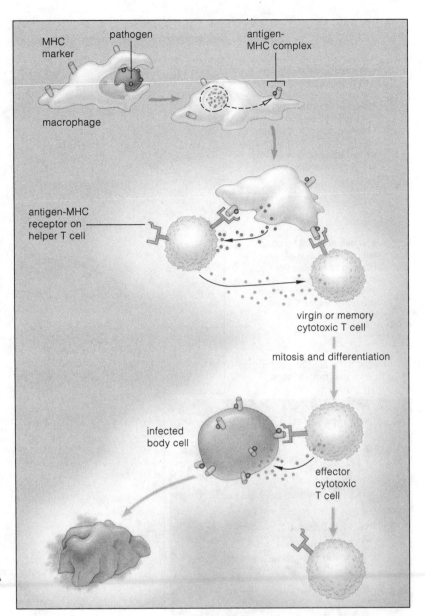

a A macrophage engulfs and digests a pathogen, then cleaves its antigen into fragments that bind to MHC markers. The macrophage becomes an antigen-presenting cell; it displays processed antigen-MHC complexes at its surface.

b Receptors on T cells bind to the complexes. Binding stimulates the macrophage to secrete interleukin-1 (pink dots). This stimulates helper T cells to secrete other interleukins (blue dots). These stimulate virgin or memory cytotoxic T cells to divide and differentiate into large populations of effector and memory cells. Only the effector cytotoxic T cells have cell-killing abilities.

c An effector encounters a target: an infected body cell that has the processed antigens bound with MHC markers at its surface. And it delivers a lethal hit. It releases perforins and toxic substances (green dots) onto its target and so programs it for death.

d The effector disengages from the doomed cell and reconnoiters for new targets. Meanwhile, perforins make holes in the target's plasma membrane. Toxins move into the cell, disrupt organelles, and make the DNA disassemble. The infected cell dies.

MHC marker
pathogen
antigen-MHC complex
macrophage
antigen-MHC receptor on helper T cell
virgin or memory cytotoxic T cell
mitosis and differentiation
infected body cell
effector cytotoxic T cell

a A macrophage engulfs and digests bacterial cells. Bacterial antigens are cleaved into fragments that bind to MHC markers. The macrophage displays antigen-MHC complexes at its surface; it serves as an antigen-presenting cell.

b A helper T cell binds to antigen-MHC complexes. Binding stimulates the macrophage to secrete interleukin-1 (dots). These stimulate helper T cells to interact with B cells.

c A virgin or memory B cell has membrane-bound antibodies that can bind to the bacterial antigens.

d Helper T cell secretes interleukins (dots) that stimulate the B cell and its descendants to divide repeatedly. Part of the resulting B cell population differentiates into antibody-secreting effector B cells. Part is reserved as memory B cells for future battles with this specific invader.

e Circulating antibody molecules bind to antigen on other bacterial cells, tagging them for destruction.

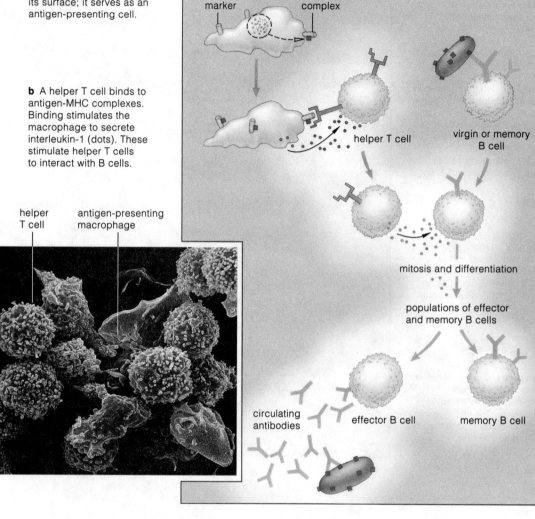

antigen

macrophage

MHC marker

antigen-MHC complex

helper T cell

virgin or memory B cell

mitosis and differentiation

populations of effector and memory B cells

circulating antibodies

effector B cell

memory B cell

helper T cell

antigen-presenting macrophage

Figure C.3 A bacterial invasion elicits an antibody-mediated response.

Organic Chemical	Microbial Sources	Selected Uses
Acetic acid	*Acetobacter*	Industrial solvent and intermediate for many organic chemicals, food acidulant
Acetone	*Clostridium*	Industrial solvent and intermediate for many organic chemicals
Acrylic acid	*Bacillus*	Industrial intermediate for plastics
Butanol	*Clostridium*	Industrial solvent and intermediate for many organic chemicals
2,3-Butanediol	*Aerobacter, Bacillus*	Intermediate for synthetic rubber manufacture, plastics and antifreeze
Ethanol	*Saccharomyces*	Industrial solvent, intermediate for vinegar, esters and ethers, beverages
Formic acid	*Aspergillus*	Textile dyeing, leather treatment, electroplating, rubber manufacture
Fumaric acid	*Rhizopus*	Intermediate for synthetic resins, dyeing, acidulant, antioxidant
Glycerol	*Saccharomyces*	Solvent, plasticizer, sweetener, explosives manufacture, printing, cosmetics, soaps, antifreeze
Glycolic acid	*Aspergillus*	Textile processing, pH control, adhesives, cleaners
Isopropanol	*Clostridium*	Industrial solvent, cosmetic preparations, antifreeze, inks
Lactic acid	*Lactobacillus, Streptococcus*	Food acidulant, dyeing, intermediate for lactates, leather treatment
Methylethyl ketone	*Chlamydomonas*	Industrial solvent, intermediate for explosives and synthetic resins
Oxalic acid	*Aspergillus*	Printing and dyeing, bleaching agent, cleaner, reducing agent
Propylene glycol	*Bacillus*	Antifreeze, solvent, synthetic resin manufacture, mold inhibitor
Succinic acid	*Rhizopus*	Manufacture of lacquers, dyes and esters for perfumes

NATIONAL INSTITUTES OF HEALTH (NIH)

The Recombinant DNA Molecule Program Advisory Committee (RAC) of the National Institutes of Health (NIH) established voluntary guidelines for research involving recombinant DNA in 1976. These guidelines have been modified and relaxed over the years. The NIH lacks statutory authority to regulate research, and the guidelines have real significance only for recipients of federal grants; noncompliance resulted in loss of federal funding. Institutional Biosafety Committees are now responsible for reviewing, approving, and monitoring the research of principal investigators. The principal investigator is responsible for following appropriate safety standards when conducting experiments.

In the late 1970s the NIH, USDA, FDA, and NSF effected informal coordination of laboratory research; though implementing no new rules, they required research to comply with existing NIH guidelines. Coordination became more difficult when recombinant DNA technology reached the stage of commercial product development, presenting new questions regarding safety. With large-scale commercial production and the potential for deliberate release into the environment, additional regulatory agencies were required. Although existing federal statutes protected both human health and the environment, there was confusion about agency jurisdiction over various products. In fact, products were often regulated by more than one statute through more than one agency during development, testing, and marketing. Consequently, the NIH no longer had sole responsibility for the regulation of biotechnology products. Today, rather than overseeing the release of recombinant organisms, the NIH focuses almost exclusively on human gene therapy research.

ENVIRONMENTAL PROTECTION AGENCY (EPA)

The EPA was created in 1970 to administer executive authority over activities that have the potential to pollute the environment (air, water, and land) in the United States. At the time the USDA was already regulating biological pesticides. Before 1979 each microbial pesticide was registered on an *ad hoc* basis since a policy had not been developed. The EPA has jurisdiction over genetically engineered organisms (GEOs) as designated within the Toxic Substances Control Act (TSCA) and the Federal Insecticide, Fungicide, and Rodenticide Act (FIFRA). In 1986, the EPA stated that the FIFRA applies to all microbial pesticides, not just genetically engineered ones, and to small-scale releases of genetically modified organisms. Also in 1986 the TSCA was extended to cover genetically engineered microorganisms with a variety of uses (such as environmental cleanup and industrial uses). New regulations also applied to small-scale field testing. The EPA review and regulatory process is evolving to address current needs of modern biotechnology. The EPA proposed new regulations for plants that have been genetically modified to resist pests (plant pesticides) and disease (e.g., viral resistance). Recommended regulations were published on November 21, 1994, in the *Federal Register*.

Federal Insecticide, Fungicide, and Rodenticide Act (FIFRA)

The FIFRA regulates pesticides such as bacteria, viruses, fungi, algae, and protozoa and derivatives of any of these. The FIFRA prohibits the distribution, sale, and use of pesticides that have not been registered with the EPA. The EPA reviews all data submitted for microbial pesticides and determines whether registration is appropriate. For registration, applicants must cite data on product composition, human health effects, environmental fate, and effects on nontarget organisms. The EPA evaluates this information in making a decision about safety. An experimental use permit (EUP) is required to conduct initial tests, even field studies. FIFRA has generally not required an EUP for small-scale testing in an area of 10 acres or less; however, in the case of genetically engineered microbial pesticides, the EPA does require an evaluation for potential risks to determine whether an EUP is required. In all other respects, microbial pesticides are afforded the same review as conventional pesticides.

Toxic Substances Control Act (TSCA)

The TSCA (1976) gives EPA the authority to regulate chemicals in research and commerce that may pose a threat to human health and the environment, and requires that the EPA review test data within a specific period and determine whether the product is safe or poses a risk. TSCA was enacted to regulate organic and inorganic substances or a combination of substances, and applies to all microorganisms produced for industrial, consumer, and environmental uses with the exception of their manufacture, processing, or distribution as foods, food additives, cosmetics, medicines, or pesticides.

National Environmental Policy Act (NEPA)

The NEPA (1970) requires that all federal agencies prepare an environmental impact statement or an environmental assessment of any major federal action that might adversely affect the human environment. The impact statement must describe the predicted effect of the action, any adverse impacts, and alternatives for the proposed action. NEPA is not a regulatory statute; it simply ensures that agencies evaluate risks and environmental impact of genetically engineered organisms. Other federal statutes regulate the release of GEOs.

FOOD AND DRUG ADMINISTRATION (FDA)

The FDA, residing within the Department of Health and Human Services, regulates the manufacture of food, food additives, drugs, and cosmetics. This agency is charged with monitoring our food and drugs and ensuring that they are not contaminated. The Public Health Service Act (PHSA) ensures that the manufacturer is licensed to produce the product, and the Federal Food, Drug, and Cosmetic Act (FFDCA) ensures

that products are manufactured within quality-control guidelines. The FFDCA gives the FDA jurisdiction over foods containing recombinant DNA. Products of biotechnology are evaluated like any other product; they must be approved, for example, under the New Drug Application (NDA), the New Animal Drug Application (NADA), or Product License Application (PLA). The FDA evaluates the product rather than the process, as demonstrated by the Flavr Savr tomato by Calgene, Inc.

UNITED STATES DEPARTMENT OF AGRICULTURE (USDA)

The USDA protects agriculture and forestry in the United States through research oversight and product regulation. This agency also encourages development of agriculture. The Animal and Plant Health Inspection Service (APHIS) regulates genetically engineered organisms within the USDA and administers several statutes to prevent the introduction of plant and animal diseases. APHIS reviews proposals for the release of genetically engineered organisms into the environment and prepares environmental impact statements as required by NEPA. In 1976, the Agriculture Recombinant DNA Research Committee (ARRC) was established to support the NIH-RAC on agricultural issues and to coordinate biotechnology regulation within the USDA. In 1985, the Committee on Biotechnology in Agriculture (CBA) replaced ARRC.

In 1986, research using genetically engineered meat became covered under the same regulations that apply to other experimental meat products by the Food Safety Inspection Service (FSIS). FSIS also regulates the commercial labeling and sale of meat products. The commercial development, sale, and labeling of transgenic meat must be addressed in the future.

Federal Plant Pest Act (FPPA) and Plant Quarantine Act (PQA)

The Federal Plant Pest Act (FPPA) and the Plant Quarantine Act (PQA) provide authority for regulating the movement into and within the United States of genetically engineered organisms that are potential plant pests (also of importance is the Federal Noxious Weed Act of 1974). Included in the 1986 biotechnology framework was a statement which applied the Federal Plant Pest Act (FPPA) to field testing of genetically engineered crops and plant pests (regulations were released in 1987 but relaxed by the Clinton Administration). Currently the role of the FPPA in the commercialization of transgenic crops is unclear.

Virus-Serum-Toxin Act

Within the Virus-Serum-Toxin Act, the USDA has authority over the exportation, importation, production, and distribution of veterinary biological products. Data that demonstrate the safety, purity, and efficacy of the product must be submitted to obtain a product license. If field testing is required to obtain these data, shipment of unlicensed products for experimental purpose is allowed if APHIS has determined that field trials carry little risk of spreading disease and the field trials are conducted under controlled conditions.

GLOSSARY

actinomycetes Aerobic Gram-positive soil-dwelling bacteria that break down plant and animal remains and help keep the soil friable.

airlift fermenter A fermenter in which cells are aerated and circulated by air mixing with the media. Air enters the column at the bottom of the vessel and rises as bubbles.

amino acid A small organic molecule that forms the unit of a polypeptide. A hydrogen atom, amino and carboxyl groups, and an R side group are covalently bonded to a central carbon atom.

androgen A male steroid hormone, such as testosterone, that is involved in the male reproductive system.

anther The pollen-containing structure of the stamen in the male reproductive system of flowering plants.

antibiotic A secondary metabolite, produced by some microorganisms, that kills or inhibits the growth of other microorganisms, often through transcriptional or translational regulation.

antibody An antigen-binding immunoglobulin produced by the B cells of the immune system.

anticodon The three-base-pair region of a tRNA molecule that base pairs with the complementary codon of the mRNA.

antigen A molecule, often a protein, that is recognized as foreign by the body and elicits an immune response.

antisense A sequence of DNA or RNA that is complementary to a functional RNA, often mRNA; the DNA or gene that is transcribed.

aquaculture The farming, often commercial, of a variety of shellfish, crustaceans, finfish, and algae. Aquaculture is a general term, but often refers to freshwater culture (*see* mariculture).

autoimmune disease An immunological disorder resulting from the attack on normal body cells and tissues by the body's lymphocytes.

autonomously replicating sequence (ARS) A sequence that can be cloned to allow extrachromosomal replication of a DNA molecule within an appropriate host cell (e.g., yeast cells); a component of yeast artificial chromosomes.

autoradiography A method by which a radiolabeled molecule such as DNA or RNA or a structure such as a chromosome produces an image on photographic film. The image is called an autoradiogram or autoradiograph.

autosomal chromosome Any chromosome other than a sex chromosome; present in both males and females.

auxin A class of plant growth-regulating hormones; an example is indoleacetic acid (IAA), which stimulates cell elongation, secondary growth, and fruit development.

bacteriophage A virus that infects bacteria; often called phage.

baculovirus A virus that specifically infects arthropods, most often insects.

baker's yeast A strain of *Saccharomyces cerevisiae* used in bread making.

batch culturing A method in which cells are cultured for a specific period of time after inoculation into the culture medium and then collected for processing.

B lymphocyte cell A type of lymphocyte or white blood cell, originating in the bone marrow, that synthesizes antibodies and secretes them as part of the immune response to a foreign antigen.

bioconversion The conversion of a particular organic compound or molecule to an important product through the enzymatic reactions provided by a specific microorganism(s). This process minimizes the synthetic chemical steps required for the synthesis of a commercially important compound.

biodegradable A substance that can be broken down into organic molecules by organisms, usually microorganisms.

biodegradation The breakdown of compounds by living organisms, usually microorganisms and plants.

biolistics A method of delivering DNA into animal and plant cells by DNA-coated microprojectiles discharged under pressure at high velocity.

bioluminescence The ability of an organism to generate light, frequently through a symbiotic association with bioluminescent bacteria. Bioluminescent bacteria produce light when proteins called luciferins, in the presence of oxygen and the enzyme luciferase, are converted to oxyluciferins.

biomass The combined cell mass of a particular group of organisms, often at a particular trophic level in an ecosystem; can be used as an energy source, food supplement, or source of chemical compounds.

bioremediation A technology that uses microorganisms and sometimes plants (phytoremediation) to convert contaminates in soil and water to nontoxic by-products.

biotechnology The commercial use of living organisms or their components to improve animal and human health, agriculture, and the environment.

bivalve A member of a class of burrowing or sessile mollusks that includes oysters, clams, and mussels.

blastodisc A disklike area on the surface of the yolky eggs of birds that divides and gives rise to the embryo.

blunt end The DNA terminus produced, without overhanging 5' or 3' ends, after restriction endonuclease digestion.

callus A term used in tissue culture for undifferentiated plant tissue; usually derived from stem or leaf tissue.

CAP *See* catabolite activator protein.

catabolite activator protein (CAP) A protein involved in regulating the *lac* operon; in the absence of glucose, enhances transcription of the genes in the operon by interacting with cyclic AMP; the CAP-cAMP complex binds to the operon CAP site.

5'-cap A chemical modification of the 5' end of eukaryotic mRNA during post-transcriptional processing of the primary transcript (hnRNA).

cDNA Complementary double-stranded DNA synthesized *in vitro* by reverse transcriptase and DNA polymerase with mRNA as the template.

cDNA library Collection of expressed mRNAs, usually from specific cells in an organism, cloned as cDNAs in a vector.

centimorgan A unit of genetic distance that represents how frequently two genes or loci on the same chromosome can be separated by crossing over. Two genes have a one percent chance of being separated (i.e. on different chromosomes) during recombination if they are located one centimorgan apart. One centimorgan of human DNA is approximately one million base pairs.

centromere A region of DNA on eukaryotic chromosomes to which the mitotic spindle fibers attach during mitosis and meiosis; the region where two chromosomes are attached after replication.

chimera A recombinant DNA molecule or organism that contains sequences from more than one organism.

chromatid One of the two daughter strands of a replicated chromosome joined together by the centromere.

chromosome Structure composed of DNA, RNA, and protein; mechanism by which hereditary information is transmitted from generation to generation. In eukaryotic cells the chromosome resides in the nucleus.

clone Used for an individual or cell from a genetically identical population derived by asexual division. Also a cell harboring a recombinant DNA; a foreign DNA fragment contained within a recombinant clone.

codon The triplet bases of an mRNA that translates into an amino acid or stop signal; the tRNA anticodon base pairs with the codon during translation.

codon engineering Modification of codons within a cloned gene to ones preferred by the host organism (codon bias); used to optimize translation within host cells.

cohesive ends The cohesive DNA termini produced, with overhanging 5' or 3' ends, after restriction endonuclease digestion; sometimes called sticky ends.

cohesive termini (cos) *cos* sites; the 12-base single-stranded complementary ends of bacteriophage λ DNA that are formed during packaging of concatemeric DNA into viral head particles.

colonies Clumps of bacterial cells on solid medium; each colony is composed of genetically identical cells that arose by cell division from a single cell; each colony is called a clone.

competent Describes cells, most often bacterial, that have the ability to take up DNA.

complementary Describes a pair of nucleotides that can base pair through hydrogen bonding; an example is cytosine (C) and guanine (G); strands of DNA also can base pair along their nucleotides.

complementation A method by which two DNA fragments (e.g., from two different regions of a gene) in a cell encode nonfunctional polypeptides that, when synthesized together in the cell, become a functional protein.

consensus sequence A highly conserved sequence, either DNA (e.g., promoter) or protein that shows the nucleotides or amino acids, respectively, most frequently found at each position.

contig A collection of overlapping cloned DNA fragments that are arranged in the appropriate order as found along the chromosome; a contig map refers to the physical location of contigs in a specific segment of DNA.

continuous fermentation A method in which culture medium is added to the fermenter continuously as medium with cells is removed to maintain exponential growth of cells or microorganisms.

cosmid A hybrid vector comprising plasmid sequences and the *cos* sites of lambda bacteriophage, enabling the vector to be packaged *in vitro* with large DNA inserts.

cotyledon A seed leaf that develops as a part of the plant embryo and provides nutrients to the germinating seedling.

crossing over A process that generates new genetic combinations: Homologous chromosomes line up and exchange segments of corresponding DNA during meiosis.

crustacean A member of the class Crustacea that harbors gills and has an exoskeleton or shell; members include lobsters, shrimp, crabs, and barnacles.

cyanobacteria Oxygenic photosynthetic, Gram-negative prokaryotes; some are able to fix dinitrogen.

cyclic AMP (cAMP) 3', 5'-cyclic adenosine monophosphate is a small ring-shaped molecule; a second messenger in cells that helps mediate a cell's response to an external signal; regulator of the *lac* operon.

cytokines Extracellular signaling proteins or peptides produced by the immune system and other tissues that are involved in cell-cell communication; includes interferons, colony-stimulating factors, interleukins, and tumor necrosis factors.

cytokinin A member of a class of plant growth regulators that promote cell division and leaf expansion.

deoxyribonucleic acid (DNA) The nucleic acid encodes genetic information that is passed to progeny; most frequently a double-stranded helical molecule composed of four nucleotides. The nucleotide unit of DNA has one of four nitrogenous organic bases—adenine, guanine, cytosine, or thymine—attached to a deoxyribose sugar with a phosphate group (called a deoxyribonucleotide).

deoxyribonucleotide *See* deoxyribonucleic acid.

diagnostics Sensitive assays used to determine the presence of a disease, DNA mutation, or pathogen.

dideoxyribonucleotide A nucleotide that is missing a 3′ hydroxyl group on the deoxyribose sugar; used in Sanger dideoxy sequencing for chain termination.

differential interference contrast microscopy A type of light microscopy that takes advantage of phase differences in light that has passed through a cell to produce a detailed image of the cell.

dinoflagellates A highly diverse group of single-cell photosynthetic or heterotrophic marine protists that often are flagellated.

diploid Having two sets of chromosomes (2 n); chromosomes of each pair are referred to as homologous.

DNA *See* deoxyribonucleic acid.

DNA helicase Protein that hydrolyzes ATP while moving along the DNA molecule and melting the double strand.

DNA library A collection of cloned DNA fragments often from a specific genome.

DNA ligase An enzyme that re-forms the phosphodiester bond between two nucleotides of DNA (the 5′ phosphate of one nucleotide and the 3′-OH group of the other).

DNA linker A synthetic oligonucleotide containing a restriction site ligated to the end of DNA fragments to create a restriction site for cloning.

DNA polymerase An enzyme that synthesizes a copy of a DNA template. DNA is synthesized in the 5′ to 3′ direction by adding nucleotides to the 3′ hydroxyl group of the newly synthesized strand.

DNA repair Replacement of damaged or incorrect nucleotides with the original sequence; DNA polymerase, DNA ligase, and other enzymes are involved in repair.

DNA replication The process of duplicating DNA for distribution to daughter cells.

DNA topoisomerase An enzyme that during DNA replication removes twists to relieve tension in the double-stranded molecule by breaking phosphodiester bonds and then reseals the breaks.

echinoderm An invertebrate with spines, plates, or needles that exhibits radial symmetry with some bilateral characterisics; examples include sea urchins, sand dollars, sea stars, and sea cucumbers.

electroporation A method for introducing DNA into cells; exposure to a brief electric pulse is thought to open transient pores through which DNA enters.

embryo A multicellular structure (such as a blastula and gastrula in animals, sporophyte in plants) formed by subsequent cell divisions after fertilization.

embryonic stem cells (ES) Cells from an early embryo that are not terminally differentiated but can divide to give rise to differentiated cells, including germ line cells.

endospore A dormant structure that forms within cells of certain bacteria during stressful conditions and can germinate and give rise to new cells when conditions improve.

enhancer A regulatory DNA sequence element that enhances transcription of a eukaryotic gene; functional from thousands of base pairs away and in either orientation.

enzyme A protein that catalyzes a specific chemical reaction.

ethidium bromide A molecule that binds to DNA by intercalating between the bases and fluoresces in the presence of ultraviolet light; used as a stain for DNA, especially in agarose gels.

eukaryotic cell A cell that is compartmentalized with a membrane-bound nucleus, other membrane-bound organelles, and a distinct cytoplasm; eukaryotic organisms are composed of one or more cells.

explant Excised plant tissue from which plants can be regenerated; a term used in tissue culture.

expressed Refers to a gene that is transcribed and translated to yield a product.

expression library A collection of DNA fragments inserted into a vector that has a promoter sequence adjacent to the insertion site; gene expression is controlled by the regulatory elements of the expression vector.

extrachromosomal Not part of the host cell chromosome; usually refers to DNA such as a plasmid.

fermentation The breakdown of organic compounds by cells or organisms in the absence of oxygen to generate ATP; in industrial applications, cells or microorganisms are cultured in fermenters (i.e., bioreactors) with or without oxygen to produce desirable products or to increase biomass.

fermenter A growth chamber used for cultivating cells and microorganisms; used for the production of important compounds; also called a bioreactor.

fingerprinting A method used to identify individual DNA banding patterns derived from hypervariable regions of DNA; used in forensics, to establish paternity, and in conservation biology.

flow cytometry A method for sorting cells and metaphase chromosomes.

gastropod A member of the phylum Mollusca; characteristics include a soft body that is supported by a flat, muscular foot and protected by a cap-shaped shell; examples are snails, abalones, and sea slugs

gel electrophoresis A method used to separate DNA or RNA fragments by length; proteins can be separated by size and/or charge.

gene A region of DNA that encodes information for a discrete product that can be protein or RNA; includes coding sequences, introns, and noncoding regulatory sequences.

gene bank A facility where plant material is stored usually as seeds, tubers, or cultured tissue; *see* germplasm.

gene cloning The insertion of a DNA into a vector and the transfer of the recombinant molecule into a host for propagation; also called recombinant DNA technology, gene splicing.

gene therapy The use of genes to correct a genetic or acquired disorder.

genetic code The 64 triplet codons that encode the 20 amino acids and three stop codons used in protein synthesis.

genetic linkage map A map that indicates the genetic distances between pairs of linked polymorphic DNA markers or genes that have alternative forms; determined by recombination frequencies.

genome The entire content of genetic material within an organism (including bacteria and viruses); sometimes, the amount of DNA in a haploid set of chromosomes in a eukaryotic organism.

genotype The genetic composition of an organism; can include any number of gene pairs from one up to all the genes.

germ cell An animal sperm or egg cell (i.e., gamete) or a precursor cell that gives rise to gametes.

germplasm The genetic material within an organism; usually meaning plant genetic material.

haploid Having only one set of chromosomes (n); found in gametes, some fungi and protists, and plant gameotphytes.

helper virus A virus that provides functional properties to another virus in the same cell that lacks specific functional proteins.

heterogeneous nuclear RNA (hnRNA) The primary product of transcription in eukaryotes that has not been processed to remove introns or to add a poly-A tail or a 5′ cap; after processing in the nucleus, the mRNA is transported to the cytoplasm for translation.

heterologous probe A DNA sequence from one organism that is used to hybridize to DNA sequences from a different organism; a DNA sequence used for hybridization that is not identical to the target sequence but has some degree of similarity.

heterozygous Having different forms or alleles of a gene at a specific locus on homologous chromosomes.

homeobox A 180 base-pair DNA sequence found in many organisms that encodes a DNA-binding motif in genes that are involved in development; *see* homeodomain.

homeodomain A DNA-binding motif comprised of 60 amino acids that is encoded by the homeobox.

homeotic gene A gene that controls a major developmental pathway.

homologous recombination The reciprocal exchange of genetic information between homologous chromosomes during meiosis; *see* crossing over.

homozygous Having two identical forms of alleles of a gene at a specific locus on homologous chromosomes.

hybridization Hydrogen bonding between two complementary single-stranded DNA sequences or between DNA and RNA sequences.

hybridoma An immortal cell line produced from the fusion of antibody-secreting B lymphocytes to lymphocyte tumor cells and used in the production of monoclonal antibodies.

hypocotyl The region of the embryo or seedling between the cotyledons and the embryonic root.

immunoassay A method that uses the specificity of antibodies to detect the presence of a specific antigen (usually protein or glycoprotein) in a biological sample, such as blood.

immunoglobulin *See* antibody.

inflammation Redness, swelling, and pain due to a reaction to injury or infection of tissue.

insulin A hormone, secreted by β cells of the pancreas, that reduces the glucose levels in the blood by regulating glucose metabolism in animals.

integrins Members of a family of transmembrane linker proteins that are involved in the adhesion of animal cells to the extracellular matrix proteins such as collagen and fibronectin.

intellectual property Once a traditional area of law that encompassed patents, trade secrets, trademarks, copy rights, and plant variety protection; intellectual property now can be living organisms.

interleukins *See* cytokines.

intron A noncoding region of a eukaryotic gene that is removed during RNA processing after transcription.

karyotype A complete set of chromosomes from a somatic cell that is arrested in metaphase and arranged in pairs of homologous chromosomes and sex chromosomes; used to identify chromosomal abnormalities.

lagging strand The DNA strand that during replication is synthesized in short fragments that then are covalently joined by DNA ligase.

leading strand The DNA strand that during replication is synthesized continuously in the 5′ to 3′ direction.

liposome transfer The transfer into eukaryotic cells of DNA, RNA, or therapeutic agent encapsulated in an artificial lipid bilayer vesicle formed from an aqueous suspension of phospholipid molecules.

leucine-zipper A structural motif found in some DNA-binding proteins that is comprised of two α-helices each from separate polypeptides held together by hydrophobic amino acid side chains, usually on leucines, thus forming a protein dimer.

linked Describes two or more genes or markers that are located on the same eukaryotic chromosome.

lysogenic Describes a bacteriophage DNA that integrates into a bacterial genome and does not become lytic; called a prophage.

lytic Describes a bacteriophage that replicates and lyses the host cell.

map *See* genetic linkage map and physical map.

mariculture The farming, often commercial, of a variety of marine shellfish, crustaceans, finfish, and algae.

meiosis Two successive nuclear cell divisions and one replication event leading to the formation of haploid gametes (eggs and sperm).

meristematic tissue A group of dividing cells, such as shoot and root apical meristems, that give rise to tissues and organs of flowering plants.

messenger RNA (mRNA) An RNA molecule that encodes the amino acid sequence of a polypeptide; translated into protein during the process of translation.

metabolite A small organic compound that is produced by an enzymatic reaction(s) or is required for metabolism.

metaphase chromosome A condensed chromosome that has replicated and is composed of two sister chromatids.

microinjection Direct injection of DNA or a therapeutic agent into a cell with a microcapillary.

microorganism A prokaryotic organism lacking membrane-bounded organelles and a true nucleus.

mismatch The lack of complementarity of two DNA sequences at one or more nucleotides leading to the inability of the two sequences to base pair or hybridize in that region.

molecular biology The study of molecular processes that encompasses DNA, RNA, and protein.

mollusk A member of the phylum Mollusca, which comprises seven classes and includes mussels, clams, snails, squid, and octopuses; members of the phylum have a tissue fold called a mantle around a soft, fleshy body.

monoclonal antibody (MAB) A single type of antibody specific for one antigenic determinant (portion of an antigen that binds to an antibody) that is secreted by a hybridoma clone derived from a single B cell.

mutation A change in the nucleotide sequence of a chromosome that is inherited.

nucleic acid A polymer of nucleotides joined by phosphodiester bonds; a general name given to DNA or RNA.

nucleotide The basic unit of DNA or RNA; a pentose sugar with a bound nitrogenous base and phosphate group.

oligonucleotide A short, single-stranded DNA that usually is synthesized *in vitro* and often used as a probe in hybridizations or as primers for the polymerase chain reaction.

operator A short region of DNA in bacteria that controls transcription of an operon (i.e., an adjacent gene or genes).

operon Several contiguous genes regulated together and transcribed into a single mRNA.

ovary The reproductive organ in female animals where the eggs are made; where seeds are produced after fertilization in plants, a mature ovary comprises all or a portion of the fruit.

patent A legal document that grants exclusive rights to sell or use an invention for a defined period of time (17 years in the United States); *see* intellectual property.

pathogen An organism or virus that causes disease.

peptide A short sequence of amino acids.

peptide bond A covalent bond between the carboxyl group of one amino acid and the amino group of an adjacent one; formed during protein synthesis.

phenotypic trait An observable characteristic of an organism or cell that is determined by the genotype or genetic composition, often in conjunction with the environment.

phycobiliproteins The primary components of the phycobilisomes, which are the primary light-harvesting antennae for photosystem II in cyanobacteria and red algae; water-soluble proteins having covalently bound phycobilins (accessory pigments) that absorb and transfer light energy to chlorophylls; examples of phycobilins are phycocyanins and phycoerythrins.

physical map A map that indicates the distance in base pairs between markers or genes along the chromosome; markers can be restriction recognition sites, sequence-tagged sites, and restriction fragment length polymorphisms (RFLP).

plaque A clear, circular area that forms after a bacterial colony cultured on solid medium has lysed by bacteriophage.

plasmid An extrachromosomal circular double-stranded DNA molecule that replicates independently from the chromosome; frequently modified for use as a cloning vector.

pluripotent Describes the ability of a cell to develop into a differentiated cell type; progenitor of other cells; often used in conjunction with stem cells (bone marrow, embryonic) in animals.

polymer a macromolecule composed of repeating units; examples include DNA, RNA, starch.

polymerase chain reaction (PCR) A method for selectively amplifying regions of DNA by *in vitro* replication involving repeated denaturation and renaturation of the DNA template.

polymorphic Describes variation within a population; frequently refers to DNA sequences. *See* polymorphic DNA marker.

polymorphic DNA marker A DNA sequence along a chromosome that varies in a population of the same species. Markers are inherited in a Mendelian fashion.

polypeptide A long linear sequence of amino acids; sometimes called a protein.

primer A short oligonucleotide sequence that is hybridized to a complementary DNA template and to which a nucleotide can be added at the 3' end to initiate replication; used in the polymerase chain reaction (PCR).

probe A labeled molecule, frequently DNA or RNA, that is used for hybridization.

prokaryotes Bacteria; organisms lacking a membrane-bound nucleus and other organelles. *See* eukaryotic cell.

promoter A region of DNA where RNA polymerase and other proteins involved in the regulation of transcription bind.

pronuclear microinjection A method by which one of the nuclei in a fertilized egg is injected with DNA prior to fusion of the gamete nuclei. *See* microinjection.

protease An enzyme that degrades proteins by hydrolyzing the peptide bonds; an example is trypsin.

protein A linear macromolecule that is composed of amino acids that are connected by peptide bonds; sometimes used interchangeably with polypeptide, although a protein can consist of more than one polypeptide (e.g., dimeric protein is composed of two polypeptides).

protoplasts Plant, bacterial, or yeast cells that have no cell wall; cell walls usually are removed enzymatically.

protozoan A single-cell, free-living or parasitic, nonphoto-synthetic, and motile eukaryotic organism; examples are amoebas and paramecia; a type of protist.

purine One of two classes of nitrogen-containing ring structures in DNA and RNA; adenine and guanine have double-ring structures.

pyrimidine One of two classes of nitrogen-containing ring structures in DNA and RNA; cytosine, thymine (DNA), and uracil (RNA) have single-ring structures.

reading frame Triplet bases in a nucleotide sequence; a messenger RNA can be read in any of three reading frames, although only one will generate the correct amino acid sequence of the polypeptide.

receptor A protein, located on the cell surface or intracellularly, that binds to a specific type of molecule (ligand) and elicits a cellular reaction or response.

recombinant DNA A sequence comprising DNA from different sources that have been joined together.

replica plating The transfer of cells in bacterial colonies (or plaques) from one petri dish, or plate, to another; used for screening libraries so that a master plate is retained for the isolation of positive clones.

replication fork The Y-shaped area of a replicating DNA molecule where the DNA strands are separated and the two daughter strands are synthesized.

replicative form (RF) A viral nucleic acid that serves as the template for replication in a host cell.

reporter A gene that encodes a product that can be assayed to determine whether a specific DNA construct is present and functional in a host organism or cell.

repressor protein A protein that binds to a gene promoter or operator site and blocks the binding of RNA polymerase, thereby preventing transcription.

restriction endonuclease A member of several classes of enzymes that cut DNA internally; each recognizes a specific short double-stranded DNA sequence and hydrolyzes phosphodiester bonds on both strands (type II enzymes are used in recombinant DNA technology).

restriction fragment length polymorphism (RFLP) Nucleotide sequence variations in a region of DNA that generates fragment length differences according to the presence or absence of type II restriction endonuclease recognition site(s).

retrovirus A virus with an RNA genome that replicates in a eukaryotic host cell by synthesizing a double-stranded DNA intermediate; the double-stranded form can integrate into the host chromosome.

reverse transcriptase An RNA-dependent DNA polymerase that uses an RNA template for the synthesis of a complementary DNA (cDNA); present in viruses with RNA genomes.

ribonuclease An enzyme that hydrolyzes RNA by hydrolyzing the phosphodiester bonds between the nucleotides (RNase).

ribonucleic acid (RNA) A heteropolymer of ribonucleotides. The nucleotide unit of RNA has one of four nitrogenous organic bases—adenine, guanine, cytosine, or uracil—attached to a deoxyribose sugar with a phosphate group.

ribosome A two-subunit structure that associates with mRNA and catalyzes the synthesis of protein; composed of rRNA and ribosomal proteins.

ribosomal RNA (rRNA) RNA molecules that associate with proteins to form the ribosome; types are distinguished by their sedimentation coefficient (e.g. 28S rRNA).

ribozyme A catalytic RNA molecule.

RNA *See* ribonucleic acid.

RNA polymerase The enzyme that catalyzes the synthesis of RNA on a DNA template.

sequence-tagged sites Two hundred to 300-base pair regions of known DNA sequences determined to be unique and that can be amplified by the polymerase chain reaction; chromosomal DNA sequences used to integrate genetic linkage maps with physical maps.

shellfish Marine and freshwater invertebrates; usually meaning crustaceans and mollusks.

single-cell protein Dried cells of microorganisms that can be ingested by animals and humans as a source of protein.

somaclonal variation Random mutations in the mature plant that are not genetically identical to those in the parent plant; one way that new, potentially desirable traits such as drought resistance are obtained if they can be stably maintained and passed on to progeny plants.

somatic cell Any plant or animal cell that is not a germ cell.

Southern blotting A method of transferring denatured DNA from an agarose gel after gel electrophoresis to a membrane for hybridization to detect specific sequences.

sponges Primarily marine invertebrates that exist in many forms, such as fingerlike branches, vases, and hollow columns; surfaces have tiny pores leading to interior chambers lined with ciliated cells that help flush water through the chambers where bacteria and food particles are trapped; from the phylum Porifera.

sporangium A hollow unicellular or multicellular structure in which spores form.

spore *See* endospore. (Spores are present in fungi and plants, but have a different function from that of endospores.)

stem cell A precursor cell that divides and whose progeny give rise to differentiated cells.

sticky ends *See* cohesive ends.

stirred tank reactor A fermentor that uses internal mechanical agitation of the culture medium by a series of rotor blades.

substrate An energy source for cells and microorganisms that can be transformed into valuable products; also a compound that is modified to a product by an enzyme in a reaction.

T-DNA A region of DNA within the Ti plasmid of *Agrobacterium tumefaciens* that is transferred to and integrated into the chromosome of plant cells.

T lymphocyte A type of cell that is responsible for cell-mediated immunity; examples are helper T cells and cytotoxic T cells.

telomere The end of a eukaryotic chromosome having a special sequence that enables it to replicate to prevent shortening of the chromosomal end with each round of replication.

therapeutic Any agent (DNA, protein, synthetic compound, etc.) that is used in the treatment of disease.

Ti plasmid A large plasmid in *Agrobacterium tumefaciens*; harbors tumor inducing genes that lead to the formation of crown galls.

tissue culture *In vitro* propagation of plant and animal cells or tissues; used to regenerate whole plants from cultured cells and tissues.

totipotency Describes plant cells or tissues able to develop into any differentiated cell or tissue type.

transcription The synthesis of RNA that is catalyzed by RNA polymerase and requires a DNA template.

transduction The transfer of DNA from one bacterial cell to another by a virus.

transfection The transfer of DNA into a eukaryotic cell.

transfer RNA (tRNA) Small RNA molecules that are involved in protein synthesis by serving as an adaptor between messenger RNA and amino acids; the tRNA molecules bring their amino acids to the site of protein synthesis at the ribosome.

transformation The uptake of DNA into a cell, usually a bacterium or yeast, and its expression; also refers to the conversion of a normal cell to a cancerous one.

transgene A gene that has been introduced into the genome of a eukaryotic cell(s) or organism that originated from another source.

transgenic organism A living organism that harbors an inserted functional gene in its germ line.

translation Protein (polypeptide) synthesis by the formation of peptide bonds between amino acids encoded by a messenger RNA and translated with the aid of ribosomes and transfer RNA molecules.

triploid Three copies of homologous chromosomes (3n).

tunicates Marine animals from the phylum Chordata, mostly sessile, that are anchored on rocks or other substrates (some are planktonic); also called sea squirts. Although adults do not have a notochord, nerve cord, or tail, the larvae of some tunicates exhibit these characteristics.

vaccine An agent that is used to promote an antibody response that protects the organism against infection; confers immunity to disease. Examples include protein, DNA, living or nonliving virus or microorganism.

vector A DNA molecule used to transfer foreign DNA into a host cell or organism; examples include plasmids, bacteriophage, and cosmids.

virulent Describes a microorganism or virus that causes disease and even death of an organism; also a bacteriophage that causes host cell lysis.

Western blotting A method to transfer electrophoretically separated proteins to a membrane for antibody binding to detect specific proteins.

xenobiotics Chemicals or compounds that are not produced by living organisms.

xenotransplant A tissue or organ removed from one species and transplanted into another species.

x-ray diffraction pattern The three-dimensional arrangement of atoms in a molecule (such as protein and DNA) that is generated by the diffraction pattern of x-rays passing through a crystal molecule. The method of generating the patterns is called x-ray crystallography.

yeast artificial chromosome A linear recombinant DNA molecule that replicates in a yeast host cell; important components include a large fragment of foreign DNA (up to 1 million base pairs), a centromere, a yeast origin of replication, and yeast telomeres at the ends of the chromosome.

zinc-finger A structural motif in many DNA-binding proteins that is composed of a loop of polypeptide stabilized by a zinc atom bound to the side chains of cysteine and histidine amino acids; proteins often have more than one zinc finger; the end of the fingerlike loop of polypeptide recognizes specific bases of the DNA.

zygote A diploid cell that is produced by the fusion of nuclei from a male and female gamete (i.e., sperm and egg); also called a fertilized egg.

CREDITS AND ACKNOWLEDGMENTS

CHAPTER 1

Figure 1.1: Photo of samples found during excavations in the Tehuacan Valley, Mexico, under the auspices of the Robert S. Peabody Foundation for Archaeology. © Robert S. Peabody Museum of Archaeology, Phillips Academy, Andover, MA. All Rights Reserved.

Figures 1.2, 1.3: Courtesy of The Oriental Institute of The University of Chicago.

Figure 1.4: J.-P. Amet/Sygma.

Figure 1.5: Courtesy of The Oriental Institute of The University of Chicago.

Figure 1.6: J. Ingraham and C. Ingraham, *Introduction to Microbiology*, Fig. 29.6. Copyright © 1995 Wadsworth Publishing Co.

Figure 1.7: Photos (left) National Library of Medicine; (right) Armed Forces Institute of Pathology.

Figure 1.8: Photos (a) Historical Collections, National Museum of Health and Medicine, Armed Forces Institute of Pathology; (b) Reproduced by permission of the President and Council of the Royal Society, London.

Figure 1.10: C. Starr and R. Taggart, *Biology: The Unity and Diversity of Life*, 7th ed., Fig. 13.2. Copyright © 1995 Wadsworth Publishing Co.

Figure 1.11: S. L. Wolfe, *Molecular and Cellular Biology*, Fig. 1-37. Copyright © 1993 Wadsworth, Inc.

Figure 1.12: S. L. Wolfe, *Introduction to Cell and Molecular Biology*, Fig. 1-30. Copyright © 1993 Wadsworth, Inc.

CHAPTER 2

Figure 2.2: Fig. 2.2 J. Ingraham and C. Ingraham, *Introduction to Microbiology*, Fig. 6.8. Copyright © 1995 Wadsworth Publishing Co.

Figure 2.3: D. A. Jackson, R. H. Symons, and P. Berg. *Proc. Nat. Acad. Sci. USA* 69:2904–2909 (1972). Used by permission.

Figure 2.5: S. L. Wolfe, *Molecular and Cellular Biology*, Fig. 4-29. Copyright © 1993 Wadsworth, Inc.

Figure 2.6: Photo Michael Maloney/San Francisco Chronicle.

CHAPTER 3

Figures 3.2, 3.5, 3.10: J. Ingraham and C. Ingraham, *Introduction to Microbiology*, Figs. 6.1, 6.5, 6.6. Copyright © 1995 Wadsworth Publishing Co.

Figure 3.4: C. Starr and R. Taggart, *Biology: The Unity and Diversity of Life*, 7th ed., Fig. 13.7. Copyright © 1995 Wadsworth Publishing Co.

Figures 3.6, 3.9: S. L. Wolfe, *Molecular and Cellular Biology*, Fig. 23-16. Copyright © 1993 Wadsworth, Inc.

Figure 3.11: From data presented in P. H. von Hippel et al. 1984. *Annu`. Rev. Biochem.* 53:389 as used in S. L. Wolfe, *Molecular and Cellular Biology*, Fig. 15.22. Copyright © 1993 Wadsworth, Inc.

Figures 3.12, 3.13, 3.15 to 3.24: S. L. Wolfe, *Molecular and Cellular Biology*. Copyright © 1993 Wadsworth, Inc.

CHAPTER 4

Figure 4.1: Photo courtesy of Monsanto Company, St. Louis, MO.

Figure 4.2: Photo courtesy of Robert Hammer, Howard Hughes Medical Institute, Dallas.

Figure 4.3: Based on F. Bolivar et al., Gene 2:95 (1977) and J. G. Sutcliffe, *Cold Spring Harbor Symp. Quant. Biol.* 43:77 (1979).

Figure 4.5: Based on J. Messing, *Methods Enzymol.* 101:20 (1983); J. Norrander, T. Kemps, and J. Messing, *Gene* 26:101 (1983); C. Yanisch-Perron, J. Viera, and J. Messing, *Gene* 33:26 (1985).

Figure 4.7: Photo courtesy of Keith Wood.

Figure 4.10: Adapted from C. Starr and R. Taggart, *Biology: The Unity and Diversity of Life*, 7th ed., Fig. 16.5. Copyright © 1995 Wadsworth Publishing Co.

Figure 4.14: Photo courtesy of W. F. Reynolds, University of Southern California.

Figure 4.15: Photo courtesy of Roger Duncan.

CHAPTER 5

Figure 5.1: J. Ingraham and C. Ingraham, *Introduction to Microbiology*, Fig. 29.8. Copyright © 1995 Wadsworth Publishing Co.

Figure 5.2: Photo courtesy of Steven E. Lindow, University of California at Berkeley.

Figures 5.3, 5.4: Photos courtesy of Mycogen Corporation, San Diego, CA.

Figure 5.5: Photos courtesy of Monsanto Company, St. Louis, MO.

Figure 5.6: Photo courtesy of Malcolm Fraser, University of Notre Dame.

Figure 5.7: Photo courtesy of R. C. Fuller, University of Massachusetts, Amherst.

Figure 5.8: Photo courtesy of Exxon Company, U.S.A., Houston, TX.

Figures 5.9, 5.10, and 5.11: Adapted from B. R. Glick and J. J. Pasternak, *Molecular Biotechnology*, Figs. 10.1, 10.2, 10.3, and 10.4, ASM Press, 1994. Used by permission of the publisher.

CHAPTER 6

Figure 6.1: Photo Susan Barnum.

Figures 6.2, 6.3: Photos courtesy of CREW, The Cincinnati Zoo and Botanical Garden, Cincinnati, OH.

Figure 6.4: Photos courtesy of H.-U. Koop and Schweiger, *Eur. J. Cell Biol.* 39:46–49 (1985). Used by permission.

Figure 6.5: Photos courtesy of S. Austin, M. A. Baer, and J. P. Helgeson, *Plant Sci.* 39:75–82 (1985). Used by permission from Elsevier Science Ireland Ltd and the authors.

Figure 6.6: Photo Susan Barnum.

Figure 6.7: Photo used by permission from Y. Poirier et al., *Science* 256:520–523. Copyright 1992 American Association for the Advancement of Science.

Figure 6.8: Photos courtesy of Monsanto Company, St. Louis, MO.

Figure 6.9: Photo courtesy of Clarence Ryan, Washington State University, Pullman.

Figure 6.10: Photo courtesy of R. E. Shade, Purdue University, West Lafayette, IN.

Figure 6.11: Photo courtesy of Calgene, Davis, CA.

CHAPTER 7

Figure 7.1: Photo courtesy of R. L. Brinster, School of Veterinary Medicine, University of Pennsylvania, Philadelphia.

Figures 7.2, 7.7: N. L. First, *J. Reprod. Fert. Suppl.* 41:3–14 (1990). Used by permission of the publisher.

Figures 7.3, 7.6: Adapted from B. R. Glick and J. J. Pasternak, *Molecular Biotechnology*, Figs. 15.4 and 15.10, ASM Press, 1994. Used by permission of the publisher.

Figures 7.4, 7.5: Adapted from art by Jared Schneidman in M. R. Capecchi, *Scientific American*, March 1994, pp. 54–55, 56–57. Copyright © 1994 Scientific American, Inc. All rights reserved. Used by permission of the publisher.

Figures 7.8, 7.9: Photos courtesy of CREW, Cincinnati Zoo and Botanical Garden, Cincinnati, OH.

CHAPTER 8

Figure 8.1 (above): Adapted from A. Fuji, *Oceanus*, 30:19–23 (1987).

Figure 8.1 (below): Photo courtesy of Jasper S. Lee, Demorest, GA.

Figure 8.2: Photos Dale Glantz, San Diego, CA.

Figure 8.4: Photo courtesy of Tom Mumford.

Figures 8.6 and 8.7: Photos courtesy of E. W. Becker, *Microalgae Biotechnology and Microbiology*, 1994. Figs. 10.10, 10.11, 10.22, 10.30, 10.49. Reprinted with the permission of Cambridge University Press and the author.

Figure 8.10: Redrawn from M. A. Shears et al. "Gene transfer in salmonids by injection through the micropyle" in C. L. Hew and G. L. Fletcher, eds., *Transgenic Fish*, 1992,

World Scientific Publishing Co. Ptc. Ltd, Singapore. Used by permission of the publisher.

Figure 8.11: Photo courtesy of R. H. Devlin, Fisheries and Oceans Canada, British Columbia.

CHAPTER 9

Figures 9.1, 9.3: Dept. of Energy, *Human Genome 1991–1992 Program Report, Primer on Molecular Genetics*, pp. 198 and 216. Prepared by the Human Genome Management Information System, Oak Ridge National Laboratory.

Figure 9.2: Photo Dept. of Energy, prepared by the Human Genome Management Information System, Oak Ridge National Laboratory.

Figures 9.4, 9.11, 9.12b: Courtesy of Human Genome Center, Lawrence Livermore National Laboratory, Livermore, CA.

Figures 9.6, 9.9, 9.13: Courtesy of Norman Doggett, Center for Human Genome Studies, Los Alamos National Laboratory, Los Alamos, New Mexico.

Figure 9.10: J. Yu, S. Tong, J. Qi, and F.-T. Kao, *Somatic Cell and Mol. Gen.* 20:353–357 (1994). Reprinted by permission of Plenum Publishing Corp. and the authors.

Figure 9.12a: D. A. Micklos and G. A Freyer, *DNA Science*, p. 155. Copyright © 1990 Carolina Biological Supply Co. and Cold Spring Harbor Laboratory Press. Used by permission.

Figure 9.14: Line art courtesy of R. A. Keller from L. M. Davis, et al., *Genet. Anal.: Tech. Appl.* (8)1:1–7 (1991). Copyright © 1991. Used by permission of Elsevier Science-NL, Amsterdam, The Netherlands.

CHAPTER 10

Figure 10.1: Adapted from A. Cuthbert, *J. Royal Soc. Med.* 87(Suppl. 21):2–4 (1994). Used by permission of the publisher.

Figure 10.3: Photo courtesy of P. L. Chang, G. Hortelano, M. Tse, and D. E. Awney, *Biotech. Bioeng.* 43:925–933. Copyright © 1994 John Wiley & Sons, Inc. Used by permission.

Figure 10.4: Adapted from M. Barinaga, *Science* 262:1512–1514. (1993). Copyright © 1993 American Association for the Advancement of Science. Used by permission.

Figure 10.6: Photo courtesy of P. X. Ma and Robert Langer, Massachusetts Institute of Technology, Cambridge, MA.

CHAPTER 11

Figures 11.1–11.5: Photos and line art courtesy of Cellmark Diagnostics, Germantown, MD.

APPENDIX A

Reprinted with permission from *Nature* 262:3. Copyright 1976 Macmillan Magazines Limited.

APPENDIX B

S. Wright, *Molecular Politics*, Table 9.2. University of Chicago Press, 1994. Used by permission of the publisher. See this paper for original sources.

APPENDIX C

Figures C.1, C.2, C.3: C. Starr and R. Taggart, *Biology: The Unity and Diversity of Life*, 7th ed., Figs. 40.3, 40.6, 40.7. Copyright © 1995 Wadsworth Publishing Co. Art by Raychel Ciemma.

APPENDIX D

Adapted from J. L. Marx, ed. 1989. *A Revolution in Biotechnology*. Cambridge University Press, Cambridge.

INDEX

EVOLUTION OF THE HUMAN GENOME PROJECT*

James Watson named the first Director of National Center for Human Genome Research (NIH).

Sequence-tagged sites (STSs) proposed as common mapping language.

The first genome sequence and analysis conference, Wolf Trap, Virginia.

NIH and DOE establish joint Ethical, Legal, and Social Issues (ELSI) Working Group.

Robert Shinsheimer holds meeting on sequencing human genome at University of California, Santa Cruz.

Renato Dulbecco proposes sequencing human genome at Washington, D.C. conference.

Charles DeLisi and David Smith of Department of Energy (DOE) propose Human Genome Initiative, a plan for mapping and sequencing the human genome.

House and Senate appropriations committee hearings on NIH and DOE budgets for fiscal year 1988, first year of government funding.

NIH genome research begins, administered by National Institute of General Medical Sciences (NIGMS) for two years.

DOE Health and Environmental Research Advisory Committee (HERAC) publishes recommendations for 15-year multidisciplinary collaboration to map and sequence human genome at designated centers.

DOE publishes first human genome program plans.

1985   **1986**   **1987**   **1988**   **1989**   **1990**

Sydney Brenner proposes to European Commission (Brussels) collaborative effort to map and sequence genomes from other organisms.

After DOE meeting of the Human Genome Initiative, Los Alamos National Laboratory, the Human Genome Initiative publicly announced.

DOE develops mapping and informatics projects for national DOE laboratories.

Cold Spring Harbor Laboratory symposium session of the human genome project, organized by James Watson, co-chaired by Paul Berg and Walter Gilbert.

At National Institutes of Health, Howard Hughes Medical Institute convenes international forum on human genome to formulate a more articulated plan that includes mapping.

Human Genome Organization (HUGO) formed to coordinate international effort to map and sequence human genome.

First annual Cold Spring Harbor Laboratory meeting on mapping and sequencing human genome.

NIH and DOE sign Memorandum of Understanding outlining cooperative agreement on genome research.

NIH establishes Program Advisory Committee on the Human Genome (PACHG).

The United States Human Genome Project officially begins in October.

NIH and DOE release joint five-year plan outlining goals for U.S. Human Genome Project.